W9-DFS-467

The Environmental Price of Energy

The Environmental Price of Energy

Edited by
Alfred J. Van Tassel
Hofstra University

Lexington Books
D.C. Heath and Company
Lexington, Massachusetts
Toronto London

Library of Congress Cataloging in Publication Data

Main entry under title:

The environmental price of energy.

Includes bibliographical references and index.
1. Power resources. 2. Pollution. 3. Energy conservation. I. Van Tassel, Alfred J., ed.
TJ153.E53 301.31 74-27746
ISBN 0-669-97766-7

Published simultaneously in Canada.

Printed in the United States of America.

International Standard Book Number: 0-669-97766-7

Library of Congress Catalog Card Number: 74-27746

Contents

Acknowledgments

The heart of a university is its library and we are fortunate at Hofstra to have a good one. We are particularly grateful to the Documents Library which enabled us to make good use of government studies.

To round out the treatment of the subject chosen for this volume, we found it desirable to go outside the university for other contributors and we were fortunate to secure a very distinguished group of contributors. We are grateful to Hans Landsberg, Director of the Resources for the Future Energy and Minerals Program, who suggested a number of those who could and did contribute significantly.

Finally, we are grateful to our colleagues: Dean Harold Lazarus of the School of Business and Frederic Stuart, who offered continuous help and encouragement; Professors Frances Sterrett and John Ullmann, and Arthur and Virginia Sugden, who read portions of the manuscript and offered helpful suggestions. The errors that remain are strictly our own.

Part I: Selected Opportunities for Savings

Since we are faced with an energy crisis, why not save wherever we can?

1 Introduction

Alfred J. Van Tassel

> Ours is a great wild country:
> If you climb to our castle's top,
> I don't see where your eye can stop
> For when you've passed the corn-field country,
> Where vineyards leave off, flocks are packed,
> And sheep-range leads to cattle-tract,
> And cattle-tract to open-chase,
> And open-chase to the very base
> Of the mountain, where, at a funeral pace,
> Round about, solemn and slow,
> One by one, row after row,
> Up and up the pine-trees go,
> So, like black priests up, and so
> Down the other side again
> To another greater, wilder country,
> That's one vast red drear burnt-up plain,
> Branched through and through with many a vein
> Whence iron's dug, and copper's dealt;
> Look right, look left, look straight before,–
> Beneath they mine, above they smelt,
> Copper-ore and iron-ore,
> And forge and furnace mould and melt,
> And so on, more and ever more,
> Till, at the last, for a bounding belt,
> Comes the salt sand hoar of the great seashore,
> –And the whole is our Duke's country!
>
> – Browning [1]

The verve of Robert Browning's lines express well the exuberance with which the Victorian English and their predecessors went at the job of putting the earth's resources to use. England became the classic model of Industrial Revolution and output expanded without sustained interruption until the Great Depression of the 1930s. The latter led another Englishman, W.H. Auden, to admonish his countrymen in a poem bereft of Browning's optimism to: "Get There If You Can and See the Land You Once Were Proud to Own" [2].

Across the ocean, the English settlers of the New World exploited the earth's resources with even more enthusiasm than their forebears. Economists argued the wisdom of using intensively the relatively abundant ones of the factors of production.

The economists' factors of production included capital, labor and land, and all the earth's natural resources were subsumed under this last. Blessed with what appeared an unending supply of land, and with precious little capital and labor, the New England settlers embraced the economists' doctrines with a fervor as profound as their fundamental ignorance of the theory on which it was based. Land was used lavishly and "cultivation . . . was extraordinarily slovenly" [4:55].

The free-handed use of land to further the economic development of the country continued to be the policy after the American Revolution. Grants of federal and state lands for farms or mineral rights were freely available, especially after the Civil War. The transcontinental railroads, in particular, were built through government land grants from whom the railroads received, directly or indirectly, 223 million acres [3:361], mostly along the right-of-way, where land values might be expected to rise most. For a variety of reasons, including the building of the railroads, the rate of economic expansion was dramatic, and by the end of World War I the United States had achieved world economic leadership.

When the United States emerged from World War II, not only unharmed but strengthened relative to other nations, the stage was set for a change in the order of magnitude of the rate of economic growth.

From time to time there had been suggestions that America might be approaching exhaustion of this or that resource. Such gloomy forecasts came to be heavily discounted because they proved wrong so frequently. Nevertheless, the rate of increase in resource use in the 1950s was so great and the prospective future increases so immense that it seemed to call for a second look at the adequacy of the resource base. To that end, Resources for the Future, Inc., an organization supported by the Ford Foundation, undertook a study in which the supply and demand of every important commodity or service in the economy were projected for ten-year intervals from a starting date of 1960 to the end of the century [4].

On the central question to which RFF addressed itself, the answer seemed fairly clear: Despite some short-term maladjustments from time to time, it was reasonable to expect that supply would expand to meet demand in virtually all situations.

At Hofstra, the RFF projections—done with great professional skill on an internally consistent basis—seemed to us to provide the start toward answering a different sort of question: If the projections proved sound, what would be the impact on the environment? For example, if we build all the automobiles as projected and burn all the gasoline, what will be the consequences in terms of air pollution? If the expected expansion of output in paper products occurs, what is the effect in terms of stream pollution? At Hofstra, we have a program for earning the degree of Master of Business Administration that calls for the preparation of a thesis as the final requirement. The RFF projections were used as the basis· for fourteen such master's essays, each covering a separate environmental question, and the collection was accepted for publication by D.C. Heath and Company as one of the first of its Lexington Books series, under the title *Environmental Side Effects of Rising Industrial Output* [5].

The general conclusion of this study was that "pollution has become a matter of such commanding urgency because of the rise in the scale of national output" [5:449]. The impact of pollution of a given type may have been trivial when output was small, but the effect of the same general sort of pollutants could be serious when discharged on a larger scale. The RFF projections pointed to continuing dramatic increases in output. Since the magnitude of pollutional effects to be overcome increases in some rough proportion to output, it seemed evident in view of the expected increase in output that we would "like Alice in Wonderland—have to run very fast to stay in the same place" [5:450].

The fact that this study necessarily relied for the most part on broad national totals robbed it of much of its persuasiveness. Pollution does not afflict some abstract nation as a whole to which national totals apply. Pollution occurs in the air basins above particular cities, in the drainage basins of specific waterways. We sought to remedy this deficiency in the next Hofstra study, which examined air pollution and problems in solid waste disposal in fifteen cities and water pollution of all varieties in nine major waterways. The fifteen cities consisted of five in the east, five in the midwest, and five in the far west, and the nine bodies of water were distributed throughout the country, each near one of the fifteen cities studied for other forms of pollution. While no claims of statistical significance were made for the sample, the wide geographical dispersion of the pollution problems studied rendered the studies conclusions of more specific relevance. These results were also published by D.C. Heath and Company under the title *Our Environment: The Outlook for 1980* [6]. Somewhat surprisingly, the conclusions of this second study were somewhat more optimistic than those of the first, particularly so far as the smaller cities were concerned, that is, those with populations less than 150,000—the outlook was for cleaner air in 1980 than in 1969 and no sign of important difficulties in disposing of solid waste. Moreover, the line of development of transportation rather favored the inter-urban transportation needs of the small cities without major adverse effects on intraurban transport in small cities [6:569].

From 1968, when the research work on the first of the Hofstra studies cited above was begun, to 1973 when the second was published, the environmental movement in the United States grew from a scattered handful of nature lovers to a highly organized body capable of exerting considerable political influence, witness the 1974 election of environmentalists as governor and U.S. senator in Colorado.

There is good reason to believe that the main driving force behind the rise in importance of the environmental movement was the increase in population and output—especially the latter—which made threats to the environment much more readily perceptible than they had been in the past. If the environmental movement has grown because of the increase in population and output it should continue to grow in the future. Most of the RFF projections point sharply upward. The RFF figure for total energy demand in 1980, for example, is 75 percent higher than the 1960 figure [4:857]. Nor is it likely that *Resources in America's Future* overstated that rise; more recent projections of RFF suggest an even greater increase by 1980 in demand for energy [7:190]. Perhaps even more persuasive is the fact that actual output or consumption was higher than called for in RFF projections when checked for 1966-1967 by Hofstra researchers [5:454]. There is no reason to suppose that environmental problems are going to go away by themselves.

Landsberg points out that the demand for gasoline increased at record rates from 1967 to 1972 and that the rate of increase in demand for other petroleum products has been even greater [8:7].

Does "Energy Crisis" Mean the End of Environmental Protection?

The upward march has been by no means an uninterrupted one. In particular the so-called "energy crisis" of the winter of 1973-1974 made it clear that the days of boundless domestic supplies of energy were over. Browning's "I don't see where your eye can stop" was no longer appropriate. Your eye stopped at the taillight of the car in front of you as you waited in what seemed endless lines at service stations. It would take us too far afield to attempt to evaluate this enormously controversial episode. Most important for our purposes were the consequences in terms of opposition to environmental control measures.

The program to improve the quality of life in America through reduction of pollution presented a handy target for those corporations and interests who were under fire due to the "energy crisis." The utilities could blame the environmentalists for delaying construction of needed additions of nuclear fueled power plants—the energy source that contributed no sulfur oxides to the atmosphere. The general public was only vaguely aware of the nature of the environmentalists' objection to expanded use of fission fuels.

Or the power industry might well argue that failure to expand sufficiently was attributable to "unreasonable" demands from environmentalists for reduction of "thermal pollution," i.e., the discharge into bodies of water of the effluent from generating stations, raised by heat absorption to higher temperatures than would have obtained naturally. There was comparatively little "hard fact" available about the effects of thermal pollution, and prescriptions for its elimination tended to be advanced with diffidence and uncertainty even by experts. The most frequently advocated solution for thermal pollution from generating stations was cooling towers, but these were expensive, unsightly and beset with localized environmental drawbacks of their own (see Chapter 2).

Nor was the technology for removal of sulfur oxides from the smokestacks of power plants any less subject to criticism. Stack gas "scrubbers" designed to remove sulfur oxides from power plant emissions have been attacked as unreliable as well as expensive, for creating a solid waste disposal problem, and for being unnecessary in any event. Philip Sporn, former president of American Electric Power Company, had long maintained that tall stacks effected sufficient dispersion of stack gases to render the sulfur oxides harmless [9]. A subsequent president of AEP, Donald Cook, has restated the case in statements and in a large-scale national advertising campaign with such vehemence as to evoke widespread criticism [10]. It must be noted, however, that analysts with no hint of self-interest in the question have characterized the stack gas scrubbing technology as not commercially viable [8:6] or unproven [8:14]. However, there is abundant evidence of solid progress in overcoming the problems of sulfur oxide removal (see Chapter 3).

The automobile industry has resisted from the start almost all efforts to make it responsible for reducing the harmful emissions to the atmosphere from its product [6:445]. It came as no surprise, then, that one of the early symptoms of the "energy crisis" was pressure from the automobile industry for deferment of the automobile emission standards that had been specified by federal and state agencies to reduce air pollution from this source [8:2]. Moreover, the industry could reasonably hope for some support from the ordinary citizen in this campaign since emission control devices were known to add to the cost of the car and were suspected as well of reducing miles per gallon. These were powerful incentives for the "man in the street" to join with Detroit in a campaign to eliminate or at least defer the imposition of strict environmental controls on the automobile.

Was the environmental price of energy to be an abandonment of legislation to improve the environment and thereby the quality of life? Were legal controls to be jettisoned in favor of laissez-faire and the voluntary measures that had proven ineffective in the past? But surely this would not be the most constructive response to the "energy crisis." Surely there were many alternatives to the wholesale abandonment of measures to improve the environment.

The American people had rarely found it necessary to conserve sources of energy, or of commodities in general, for that matter. Sound conservation practices might be expected to effect a very large and significant reduction in the size of America's energy budget. For example, average miles per gallon had declined in the years prior to 1971 in part because the percentage of cars with factory-installed air conditioning had increased from a negligible figure in 1960 to 60 percent of all new cars by the end of the decade [8:10]. Moreover, typically the air conditioned automobile carries only the driver to his place of work. This single example does not, of course, begin to suggest the extensive possibilities for conservation of energy on the American scene. A general investigation of the possibilities for energy savings in the United States was not the intent of this study and would have taken us too far afield.

Energy Saving and Environmental Protection

However, it is essential to scotch the spurious notion that there is some ineluctable conflict between measures to improve the environment and providing an adequate supply of energy. As has been frequently noted, conservation—voluntary or compulsory—is probably the most plausible route to balancing the energy budget and simultaneously improving the quality of the environment. But some environmental measures may actually boost the available supply of energy. One fairly obvious path is the use of the garbage that threatens to engulf some of our cities as a source of fuel for electric power generation.

More exciting, perhaps, is the prospect of recovering some part of the heat rejected to the environment by electric power plants. When water cooling of the electric generating equipment is employed, this reject heat is the source of "thermal pollution," producing a variety of undesirable effects in the waterway to which the heated water is discharged. The electric power industry accounts for 25 percent of all energy consumed in the United States [7:Table 5] and only 40 percent of that vast amount is converted to electric power [11:S-5]; 45 percent of energy consumed in electric power plants or about 11 percent of all energy consumed by the nation is rejected to the environment with varying degrees of harmful effect [11:S-5; 7:Table 5]. What an opportunity to kill two birds with one stone! If the reject heat can be utilized, the potential savings are tremendous and thermal pollution is eliminated as well. Such a desirable outcome is not easily realized and is limited to very special circumstances. John Bandel explores, in Chapter 2, the experiments and experiences with efforts to utilize reject heat from electric generating stations.

The amount of fuel consumed by an electric power plant is so great that it is sometimes possible to effect savings by recovery of by-products from the stack gases. In Chapter 3, A.A. Ferrara examines the cases in which utilities have sought to realize returns from such by-products as the fly ash recovered, or the sulfur or sulfuric acid resulting from the capture of sulfur oxides. Although major difficulties were encountered, there were also enough successes to encourage further experimentation in this direction.

Economies of scale may make possible some recovery of by-products from desulfurization of stack gases of power plants, but such measures are entirely infeasible with respect to sulfur oxide emissions from the millions of space heating installations and small or medium scale industrial sources. The only practical way to hold down sulfur oxide emissions from such sources is to use low sulfur fuels. Chapter 4 studies the cost of desulfurizing fuel oils at the refinery and factors affecting the price of the resultant low sulfur oil.

The disposal of solid waste has been a mounting problem for American cities. Chapter 5 reviews the costs of solid waste disposal in general and the possibilities of offsetting returns by using garbage as a fuel for power or steam generation and recovery of incombustible by-products.

Chapter 6 examines, among other things, a unique approach to energy saving, notably through the redesign of major consumer appliances. Based on a study by the Center for Policy Alternatives of the Massachusetts Institute of Technology, Chapter 6 reveals that the power demands of refrigerators have increased because of increase in average size, but also because of larger frost-free freezer compartments. Without attempting to reverse these two popular trends, the study suggests that power requirements could be reduced by better insulation and heavier wiring in the motor. Both changes would raise initial purchase costs, but would result in lower costs over the life of the machine because of power savings.

Thus, Part I of the book illustrates in a variety of ways the general thesis that there need be no abandonment of environmental protection measures because of the need for energy saving. However, while energy saving may frequently be consistent with environment protection, it seems unlikely that such savings would be sufficient to balance the energy budget. That is likely to call for expanded exploitation of old sources or initiation of new sources. Part II will consider the environmental price of new or expanded sources of energy.

The Environmental Price of New Sources of Energy

The American public, accustomed for decades to free use of relatively cheap and abundant energy, is unlikely to adjust easily and willingly to sustained constraints on the magnitude of energy supplies. The possible sources of enlarged energy supply and the environmental impact of their exploitation are analyzed in Part II, consisting of Chapters 7-12.

Nuclear energy is the source perhaps most looked to for an early increase in energy output. Chapter 7 consists of a careful and thoughtful analysis of the consequences of relying heavily in the future on atomic fuels to fill the energy gap. The analysis is performed by Dr. Alvin Weinberg, formerly a director of the Oak Ridge National Laboratory and presently director of Energy Research and Development for the Federal Energy Administration. On balance, Dr. Weinberg feels that the extraordinary virtues of nuclear fuels outweigh the risk involved in their use. But this view does not deter Dr. Weinberg from analyzing with objective care the difficulties likely to be encountered at nuclear installations in the years to come.

Dr. Hannes Alfven, a Nobel laureate in physics, who is affiliated with the Royal Institute of Technology in Sweden and the University of California at San Diego, takes a less sanguine view of heavy reliance on nuclear energy. He points out that the nuclear fuels and especially the waste products of nuclear fission are among the most poisonous elements known to man. Dr. Alfven does not wish the availability of nuclear power to divert mankind from the search for other sources of energy such as fossil fuels, atomic fusion, the sun, and geothermal sources.

In Chapter 9, Ronald Axler looks at the problems and potentialities of one of the newest sources of energy for the United States—geothermal power. The first commercial exploitation on a substantial scale is only a few years old, but geothermal energy seems likely to occupy a small but significant niche in the energy structure. Though the drill sometimes opens up gases and liquids too corrosive to handle, geothermal energy is usually substantially pollution free.

This is not true with strip mining, which has wreaked havoc on much of Appalachia, where the search for strip mined coal has been conducted with little regard for the environment. Irwin Zurkowski, in Chapter 10, shows that the cost of rehabilitation of strip mined lands in Appalachia would not have added greatly to the cost of the coal. But the major threat today is in the Far West, which has the bulk of the nation's coal reserves, much of it lying under a relatively thin overburden. The difficulty is that lack of rainfall and other natural conditions in much of the Far West are such that restoration of strip mined lands there would be a slow process with no assurance of success.

Chapter 11 deals with another subject of much concern to environmentalists—the exploitation of the oil and gas deposits of the outer continental shelf. What is presented is a summary of a very thorough study of the problems and technology involved in OCS operations. The analysis was conducted by the Science and Public Policy Program of the University of Oklahoma.

Another relatively untapped source looked to with high hopes is oil shale, which is discussed by William D. Metz in Chapter 12. Oil shale operations may leave areas as heavily damaged as strip mining without rehabilitation has done. Also, oil shale operations would be handicapped by lack of rainfall in areas where outcroppings appear.

To treat this broad range of topics adequately, it was necessary to draw on contributions from "outside" experts, that is, scholars other than those participating in the Hofstra seminar. In this endeavor, we were fortunate enough to obtain contributions from a very distinguished group of experts.

In Chapter 13, the editor has sought to summarize and consider in a common context the contributions of all the authors involved.

Notes

1. From "The Flight of the Duchess," *The Poems of Robert Browning, 1832-1864* (Oxford: Oxford University Press, 1928), p. 186.
2. Selden Rodman (ed.), *A New Anthology of Modern Poetry* (New York: Random House, 1938), p. 373.
3. Edward C. Kirkland, *A History of American Economic Life*, 3rd Edition (New York: Appleton-Century-Crofts, Inc., 1951).
4. Hans H. Landsberg, Leonard L. Fischman, and Joseph L. Fisher, *Resources in America's Future* (Baltimore, Md.: Johns Hopkins Press, 1963).

5. Alfred J. Van Tassel (ed.), *Environmental Side Effects of Rising Industrial Output* (Lexington, Mass.: Lexington Books, D.C. Heath and Company, 1970), 548 pp.
6. Alfred J. Van Tassel (ed.), *Our Environment: The Outlook For 1980*, (Lexington, Mass.: Lexington Books, D.C. Heath and Company, 1973).
7. Sam H. Schurr (ed.), *Energy, Economic Growth, and the Environment* (Baltimore, Md.: Johns Hopkins Press, 1972).
8. Hans H. Landsberg, "Low-Cost Abundant Energy: Paradise Lost?", Washington, D.C., Resources for the Future, Reprint # 112, 1973, 23 pp. (Reprinted from RFF 1973 Annual Report.)
9. Philip Sporn, "Possible Impacts of Environmental Standards on Electric Power Availability and Costs," in Schurr, *Energy, Economic Growth*, pp. 69-88 (especially p. 75).
10. "Donald Cook takes on the Environmentalists," *Business Week*, October 26, 1974, pp. 67-70, 72, 77. (Criticism of Cook, pp. 68-69.)
11. "Cooling Towers," *Power*, March 1973, 22 pp.

2 Opportunities for Energy Savings in the Beneficial Use of Waste Heat

John M. Bandel, Jr.

One of the biggest opportunities for saving would appear to be in the generation of electricity where 45 percent of the input is at present rejected to the environment, sometimes harmfully.

All of society's activities release energy in the form of heat. Some is usefully employed, and some is wasted, but most of it is eventually dissipated to the environment. In the past, the amount of energy which was released in a given area did not exceed the ability of nature to carry away energy rapidly enough to prevent environmentally damaging temperature in most cases. By the year 2000, however, the energy release from all sources in the United States is projected to reach 190,000 trillion BTUs per year. The Boston-Washington megalopolis release will exceed 30 percent of the incident annual solar energy received by the area. For the nation as a whole, waste heat from electric power production will constitute about one-third of total energy demanded in the United States [18:31]. Figure 2-1 shows the expected increase in the amount of waste heat to the end of the century.

Thermal wastes from electrical power plants are predictable and for each plant are concentrated in a single place. These characteristics make it possible to deal with them. Waste heat from homes, in contrast, is not predictable as to quantity but depends upon construction, location, and weather, and is not concentrated but is spread out over a large land area. The power industry is well developed and the engineering and economic factors affecting thermal wastes are well established. Thus, it is possible to base a study of thermal pollution control upon established facts presented in the form of national parameters and individual case studies. Finally, thermal pollution control with regard to power plants is a timely topic. The rapidly rising demand for electric power makes measures for the control of thermal pollution a matter of some urgency. At the same time, the general energy shortage confronting the nation makes it desirable to conserve as much waste heat as possible rather than simply making sure that it is dissipated with a minimum of harm to the environment.

This study will concentrate on all aspects of electric power generation in the United States which are related to waste heat generation and dissipation. Special emphasis will be placed upon alternate methods of heat dissipation to the environment, uses for waste heat, and the costs or benefits involved.

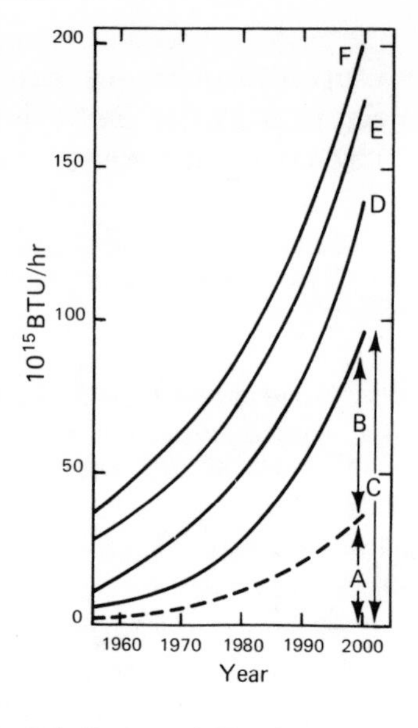

Figure 2-1. Projected Total Energy Demand for the United States, by Decades. Source: F.C. Olds, "Cooling Tower," *Power Engineering* (December) p. 31.

The first section concerns the methods of power generation and anticipated changes in the amount and methods of power generation and their effect upon the quantity of waste heat. The second section discusses the operational and environmental effects of alternate methods of dissipating waste heat to the atmosphere with special emphasis upon the costs associated with waste heat control. The third section concerns the possibilities of utilizing waste heat beneficially. An alternative heat-electric system will also be examined. The concluding section summarizes the options for the future.

Methods, Efficiencies, and Prospects of Electric Power Generation

The problem with electric power production is that the processes involved are not very efficient, which results in the release of a large quantity of heat to the environment. In industry terms, this waste heat is known as rejected heat. The total quantity of rejected heat increases as the amount of electrical power increases, and electric power production is projected to increase about two and a half times by the end of the century [8:68]. The relationship between the amount of waste heat and the amount of electric power produced is not a linear one as the efficiency of power production varies with the method employed and there is variation in the methods used. The efficiency of power production will also change in the future. The remainder of this section discusses the present and future relative importance and efficiency of alternative methods of power generation.

Methods of Base Load Power Generation

Fossil Fuel Power Generation. The most important method of power generation today is by steam-electric means utilizing fossil fuels. This method produces most of our present electric power—about 75 percent. By 1990, it is likely to account for only about 43 percent [6:9].

A steam-electric fossil system is composed of four primary elements. Oil, gas, or coal is burned in a boiler, which produces high temperature and pressure steam. This steam is expanded inside a turbine thereby transforming heat energy to mechanical energy. The turbine shaft turns a generator which converts the mechanical energy applied to electrical energy. Exhaust steam from the turbine is condensed into water in the condenser unit, and the water is (usually) pumped back into the boiler. Cooling water is circulated through the condenser, and it is this water which carries away the rejected heat.

Of the total heat energy produced by the burning fuel, about 40 percent can be converted to power, 15 percent is lost in ash and up the stack, and 45 percent is rejected to the environment [16:S-5]. Although these figures represent the result of a steady improvement year after year during this century, most of this improvement has come from higher steam temperatures and pressures and larger unit sizes. Major further improvement is not expected, however, because a major breakthrough would be necessary to produce a metal that has sufficient strength to withstand higher temperatures and pressures. Improvements in efficiency since the 1950s have been small [12:1,5,7].

Hydroelectric Power Generation. Hydroelectric power accounts for about 15 percent of present power production, but it will account for only about 6.5 percent of power production by 1990 [6:9].

Hydroelectric power is produced by causing water to flow from a higher elevation to a lower elevation through a turbine. The turbine converts the linear flow of the water through it into rotary motion which drives a generator. There is no fuel efficiency associated with this method of power generation, as no fuel is consumed, but the total amount of power available is limited by the available quantity of water and the difference between the higher elevation and the lower elevation of the water. Future increases in hydroelectric power production are limited by the number of geographical sites available, and many of these sites will probably not be developed because such development would spoil the natural beauty of many of our scenic rivers [12:I,7,21].

Nuclear Power Generation. Nuclear power accounts for a very small percentage of present power production—less than 4 percent—but it is expected to account for at least 39 percent by 1990. This figure is probably conservative, as each later estimate has increased this percentage [6:9]. It is quite possible that as much as 50 percent of our total power will be generated by this method by 1990.

There are two types of reactor systems presently in use and being built. In the pressurized water reactor system, water is pumped through the reactor core where it acts as both a moderator and a coolant at very high pressure. The heated water then passes through a heat exchanger which transfers the heat contained in the superheated water to a separate water circuit where lower pressure steam is produced. This steam drives a turbine which drives a generator, as with the fossil fuel electric system. All radioactivity is confined to the reactor and heat exchanger. The boiling water reactor is the other type system. As the name implies, water is converted directly to steam in the reactor, which then drives a turbine as in the other system. Some radioactivity is present in all parts of this system [12: I,6,9].

Of the total heat energy produced by present nuclear reactors, only about 33 percent is converted into power, 5 percent is lost within the system, and 62 percent is rejected to the environment [16:S-5], as compared to a 45 percent rejection rate for fossil fuel plants. Nuclear reactors presently in use, known as light water reactors, are only the first generation, and rapid technological advance to more sophisticated types which convert a much larger portion of their rare and expensive nuclear fuel to heat energy is imminent. Present efficiency is limited by maintenance of large safety margins which reduce steam temperature and pressure to far below optimum values. Present projections of the ultimate thermal efficiencies run to 43 percent [8:357]. Ultimately, nuclear steam-electric systems should be about as efficient as fossil steam-electric systems.

Methods of Peak Load Power Generation

Steam-electric and hydroelectric systems are generally base load systems. This means that they are designed to run as much as possible. Electric power demand varies from hour to hour and from day to day during the week. Systems which are designed to run for short periods to supply peak power demands are known as peak load power systems. Until recently, as older steam-electric equipment was replaced by newer steam-electric equipment, the older equipment was downgraded to peaking duty. As average unit size increased and efficiency leveled out, the economies became such that other methods of power generation were developed to meet peaking requirements. Ideally, these units should be low in capital cost, able to start quickly, operate satisfactorily at partial load, require little manpower, and be capable of remote operation. High efficiency is not a prime factor because of the relatively small amount of power such units generate [12:I,8,1].

Gas Turbine Power Generation. Gas turbines presently represent about 4.4 percent of total generating capacity; by 1990, they will represent about 4 percent [6:9]. There are many different types of turbines in use, but the typical installation in use by utilities is little more than an array of aircraft-type jet engines driving a turbine. Gas turbines operate by compressing air, spraying fuel into it, and igniting it. The rise in temperature expands the air which drives a second turbine. A shaft connects the compressor and turbine and, in the simple open-cycle systems, also drives the generator. In some systems, a series of gas turbine units drive an external turbine which drives a large generator. The type of turbine units favored by utilities operates at about 20 percent efficiency, but more complicated units can achieve much higher efficiencies [12:I,8,5].

Pumped Storage Power Generation. Pumped storage presently represents about 1 percent of total peak generating capacity but is projected to represent about 5.5 percent of total peak capacity by 1990 [6:9]. Pumped storage is not a method of power production, as such, but rather a method of storage. Water is pumped from a lower level to a higher level by a turbine during periods of low power demand with the generator acting as an electric motor. During peak periods the process is reversed, and the motor becomes a generator. Pumped storage is often combined with conventional hydroelectric generation to provide great versatility in meeting fluctuating power demands. The efficiency of this process is at the present time about 66 percent [12:I,7,4]. Pumped storage will increase in the future but may be limited by the number of acceptable sites and environmental considerations.

Possible Future Methods of Power Generation

Research is going on which is aimed at raising the efficiency of power production. This research could be classified into three categories with some overlap—improvement of the present steam cycle by the use of combined cycles or topping cycles, nonmechanical generators which could operate at higher temperatures, and the capture of the power of natural forces.

Methods to Improve Efficiency of Present Steam Cycle

Combined Cycle Power Generation. The most common combined cycle is that of the gas turbine and steam turbine. A gas turbine generates electricity in its normal way. The exhaust from this unit is enriched with fuel and air and blown into a steam boiler where combustion takes place, and the resulting heat produces steam which is used to produce electricity. This process raises the efficiency of either cycle alone because the exhaust of the gas turbine, which is about 900 degrees Fahrenheit, is used to help heat the boiler, thus utilizing this heat rather than rejecting it to the environment. Several small test plants have been built and operate successfully, but wide utilization of this combined cycle is unlikely as gas turbines use costly natural gas or other relatively expensive fuels such as #2 fuel oil. If coal burning gas turbines could be developed, this cycle might find wide use [12:I,5,12].

MHD Topping Cycle. A topping cycle which is being developed for use with the steam cycle is magnetohydrodynamic (MHD) power generator. Magnetohydrodynamic power generation is accomplished by passing a hot ionized gas or liquid metal through a magnetic field. The basic open-cycle system consists of a high-temperature rocket-like combustion chamber and a channel, which is surrounded by a high efficiency magnet, to transmit the hot, ionized gas. Fossil fuels are burned under high pressure with seeding material added. This produces a highly ionized gas, which is released at very high velocities through the MHD channel and magnetic field. The high velocity motion of charged particles in a magnetic field induces electricity within coils around the chamber in the same way that the movement of conductors through the fixed magnetic field of a conventional generator results in electricity. Thus, MHD is, in effect, a linear generator. MHD is combined with a steam cycle by using the exhaust gas from the MHD to produce steam for the conventional steam cycle. The combined cycle could raise overall system efficiency to 50 or 60 percent. Although there are still many problems to be solved, MHD steam generation seems very promising. The Soviet Union is constructing pilot plants, as are other European nations. Soviet engineers feel major power plants utilizing this cycle will be operating in the 1970s [12:I,7,3].

There is also a liquid metal MHD combined cycle which could be used with nuclear fueled systems when reactors are built which are capable of operating at the necessary temperatures [12:I,9,5].

Non-Mechanical Methods of Generating Electric Power

Fuel Cells. MHD by itself is an excellent example of a non-mechanical generating system. Fuel cells are another type of non-mechanical generating system. A fuel cell is similar to a battery in operation except that it is an open cycle rather than a closed one. The hydrogen-oxygen fuel cell generates electricity from the chemical process of combining these two gases to make water as a wåste product. The reaction is usually reversible (electricity is used to break down water into hydrogen and oxygen which is compressed into storage tanks) so that the fuel cell becomes a storage device as well as a generator. If present engineering difficulties are worked out, the fuel cell could find wide use as a peak load storage device [12:I,9,9-10]. Base load power generation has been installed on a very small scale, such as the home fuel cell units installed in a Connecticut suburb in 1971. Several large companies plan to develop large-scale base load units in the near future which will use natural gas as a fuel. Widespread use of large installations is questionable due to the limited supply of natural gas [6:83].

The Harnessing of Natural Forces to Produce Electric Power

Natural forces could provide all our power needs if they could be captured. As naturally generated power requires no fuel and produces no pollution, methods have long been sought to generate power from the changing level of the tides and the heat from the sun. No large amount of power has ever been generated from these sources because their energy density is too low and dispersed over too large an area. This is analogous to saying that although all sea water contains gold, no gold is extracted from sea water. Future prospects for naturally generated power are not particularly good, especially in the near future. Geothermal steam-generated power is a possible exception, which is discussed in Chapter 9 of this volume.

Projection of Methods of Power Generation into the Future

It is clear that by 1990 all significant amounts of base load power will be produced by nuclear fueled thermal, fossil fueled thermal, and hydroelectric means. Pumped storage hydroelectric and gas turbine will fill the bulk of peak load requirements. Even if an important breakthrough should occur with regard to one of the other methods of power generation, the effect would not be felt before the 1990s at the earliest. Lead times for new technology power plants are very long—the majority of new plant starts have been nuclear since the mid-1960s, but even by 1980 nuclear power will account for less than one-third of total generating facilities. Figure 2-2 summarizes the projected changes of these methods of power generation; steam-electric systems, whether using fossil or nuclear fuels, produce thermal pollution. If all projected power requirements were generated using nuclear fuels, the projected amount of heat rejected would follow the upper curve in Figure 2-3; conversely, if fossil fuels were the sole

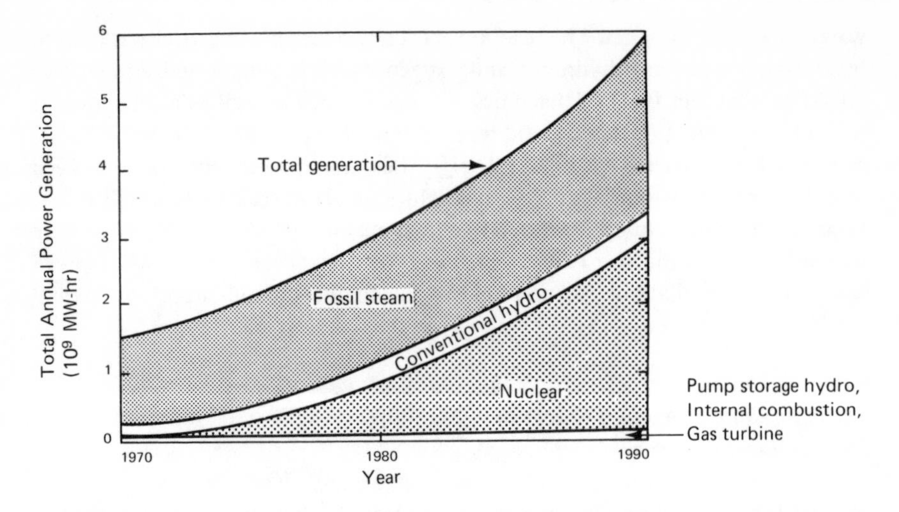

Figure 2-2. Total Annual Power Generation by Method, to the Year 1990. Source: U.S. Federal Power Commission, *The 1970 National Power Survey* (Washington, D.C.: U.S. Government Printing Office, December 1971), Vol. I, Chapter 18.

energy source, the projected amount of heat rejected would be given by the lowest curve in Figure 2-3. A projection of total future heat rejection has been made assuming a mix of fossil and nuclear fuels and this is illustrated by the intermediate curve in Figure 2-3.

Future Power Generation and Rejected Heat

The total amount of rejected heat will increase faster than demand in the short run due to the lower efficiencies associated with present nuclear power production and its increasing percentage share. As nuclear power efficiencies improve in the future the rate of increase of heat rejection will decline. Overall, during the next twenty to thirty years, the amount of heat rejected will increase roughly in proportion to the amount of power produced. The total amount of heat rejected to the environment will therefore increase rapidly.

With regard to power usage, there is one unknown, and that is the amount of conservation of electric power which may be practiced in the future. Consolidated Edison has demonstrated that a simple appeal to the public can significantly lower power usage without any serious deprivation. In May of 1971, Consolidated Edison asked its consumers to conserve power due to inadequate generating capacity. The appeal reduced load in July of 1971 by 300 megawatts from the previous year [6:55-56]. This type of experience has caused

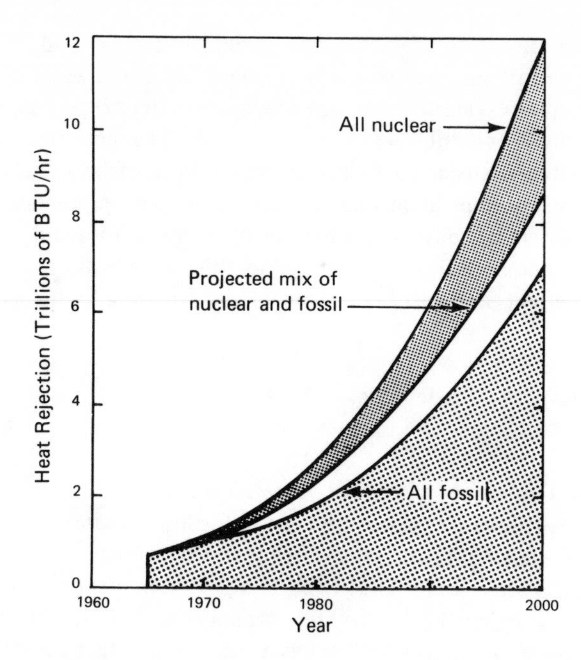

Figure 2-3. Projected Heat Rejection from Thermal Generating Plants. Source: U.S. Congress, Senate, Committee on Public Works, *Hearings Before Subcommittee on Air and Water Pollution: Thermal Pollution–1968*, 90th Congress, Second Session, 1968, p. 26.

some to advocate electric rates which would charge large users more so as to encourage conservation. The outcome of such a conservation movement is uncertain at the present time, but future price increases will almost surely encourage increased conservation on the part of the large users as well as the general public.

Operation, Cost, and Environmental Effects of Alternate Heat Disposal System

The large amount of heat produced as a necessary by-product of electric power generation must be disposed of. All rejected heat is ultimately dissipated within the environment; there is no way known to convert it into matter or send it directly out into space. When too much heat is dissipated in too small an area, the temperature of that area will increase significantly. If the local increase in air or water temperature is sufficient to have an adverse effect upon any beneficial use of that air or water, then thermal pollution can be said to have been created [9:3]. All present-day steam cycle power plants use water to carry away waste heat from the condenser unit which condenses the exhaust steam from the turbines back into water for return to the boiler. The heat from this cooling water is ultimately transferred to the air in the form of heat and/or increased water vapor content due to increased evaporation. The problem is to transfer this heat energy to the environment in such a way as to avoid local thermal pollution. There are several ways of doing this, each with its characteristic environmental effects and costs, and these will be discussed in this section.

Methods of Dissipating Waste Heat in the Environment

Once-Through Cooling Using Fresh Water. Once-through cooling is used in most present steam-electric power generation facilities where adequate supplies of water are available. As its name implies, once-through cooling consists of drawing cooling water from a local river, lake, or reservoir, pumping it through the condenser, and then back into the original source. This type of cooling is the least expensive and is usually used as a basis in equating the costs of other types of cooling.

Rivers have been used extensively as a means of waste heat disposal. One-sixth to one-fifth of the total run-off in the coterminous United States is being passed through steam condensers for cooling purposes [22:27]. By the year 2000, 100 percent of fresh water run-off will be used in most parts of the United States at low flow periods. Although cooling water is not lost for subsequent use (along the river, water may be used over and over again for cooling purposes as well as other purposes), its temperature is increased above natural levels. In areas where temperature is sufficiently increased so that any previous uses of the water are altered, thermal pollution exists.

In addition to use for cooling, river water is most often used for drinking water, production of economically valuable marine life, process water, to carry away chemical and organic waste, and for recreation. Raising river water temperature has the most marked effects upon marine life, the ability to assimilate wastes, and recreational uses. Raising water temperature affects marine organisms in two ways. Increasing water temperature increases their metabolism rate at the same time as it reduces the amount of dissolved oxygen the water can hold. The net effect is to disadvantage oxygen-using organisms, such as fish, while benefiting lower level and non-oxygen using organisms, such as plants. The result is a disruption of the natural ecological balance in the river, which causes some organisms to disappear while others increase out of control. Disruption also occurs from rapid changes in water temperature. From hour-to-hour during the day, the amount of electric power produced varies in response to demand for it. The cyclical change in power production causes a cyclical change in the amount of waste heat. The shock temperature changes which result can be very damaging to biological organisms because they may not be able to adapt to the changes [22:30]. The same effect occurs when fish and other organisms move past power plant cooling water outfalls where temperatures are suddenly increased 10 to 15 degrees. Biological impact of water temperature in this region can be minimized by proper design of the discharge structures [5:105]. A similar effect takes place with regard to waste assimilation. Raising water temperature lowers the quantity of dissolved oxygen, which reduces the ability of the river to break down wastes. One study shows that a rise in water temperature of 5 degrees centigrade from 25 degrees centigrade is equivalent to between 5,200 and 11,000 additional pounds of BOD per day [22:33]. Thus, thermal pollution tends to intensify problems caused by other types of pollution. A river which is devoid of desirable marine life and which is incapable of assimilating wastes dumped into it becomes little more than an open sewer and unusable for recreational purposes.

Once-through cooling has one other serious effect on marine life which is not related to thermal effects. Young fish and other small marine organisms are often sucked into power plant cooling water intakes thereby destroying them. Once-through cooling often withdraws most of a river's flowing water so that few slow moving organisms can escape. Various methods of diverting these organisms into bypass canals are being tried. The most successful method seems to be a traveling screen, which is a continuous, closed screened loop that moves horizontally. The screen extends from the top of the water to the bottom of the river or intake canal and is positioned in such a way that it deflects fish into a bypass or holding tank. One problem with screens is that the rate of water flow in the river or canal must be slow enough so that young fish can swim sufficiently fast horizontally upon meeting the screen to avoid impingement on it, otherwise the current may damage or kill the fish by sweeping them against the screen [5:104].

The economic impact of thermal pollution resulting from once-through cooling is impossible to judge. Even if a cost value could be established for all the fish lost, the impairment of recreational sites and other costs, the situation is so different from river to river that specific cases would have unique results. By using good engineering design with proper assessment of biological effects, once-through cooling could remain a viable mechanism for waste heat dissipation in major rivers, reservoirs, and large lakes [22:225-242]. Although thermal pollution is not a serious problem today except in isolated incidences, projected increases in power demand make it clear that this will not be true in the future in most areas.

Once-Through Cooling Using Salt Water. Once-through cooling utilizing salt water is a common practice in some parts of the country. Coastal power plants can use salt water for cooling, but the local effects of thermal pollution are much the same as those encountered with the use of fresh water. Although the ocean is a nearly infinite heat sink, the problem of elevated local temperatures remains. Power plants located at sea on man-made islands have been proposed. It is thought that thermal pollution around such a plant would be insignificant due to the depth of the water and the strong currents resulting in rapid mixing in the open sea. The use of salt water is more expensive than fresh water for once-through cooling due to the increased corrosion resistance required.

Cooling Ponds and Canals. Once-through cooling from an artificial source is the next least expensive method. Cooling ponds are artificial lakes from which cooling water is pumped and into which it is returned. Thus, cooling ponds are usually closed systems. Although cooling ponds are simple and economical to construct, they are very inefficient in dissipating heat, so that roughly one acre per million watts of generating capacity is required. Sufficient land at reasonable cost is not available in much of the country to make cooling ponds practical. Spray ponds in which water is sprayed 6 to 8 feet into the air over the pond where it is cooled by mixing it with the air with a portion of it evaporating, have become more popular due to the higher efficiency achieved. Such ponds can reduce land area requirements by a factor of 20. A cooling canal receives cooling water containing waste heat at one end and discharges it to a river or lake at the other end, with natural cooling occurring in between. Cooling canals are used in conjunction with once-through cooling to reduce the amount of waste heat the river or lake will be required to dissipate [16:S-2]. Cooling ponds and canals are simple but expensive to build, require large land areas, and in the case of spray ponds, require a higher pumping head, which makes cooling ponds a more expensive method of cooling than the once-through cooling with sea or fresh water.

Cooling Towers. Cooling towers are the most expensive method of cooling. As a group they are two to three times as expensive per million watts of generating capacity as cooling ponds, depending upon the tower type employed. Cooling towers discharge waste heat directly to the atmosphere without a body of water being used as an intermediary. A large volume of air must be passed through all cooling towers, heavier outside air enters around the base of the tower displacing the lighter, warmer air inside, forcing it out the top. Mechanical draft towers use mechanical fans to provide the air draft. Two different cooling processes are employed. The dry process transfers heat to the air from the cooling water by conduction; cooling water is not exposed to the air, so no cooling by evaporation takes place. The wet process depends upon the evaporation of the cooling water for most of the heat disposal, although some heat is also conducted to the air. Combinations of the two methods of inducing air draft and the two cooling processes produce the four basic tower types.

Types of Cooling Towers

Mechanical Draft, Wet Cooling Towers. The most common type of tower is the mechanical draft, wet cooling tower. The cooling water in this type of tower is sprayed from overhead nozzles over fill, usually redwood or treated douglas fir planks, placed so that water dripping off the edge of one plank will splatter upon the next lower plank. A wet tower is like a stall shower containing a series of boardwalks, one placed under the other. At the bottom of the tower, the remaining water is collected and returned to the steam condenser. A powerful fan sucks air between the boards of the fill so that water droplets and air come into intimate contact. Most wet towers are fitted with drift eliminators which serve to conserve water by reducing the amount which is carried away unevaporated in the form of droplets by the cooling air. There are two general types of mechanical draft, wet towers, depending on how the air flows relative to the cooling water. In both types, the draft creating fan is located on top; if the air enters on the bottom so that it travels in the opposite direction to the flow of the cooling water down the tower, the tower is of counterflow design; if the air enters on the side so that it travels across the falling cooling water to a central well where it is exhausted through the top, the tower is said to be of crossflow design. Most new towers are of crossflow design, as the characteristics of this design are such that less land area and less horsepower to drive the fan are required and less water is lost as drift and thermal performance overall is better although the counterflow design offers maximum thermal efficiency because the coolest water contacts the coolest air [5:120-121].

Natural Draft, Wet Cooling Towers. Natural draft, wet cooling towers have become very popular. Ten years ago, there was only one natural draft, wet tower in the country, but by the end of 1974 there will be 46 in service, and by 1980, at least 100 [16:S-4]. Natural draft towers consist of a tremendous hyperbolic tower made of steel reinforced concrete sitting on top of the cooling system of the forced draft wet tower. The hyperbolic tower serves to create an air draft by differences in air density in and outside of the tower and, thereby, replaces the mechanical fan. The hyperbolic shape is an aerodynamic necessity to maximize the rate of air flow up the tower. Hyperbolic structures are also strong, which reduces construction cost. Most hyperbolic towers are very large–they may be as much as 500 feet high and 400 feet in diameter [16:S-4]. Natural draft, wet towers may be of crossflow or counterflow design dependent upon the operating conditions required, although the crossflow design is usually favored.

Dry Mechanical and Natural Draft Cooling Towers. Dry mechanical and natural draft cooling towers depend purely upon conduction to dispose of waste heat. They consist of coils of tubes with fins placed so a fan or hyperbolic tower provides a draft of air across them. The automobile radiator is a dry mechanical cooling system. A variation of this scheme is air cooling the steam condenser directly rather than using a secondary cooling water system. Dry cooling has not been used in the United States for major power plants because of its low efficiency.

Hybrid Towers. Several hybrids of the basic types of towers are being designed to take advantage of the unique characteristics of each type. Fan-assisted natural wet towers and wet/dry towers are on the drawing boards. Various special cycles which use refrigeration coolants to operate at higher temperatures have been built.

Characteristics of Different Tower Types

Wet Versus Dry Towers. The unique characteristics of each tower type make one type superior, given certain circumstances of plant operation and climate. For the most part, wet towers cool through evaporation. As water evaporates, it absorbs heat. Some of the water circulating through the tower evaporates, cooling the remainder. The cooling water could be cooled, in theory, below air temperature to the wet bulb temperature, which is the temperature at which the air is saturated. At this temperature, the air cannot hold any more moisture, so evaporation stops. The temperature of the cooling water is said to "approach" this temperature. The tower design controls the degree of "approach." When the relative humidity is 100 percent, the wet and dry bulb temperatures are the same. Under this condition, a wet tower acts as if it were a dry tower, and all heat loss is by conduction. Wet towers have four important characteristics based upon these evaporative effects: (1) Wet towers are much more efficient in dissipating waste heat than dry towers. As a corollary, wet towers are capable of cooling to lower temperatures than dry towers, which improves the thermal efficiency of the steam cycle 6 to 8 percent [12:I,10,8]. If a dry tower is to dissipate the same quantity of heat as a wet tower, it must be much larger and more expensive; (2) The efficiency of wet towers depends upon the wet bulb temperature of its location. Wet towers are most efficient in areas of low humidity; (3) Wet towers consume water. Water which has evaporated must be replaced from some local source. In most areas of the country this is not a problem, but in the Southwest and similar arid areas, some utilities use sewage and industrial waste waters for replacement water, so as to conserve fresh water supplies [22:254-255]. In addition, some water is used for blowing-down. Blowing-down is a method of flushing out the cooling tower so as to remove deposited minerals and biologic growth on the fill in the cooling tower. Blow-down water is often released in a local river or lake without treatment which leads to a small amount of additional water pollution; (4) The plume of all wet towers consists of a tremendous amount of air at an elevated temperature saturated with water vapor. The result is a considerable drift loss of moisture which can cause disruptions in local meteorology. As the warm saturated air leaves the tower it comes into contact with the cooler outside air causing condensation of the water vapor into fog. Depending upon atmospheric conditions the fog eventually re-evaporates. Local fog may cause traffic accidents [22:263]. Rainfall may also be much greater than normal under the plume area [18:34].

Natural Draft Versus Mechanical Draft Towers. Natural draft towers also have different characteristics than mechanical draft towers: (1) Natural draft towers are more variable in operating performance. Optimum conditions consist of low wet bulb temperatures and high relative humidity, low wet bulb temperatures and high cooling water inlet and outlet temperatures, and heavy winter heat loads. Variance from optimum conditions reduces the chimney effect of the hyperbolic tower, thereby reducing air flow and cooling ability. Most hyperbolic towers are located in the South Atlantic and East Central sections of the United States, as these areas fulfill the climatic requirements most of the year [16:S-4]; (2) Natural draft towers are much more expensive to build than mechanical draft towers, but mechanical draft towers are more expensive to operate. The tower portion of natural draft towers accounts for one-third to one-half of the total tower cost [16:S-8], but as much as 1 percent of the total power plant output may be used by the draft fans in mechanically induced systems [12:I,10,8]; (3) Hyperbolic towers occupy a large land area and are very tall, which may destroy the pristine beauty of some areas; (4) Natural draft towers are less prone to fogging at ground level because of the higher altitude of the cooling air discharge. Mechanical draft towers often require four to five times the land area required by natural draft towers of equal capacity because they must be located away from the power plant and each other to minimize recirculation of cooling air and local fogging.

Selection of Cooling Methods to be Used and Associated Costs

Given an unlimited amount of capital, tower types are chosen on the basis of which type will have the minimum operating costs and acceptable fogging effects given the climatic and load conditions. Plants located in areas with the appropriate climatic conditions and very heavy, continuous loads favor natural draft, wet towers. Plants with variable loads, such as those supplying power mainly to individual households, favor mechanical draft, wet towers. Plants in the Northeast, where fresh water for cooling is available, favor once-through cooling with perhaps mechanical draft, wet towers to lessen the peak heat load which rivers or lakes are required to accept during periods of peak power generation or hot weather. In the West, cooling lakes are favored due to the cheapness of land, where fresh water is available in sufficient quantity. The development of dry towers is greatest in arid areas of the West where even make-up and blow-down water is difficult to come by.

When capital costs are considered, the most ecologically desirable tower, the dry tower, is also the most expensive. Utilities must raise capital for new equipment. The cost of the additional capital needed for cooling towers must be passed on to consumers as well as the operating costs. Thermal pollution control will raise the costs of electric power to the consumers considerably. The cost increase for any one unit is dependent upon local conditions, but an average projection based on 1971 costs projected to 1975 operation and assuming 32 percent thermal efficiency (the present average) is presented in Table 2-1.

Table 2-1 shows that the cost of producing electric power at the plant could be increased by as much as 19.2 percent by using natural draft dry cooling towers with investor financing. This is not the price the consumer pays but the cost of production at the plant.

Table 2-2 shows the estimated percentage cost increase the consumer would be required to pay at retail prices. These figures are abstract. The TVA system spent one million dollars per week on thermal pollution abatement in 1972 and expects to spend double that per week this year. At maximum projected expenditures the average family buying electricity from TVA could experience an annual cost increase of $360 [18:33].

The costs of waste heat disposal are not insignificant and must be weighed carefully against benefits in each case. Large-scale cooling towers are relatively new and may have hidden costs in terms of conservation. Dr. James H. Wright of Westinghouse Corporation made several cost studies of hypothetical tower applications. The results of one study concluded that if every new power plant in the state of Florida from now to the year 2000 used towers rather than once-through cooling with sea water, the reduced steam cycle efficiency would result in the additional consumption of two billion barrels of oil by the end of the century [18:33]. Single-mindedly building cooling towers at every new power plant is not likely to yield the best answer in many, probably most, instances.

This conclusion is strengthened by reference to the results of a Federal Power Commission study which sought to put the projected cost of cooling facilities for power plants in some reasonable perspective. Analyses were made on a plant-to-plant basis of the costs of possible alternative cooling systems for steam-electric plants and the results projected to 1990. Only plants over 500 megawatts were selected, as plants of this size constitute 75 percent of the total 1970 steam-electric capacity; by 1990, they are projected to constitute 97 percent of the total steam-electric capacity.

Three separate studies were made utilizing different assumptions as to the

Table 2-1
Typical Plant Cooling Systems Costs for 1000 MW Nuclear Fueled Plant With a 32% Steam Cycle Efficiency

	1971 Costs Projected to 1975 Operation							
			Typical Power Costs mil/kwh Public Financing			Typical Power Costs mil/kwh Investor Financing		
System Type	Typical Cost*	Cost Above Base Case*	Capital Charges	Operating Cost	Net Cost	Capital Charges	Operating Cost	Net Cost
Once-Through: Fresh Water								
Canal-Discharge	7.78	–	–	–	5.33	–	–	–
Deep Discharge	8.39	.61	.007	–	5.34	0.013	–	5.35
Once-Through: Sea Water								
Canal Intake & Discharge	8.53	.75	0.008	–	5.34	0.016	–	5.37
Deep Intake–Canal Discharge	11.30	3.52	0.036	–	5.37	0.073	–	5.42
Deep Intake–Deep Discharge	15.30	7.52	0.077	–	5.41	0.158	–	5.51
Evaporative Systems:								
Cooling Pond	14.50	6.72	0.071	0.040	5.44	0.154	0.040	5.54
Natural Draft Towers	22.30	14.52	0.150	0.187	5.67	0.305	0.187	5.840
Induced Draft Towers	15.80	8.02	0.084	0.250	5.68	0.172	0.250	5.77
Dry Cooling Systems:								
Natural Draft Towers	50.20	42.42	0.440	0.349	6.12	0.900	0.349	6.60
Induced Draft Towers	35.20	27.42	0.282	0.502	6.11	0.577	0.502	6.42

*All costs in millions of dollars.

Source: F.C. Olds, "Cooling Towers," *Power Engineering*, (December 1972), p. 32, who credits Battelle Pacific Northwest Laboratory.

Table 2-2
Percentage Cost Increase Because of Thermal Pollution Control for the Average U.S. Consumer

Type of Cooling System	Class of Consumer		
	Industrial	Commercial	Residential
Once-through–fresh water	No added cost; the base situation. All costs listed below are increases over this situation		
Once-through–sea water	0.34%	0.16%	0.14%
Cooling pond	0.94%	0.43%	0.39%
Wet mechanical draft tower	3.17%	1.41%	1.28%
Wet natural draft cooling tower	1.48%	0.68%	0.63%
Dry cooling tower	Very uncertain, may be in the range of 1 1/2% to over 3% for residential consumers		

Source: U.S. Congress, Joint Committee on Atomic Energy, *Selected Materials on Environmental Effects of Producing Electric Power* (Washington, D.C.: U.S. Government Printing Office, 1969), p. 280.

extent of thermal pollution control system usage. Study "A" assumes compliance with thermal criteria comparable to those suggested by the National Technical Advisory Committee. Each site was evaluated in such a way that the least expensive type of cooling system was used (including once-through) consistent with maintaining the following standards for maximum temperature rise with the condenser discharge water fully mixed with the waters of the receiving body:

Streams with cold water fisheries	– 2 degrees F.
Streams with warm water fisheries	– 5 degrees F.
Estuaries–summer months	– 1.5 degrees F.
Estuaries–winter months	– 4 degrees F.

Study "B" assumes that all new plants constructed after 1970 will require some type of pollution abatement system except where ocean cooling is possible. Study "C" assumes that all plants operating after 1980 will require some type of pollution abatement system except where ocean cooling is possible. Even in Study "C", however, the use of dry cooling towers was not considered as an option because of their present high cost and limited applicability. However, costs for study "C" were higher because it was assumed that plants "operating in 1980" and not simply those "constructed after 1970" would require pollution abatement equipment.

Table 2-3 shows the results of these studies and puts them in perspective relative to the total cost of electric power [12:I-10-15 to I-10-20].

Table 2-3

Projected Total and Incremental Costs of Thermal Pollution Abatement for Plants of 500 Megawatts or More in 1970 Dollars

Year	Annual Costs of Cooling Facilities in Mils Per Kilowatt-Hour[a]				Percentage Cost Increment Over Existing Cooling System Costs			Average Revenue Per Kilowatt-Hour Sold– Total Electric Industry in Mils Per Kilowatt-Hour[b]	Percentage Cost Increase Over 1970 Average Revenue			Total U.S. Generation Steam-Electric Systems in 10° Millionwatt Hours[c]	Annual Cost of Pollution Control Including Existing Cooling System Costs in Billions of Dollars			
	Existing Cooling System Costs	Study			Study				Study				Existing Cooling System Costs	Study		
		A	B	C	A	B	C		A	B	C			A	B	C
1970	.15							15.9				1,263	.19			
1980*		.22	.28	.39	47%	87%	160%		.44%	.81%	1.5%	2,769		1.02	1.19	1.49
1990*		.24	.33	.37	60%	120%	147%		.56%	1.1%	1.3%	5,492		2.14	2.64	2.85

*Projected Figures.

Sources:

[a]U.S. Federal Power Commission, *The 1970 National Power Survey* (Washington, D.C.: U.S. Government Printing Office, December 1971), Vol. I, Chapter 10, p. 20.

[b]Edison Electric Institute, *Statistical Yearbook of the Electric Utility Industry for 1971*, Report of the Statistical Committee of the Edison Electric Institute (New York: Edison Electric Institute, 1972), Number 39, Publication No. 72-25, p. 53.

[c]U.S. Federal Power Commission, *1970 National Power Survey*, Vol. I, Chapter 18, p. 29.

On the basis of these results, two conclusions are evident. The cost of thermal pollution control is not very great as a percentage of total electric power costs, but it is large in gross figures. There is a significant difference between the cost of reasonable and prudent control represented by Study "A" and rigid and dogmatic control represented by Study "C."

Potential Benefits and Environmental Effects of Schemes for Capitalizing Waste Heat

The disposal of waste heat in cooling towers involves great cost and may have undesirable side effects. Heat is needed and hence valuable for many uses. It is indispensable in many industrial processes, for heating buildings and homes, and through the refrigeration process can be used as an energy source for cooling. If waste heat from power plants could be used for such purposes, it could be capitalized; that is, it could be converted from waste heat to a resource.

Waste heat would still be dissipated within the environment, but it would displace heat generated expressly for these purposes. Fuel resources could be conserved, and air pollution could be reduced as well as thermal pollution of the overall environment. Savings resulting from such cheap heat might make entire new industries possible. The basic obstacle to its use is that the temperature of the cooling water is not very high, usually less than 100 degrees Fahrenheit, and accordingly the waste energy occupies a large volume of water. This precludes many possible uses for waste heat. Another difficulty is that the amount of heat available depends upon the amount of power being produced. In hot weather electric loads are high to provide power for electromechanical air conditioning, but waste heat is least useful in hot weather. The opposite is true for winter nights—power demand is low, but the demand for waste heat is at its maximum. A related problem is the reliability of waste heat production. Most uses of waste heat require a continuous, uninterrupted supply. Power plant outages for maintenance or due to mechanical or electrical failure could result in large economic losses to waste heat users.

Waste Heat Versus Low Temperature Heat

Consideration of the beneficial uses of waste heat requires a careful definition of the terms to be used. "Waste heat" is the heat associated with the difference in temperature between the cooling water entering the condenser and the water leaving the condenser multiplied by the volume of water flowing through the condenser *without* any alteration in the thermal cycle which would reduce its overall efficiency for the production of electric power. The relevant engineering principles dictate that the discharge water from the condenser will in most instances be not greater than 100 degrees Fahrenheit under such conditions.

Higher temperature steam can also be extracted from various parts of the power generation cycle, and this kind of heat resource is called "low temperature heat." Higher temperature steam is removed from the power generation cycle before all of its energy potential for the generation of electric power has been realized. As some of the heat energy is removed from the power generation cycle, the efficiency of the thermal cycle for the production of electric power *will* be degraded but the use of heat energy for other purposes in combination with the generation of electricity can result in higher overall efficiency of heat utilization than would have been achieved using all of the heat energy to generate electricity. Low temperature heat can be considered a method of thermal pollution reduction to the extent that the total system of power generation plus other uses will reject less heat to the environment than would the conventional cycle for the production of electricity alone. The important consideration in the use of low temperature heat is the total efficiency of heat utilization (that used for other purposes as well as that used to generate electricity), which will be called the "system efficiency," to differentiate it from the "cycle efficiency," which refers only to the conversion of heat energy into electrical energy without credit being given for steam uses outside the power plant.

Many schemes have been proposed to utilize thermal waste, but only a few have been practical enough to bear investigation. Of these, most are still in the early stages of development. The schemes investigated here were selected on the basis of meeting three criteria: (1) the scheme has reached a stage of development where data have been developed, which makes realistic feasibility studies possible, (2) the nationwide utilization of the scheme would use a significant amount of waste heat, and (3) the opportunity exists for nationwide use of the scheme in the United States as opposed to the opportunities which might exist in foreign countries.

Schemes for Utilizing Waste Heat. The main obstacle to the utilization of waste heat is the large volume of water at less than 100 degrees in which the heat is contained. Temperatures this low make it unsuitable for purposes other than promoting the growth of biologic organisms [13:4]. Waste heat can find use in agriculture and aquaculture by providing optimal temperature and humidity conditions for maximizing the growth rate of plants and animals. Agricultural uses of waste heat include open field use (extensive farming), greenhouse use (intensive farming), and raising animals.

Open Field Agriculture. Open field utilization of waste heat can provide several benefits: damage from temperature extremes can be prevented, the growing season can be extended, growth can be promoted, crop quality can be improved, and some diseases can be controlled [13:4]. Of these beneficial uses of waste heat, the economics of increased crop production has been studied by L. Boersma of Oregon State University. The system he describes contains many elements. Discharged cooling water is taken from a local power generation station and circulated through pipes laid 36 inches below the surface of the ground and laid in such a way that all areas of the fields are heated evenly by the water flowing through the pipes. The cooling water is ultimately discharged into an evaporative cooling basin where the remainder of the waste heat is dissipated. Cool water is pumped back to the power generating station. The system is thus closed with only that portion of the water which is lost to evaporative cooling to be made up. To gain experimental data, corn, tomatoes, soy beans, and bush beans were grown in plots heated with electrical heating cables which simulated the effects of waste heated water, while the same crops were also grown without soil warming as a control during the 1969 season in Corvallis, Oregon. The results are presented in Table 2-4.

A thorough economic analysis of all costs was made including fixed costs, operating costs, and interest on capital for a subsoil heating system of 5,000 acres with a cooling pond of 500 acres. All of the waste heat produced by a 1,000 megawatt power plant operating at 34 percent cycle efficiency would be dissipated by the system with the exact heat requirements determined by a complicated modeling study. The total capital investment would be $24 million not including working capital. Combining these factors, if the 5,000 acres were

Table 2-4
Crop Production at the Corvallis, Oregon, Agricultural Experiment During 1969

Crop	Temperature Treatment		Yield Increase
	Natural	Heated	
	T/A	T/A	%
Corn (silage)	5.5	8.0	45
Corn (grain)	3.2	4.3	34
Tomatoes	32.1	48.3	50
Soybeans (silage)	2.25	3.74	66
Bush beans (first planting)	6.44	7.80	21
Bush beans (second planting)	(3.30)	5.70	73
Bush beans (total)	(9.74)	13.50	39

T/A = Tons per acre

( ) Indicates crop did not reach marketable size before growth was terminated by the approach of winter.

Source: S.P. Mathur and R. Steward (eds.), *Proceedings of the Conference on the Beneficial Uses of Thermal Discharges*, New York State Department of Environmental Conservation, September 17-18, 1970 (Albany, New York: Thruway Hyatt House, 1970), p. 94.

planted in bush beans, the break even point for one crop per year would be 7.9 tons per acre; for two crops per year, it would be 9.9 tons per acre. For the test sample, one crop per year would have lost $34,500; two crops per year would have made $1,815,500. If corn (for silage) had been planted, the operation would have lost money in any event. This study does not take account of any credit for the cost of cooling towers as an alternate method of dissipating the waste heat [23:74-107].

Frost Prevention. The use of waste heat to protect high value fruit and nut trees from frost damage was investigated by the Thermal Water Horticultural Demonstration Project at Springfield, Oregon. Cooling water heated by waste heat was obtained from a local pulp mill ranging at temperatures from 100 degrees Fahrenheit to 140 degrees Fahrenheit. The water was piped over two miles to the project which consisted of 170 acres of land containing a variety of fruit and nut trees plus some field crops and berry bushes. It was sprayed into the air in the groves by spray nozzles located 10 to 15 feet above the ground. The sprayed water temperature varied from 90 degrees Fahrenheit to 130 degrees Fahrenheit. One objective of this project was to evaluate the effect of waste heat in preventing frost damage to buds and trees in winter and "sunburning" in summer. On cold winter nights, the warm water sprayed on the trees freezes, but as it freezes, it liberates its heat of fusion (the heat required for water to change state from a liquid at 32 degrees Fahrenheit to ice at 32 degrees Fahrenheit), which is sufficient to maintain buds and trees at the freezing point. In summer, the rapid evaporation of the warm spray water from the fruit trees cools them by a similar process as well as raising humidity. The demonstration project was a success insofar as no frost damage in winter or wilting in summer was observed on the protected trees, while in adjacent, unprotected groves, up to 50 percent losses were sustained during the period. The only problems were that ice formation became so heavy on some of the trees in winter that limbs were broken off. Peach trees reached fruit-bearing maturity two years earlier than normal, requiring special pruning to prevent limb and trunk damage. The fruit yield from the sprayed trees was thought to be greater and the quality much better [24:166-182; 23:62-66].

Economic data adequate for evaluation do not exist for this project as that was not the purpose of this first phase. Any economic conclusion would necessitate consideration of certain factors. In this case, the local pulp mill had been pumping its waste cooling water into the river on which the project was sited. The mill was looking for a way to dispose of its waste heat without dissipating all of it in the river. One could assume that the main line pipe and pumps necessary to get the water to the project would have to be constructed by the plant in any event and that these are "free" from the project standpoint. Some of the distribution lines and pumps within the project would have been needed for irrigation if the project were on a larger scale and, therefore, represent a moderate additional cost. However, the value of fruit and nut crops is sufficiently high so that 50 percent losses should permit such a project to at least break even on a larger scale.

Greenhouses. Waste heat can be used in greenhouses to maximize production and improve on the quality of vegetables. The use of heated greenhouses to produce premium quality (and premium price) vegetables all year round is widespread as commercial greenhouses can be found in most states, but waste heat had not been used in large installations to displace fossil fuel in greenhouses as late as 1972 [13:12-15]. Several test projects have been started to evaluate the economic value of waste heat where optimal temperatures and humidities are maintained rather than using fossil fuel to maintain "safe" temperatures as is now done. The oldest and most complete project of this type to date is the one jointly operated by the Universities of Arizona and Sonora at Puerto Peñasco, a small Mexican village on the Gulf of California. In 1963, the universities set up a pilot plant to desalt water for agricultural use. The project proved impractical and experiments with greenhouses to reduce the loss of moisture were begun in 1966. A closed system eventually evolved which utilizes waste heat to desalt sea water that is used to irrigate plants in a greenhouse. The greenhouse is heated and cooled by utilizing salt water pumped from a salt water well close to the ocean. The water in the well is maintained at 80 degrees Fahrenheit year round by close contact with the sea. The salt water is sprayed through a direct contact heat exchanger at one end of the greenhouse which tends to maintain the greenhouse at 80 degrees Fahrenheit year round. A small amount of fresh water evaporates from the salt water, maintaining the humidity at nearly 100 percent. A fan circulates the air through the greenhouse areas in a closed loop. The plants grow in sand and are fed liquid nutrients which are distributed by an extensive irrigation system. The increases in yields are impressive, as shown in Table 2-5.

As an additional benefit, up to 1,500 gallons of fresh water a day condensed on the greenhouse wall in winter. If this fresh water were stored, no additional input of fresh water would be required [23:108-116; 7:31-40; 13:9-18; 24:21-38].

The results of this project have been applied to a study of the feasibility of utilizing all of the waste heat from the Fort St. Vrain power plant near Denver, Colorado. The St. Vrain power plant, a 330 megawatt generating unit of 40 percent cycle efficiency, could supply sufficient waste heat at 102 degrees Fahrenheit for 200 acres of greenhouses (500 acres would be sufficient to fulfill all the vegetable needs of the city of Denver). Greenhouses in the area cost $175,000 an acre for one or two acres. The capital cost of conventional greenhouses is about $20,000 an acre, and fuel bills run $4,000 to $8,000 an acre. The cost of all necessary hardware for utilizing waste heat would be

Table 2-5
Vegetable Crops Grown in Greenhouses in Puerto Peñasco, Sonora, Mexico (1969-1970)

Puerto Peñasco greenhouses		Comparative data for U.S. (acres)		
Kind of vegetable	Marketable yield/ac.	Approx. av. yield from greenhouses	Approx. av. yield outdoors/yr.	Good yield outdoors/yr.
Cucumber (European type)***				
Fall crop	6,600 bu @ 48 lb/bu	–	185 bu*	500 bu**
Spring crop	7,290 bu***			
Eggplant				
Fall crop	4,000 bu***	–	433 bu*	500 bu**
Spring crop	4,000 bu*** @ 33 lb			
Lettuce				
Bibb & Leaf				
Winter crop	3,500 ctn @ 2 doz.	3,500 ctn	–	–
Okra				
Winter crop	40 ton	–	–	5 ton**
Peppers				
Bell				
Winter crop	1,200 bu @ 25 lb	–	372 bu*	500 bu**
Radish				
Winter crop	40,000 bnch @ 12 lb	40,000 bnch	–	20,000 bnch**
Squash				
Zucchini				
Spring crop	2,000 bu*** @ 45 lb	–	–	400 bu**
Tomato				
Fall crop	75 ton***	40 ton***	6.8 ton*	30 ton
Spring crop	65 ton	60 ton***	–	–

*United States Department of Agriculture, *Agricultural Statistics, 1969.*

**James E. Knot, *Handbook for Vegetable Growers* (New York: John Wiley & Sons, Inc., 1962).

***Based on a harvest period of 90 days.

Source: M.M. Yarosh (ed.), *Waste Heat Utilization Proceedings of the National Conference, October 27-29, 1971, Gatlinburg, Tennessee*, Electric Power Council on the Environment and Oak Ridge National Laboratory (Springfield, Virginia: U.S. Department of Commerce, May 1972), CONF-711031, p. 35.

$28,000 an acre (assuming no credit for cooling towers not constructed—with towers considered, the cost would be reduced $10,000 an acre). The additional capital cost of using waste heat could be amortized the first year of operation, and increased profits of $4,000-$8,000 an acre could be realized thereafter [23: 191, 195].

This analysis is based on a mixture of nine crops in demand in the Denver area. The gross value of these crops is $22,146 an acre using productivity data from the Puerto Peñasco project [13:17]. These figures point up two sources of miscalculations which can be made if the data are extrapolated to other areas. The figures apply for locations with very high solar input due to the Southern location of Puerto Peñasco and the near total absence of cloud or air pollution cover there. Productivity may be much lower further north and east. Productivity increases in greenhouses vary greatly with the crop grown—the best crop is tomatoes, which could produce gross returns as high as $33,000 per acre; lettuce and cucumbers are also attractive, but other crops which are in demand locally might represent a financial disaster if greenhouse grown [13:16]. Greenhouse pilot and experimental projects are being started in different parts of the country to evaluate various local situations.

Animal Raising. The efficiency of feed conversion into flesh in farm animals is also sensitive to temperature. Waste heat could be used to heat and cool animal shelters and achieve results analogous to those in greenhouses. Assuming that the two main food animals, chickens and hogs, which are grown indoors, are maintained at optimum temperatures, $8 million could be saved in fuel costs through use of waste heat and such saving might be valued at $9.5 million insofar as the use of waste heat permitted the practical maintenance of higher temperatures at which feed efficiency is higher [13:18-24]. This potential use for waste heat is under research at the present time.

Aquaculture. Aquaculture is the raising of fish and bears the same relationship to fishing as agriculture does to gathering. Although aquaculture is an ancient art in the Orient, it has been practiced only a few years in the United States. Of all of the 2,500 known fish species, less than 1 percent have been successfully cultured and less than .5 percent intensively cultured [24:43]. Fish and shellfish can be cultured in ponds, channels or rivers, or in the ocean. They may be free swimming or confined in pens or small cages. As fish and shellfish are cold blooded, their metabolism rate is determined by the temperature of their environment to a much greater extent than warm blooded animals like chickens and hogs. Raising the temperature of catfish from 20 degrees Centigrade to 32 degrees Centigrade increases feed efficiency from 25 percent to 65 percent [23: 191]. Artificial heating of the tremendous amounts of water necessary for intensive aquaculture is impractical except by the use of waste heat or geothermal heat in the form of warm springs or warmed ground water. Only a couple of types of fish are being commercially grown, and a few types of shellfish are being experimented with in the United States at this time.

John E. Tilton and James F. Kelley have been raising channel catfish in cages in the discharge canal of the Texas Electric Service Company at Lake Colorado City, Texas. Mid-winter temperatures in the discharge channel averaged 67 degrees to 74 degrees Fahrenheit, which was sufficient to result in a 45.6 percent increase in weight compared with fish not kept in warmed water which failed to show any weight gain over the same period. Based on these data, a 1,000 acre reservoir located in Texas, which is part of a closed cooling system with a power plant, could produce two crops of two million pounds each worth $1.6 million at a market price of $.40 a pound. Capital costs would be $894,850, and profit for the first year's operation would be $705,150 less interest and taxes. Even if the price dropped to $.30 a pound, a profit of $305,150 less interest and taxes would be realized [3:148-153].

A feasibility study of sophisticated channel culture for shrimp has been made. The study envisions a series of channels in a closed cooling system with a thermal nuclear power plant. Shrimp would be confined in pens along the channel based upon maturity so that the number of pounds of live shrimp would remain constant relative to the water volume of that pen. Shrimp would be harvested weekly and processed on the site. Ten tons an acre per year would be produced, or ten million pounds a year in total. The wholesale price of shrimp in 1970 was $1 a pound, while total production costs would have been $.80 a pound [24:49-50].

Aquaculture may not be able to utilize large amounts of waste heat because of several problems. Fish and shellfish are much more sensitive to sudden changes in temperature than are plants. A power plant shutdown would result in a rapid drop in temperature which could not as a practical matter be ameliorated by standby boilers, and such a temperature drop in mid-winter could kill most of the fish or shellfish under cultivation. Variations in electrical load could also present temperature change problems. Trace amounts of chemicals such as copper, or radioactive elements that may enter the cooling water from pipes or the condenser tubes, could slowly poison the fish or shellfish or make them toxic to consumers. For most locations and most fish types, much of the waste heat would have to be dissipated by alternate means in the summer because aquacultural use of waste heat would not be consistent with summer cooling by evaporation. Conversely, during very cold winters, less than optimum growth may be obtained due to an inadequate supply of waste heat. With open cooling systems, a large amount of water would be returned to the river with a large BOD load from the organic fish wastes and with depleted oxygen. Open systems do not help dissipate waste heat either although the heat may still be considered a resource [24:60].

Prospects for the Use of Waste Heat. The projects which have been presented have an excellent probability of breaking even economically, especially since a credit for the alternative cooling tower costs has not been taken. These projects are considered very risky, however, because they have not been tried on a large enough scale to validate the economic expectations. The largest problem is the market price equilibrium in the case of agriculture and market acceptance of the products in the case of aquaculture. Intensive agriculture requires much greater capital investment than does extensive agriculture and would be more sensitive to extremely low prices caused by over supply in the market place. In the event of overproduction of one crop, extensive producers may choose to plant a cover crop to renew the soil or shift to a different crop, but greenhouses may be limited to a narrow range of crops—those which are sufficiently temperature sensitive. Greenhouses may eventually be limited to the mass production of a few economically attractive crops and limited production of others for off season, premium quality, or local markets.

With regard to aquaculture, the supply of fish available from fishing is simply running out. It is estimated that the maximum sustainable limit has been reached for fin fish and will be reached by 1980 for tuna and crabs. By the turn of the century, it will have been reached for all types of seafood we now consume [24: 243-244]. The question is whether intensive fish farming can produce economically desirable fish or if not, whether public tastes will shift to fish like catfish, which can be produced economically on an intensive basis. Extensive cultivation of fish appears impractical, as not enough fish could be produced and harvested. Gourmet seafood like shrimp have shown relative price inelasticity compared to beef and pork and are a good bet for intensive farming [13:37].

In light of this discussion, it is impossible to predict how much waste heat may be used beneficially. It has been estimated that of the total amount of waste heat resulting from power generation in 1970, 1 to 5 percent could be utilized by greenhouses, 2 percent for animal enclosures, and 14 percent for thermal aquaculture. Unfortunately, all these activities are growing much less rapidly than the demand for power, so that by the turn of the century, all these uses are estimated to constitute less than 5 percent of the total waste heat available. In the case of aquaculture, this estimate is based on an assumed 10 percent market share for cultured fish relative to alternative foods [13:17, 24, 38]. If fish farming does succeed in producing large amounts of desirable fish at low market prices, then consumption patterns may change drastically, as happened after the Second World War when intensive chicken farming resulted in lower prices and consumer preferences shifted. It is also true that the natural supply of fish will be rapidly diminishing by this time. In light of these factors, it is likely that the demand for cultured fish may increase rapidly so that aquaculture would utilize as much as 10 percent of the total amount of waste heat available by the year 2000.

Schemes for Utilizing Low Temperature Heat. The other approach to reducing thermal pollution is the generation of low temperature heat combined with electric power. Low temperature heat is generated by altering the standard closed cycle steam turbine. A steam turbine contains a number of stages. In each stage, steam is expanded which reduces its pressure and temperature from inlet to outlet. The resulting flow of steam from the higher pressure inlet to the lower pressure outlet through the turbine blades converts the heat energy, represented by the flow of steam, into mechanical energy using the same principle as the windmill. There are several stages to a steam turbine with distinct pressure-temperature drops associated with each. The final stage exhausts into a vacuum caused by the rapid condensation of the remaining steam into a much smaller volume of water in the condenser. Steam can be tapped from various stages to achieve different temperatures and pressures. Low temperature heat generated in this manner is said to be "extracted" from the turbine, and the specially designed turbine is called an extraction turbine. Low temperature heat can also be generated by using fewer total stages and/or not condensing the remaining steam after the last stage. Since there will not be a vacuum created for the last stage outlet, the "back pressure" has been increased. The "back pressure" turbine rejects hot water steam rather than waste heat. With either type of turbine, some of the energy contained in the steam has been diverted to other uses so that the cycle efficiency for the production of electric power has been reduced. For instance, with a conventional turbine of 40 percent efficiency, 40 percent of the steam input produces electricity while the remainder is waste heat. With an extraction type turbine, 35 percent of the steam might produce electricity, 35 percent be extracted at 250 degrees Fahrenheit, and the remainder be waste heat. With the back pressure type, 30 percent of the steam would produce electricity and 70 percent be rejected at 250 degrees Fahrenheit with no waste heat. Thus, the extraction and back pressure type turbines produce a low temperature heat at the expense of some loss of output of electricity. If such low temperature heat can be used in such a way that its economic value exceeds that of the lost electricity, the economic efficiency of the system as a whole is increased [24:68-69]. Herein lies the potential value of use of low temperature for reducing the amount of waste heat. However, it must be noted that system efficiency is dependent upon maintaining the particular balance between steam and mechanical energy for which the system was designed, and any imbalance may greatly reduce both cycle and system efficiency.

Process Heat. The two principal uses for low temperature heat have been employed for many years on a very limited basis. Process steam is used in industry to supply heat for production operations. It can be supplied by a power plant by use of either an extractor or back pressure type turbine. Industry and electric utilities have often cooperated to mutual advantage in this area. The new Midland, Michigan, Nuclear Plant will supply process steam to a Dow Chemical Company Plant across the Tittabawassee river. In addition to generating 1,300 megawatts of electricity, the plant will supply 4,050,000 pounds per hour of process steam to Dow. Two turbines are to be used—one supplying 800 megawatts of electric power, the other producing 500 megawatts of power and the extracted steam. The system is said to reduce costs and thermal pollution relative to Dow's producing its own steam in a fossil fuel boiler [24:77-82].

A study of the potential for the utilization of process steam taken from power plants reveals that the six industries with the largest demand (chemicals, petroleum, paper, food, textiles, and rubber) will use heat roughly equal in amount to the total heat equivalent of electrical energy required by 1980 [7:68]. To the degree that process steam from heat-electric systems is used, thermal pollution may be abated. Predictions on potential use nationwide are impossible as careful matching of suppliers' equipment and users' requirements are necessary for each case. Provision must be made for outages and the operating cycle of the electric plant, and process steam users must interface so that the correct amount and temperature of steam can be supplied without further reducing the cycle efficiency for the production of electricity. The user and supplier of steam must be in close proximity as transportation of steam over long distances is expensive. Many electrical generation stations are located in urban or suburban areas where the heavy industry represented by the major users of process steam would be unwelcome. These factors make it likely that increased process steam use supplied by heat-electric systems will be most practical where new generating stations are located next to a steam customer's plant and are specifically designed to supply his requirements.

Space Heating and Cooling. The other major use for low temperature heat is for space heating and cooling. This is a larger category which would include process steam plus several other uses which would be made possible by the distribution costs of a large system being shared by many users. Heat-electric systems which fill most of the energy needs of the communities in which they are located have been described as "energy utilization systems" or EUS [24:67]. EUS could supply all the electricity and heat energy required for space heating (and cooling), domestic hot water, water distillation, process steam, snow and ice melting; EUS would also make waste heat uses less dependent upon close location to the central power facility for urban centers. This would be accomplished by distributing steam throughout the city to substations where heat exchange systems and controls would distribute hot water, steam, or cooled water to local homes, offices, and businesses, depending upon their requirements.

There are two precedents for the mass use of space heating with steam or hot water supplied from a central source. District heating was first used in

Reykjavik, Iceland, in 1928. Hot water was piped from wells where it was heated by geothermal processes. Today, more than 40 percent of the population of Iceland lives in houses heated by geothermal energy. The total cost of space heating with hot water averages only about 60 percent as high as similar heating with oil, and the large amount of inexpensive heat has spawned several new industries as well [3:418]. Iceland's experience varies from the situation in the United States in important ways. In Iceland, the heat is free with little fuel being required, and oil is more expensive, as it is imported. The supply of steam is independent of electricity production, so there is very little problem with matching supply and demand. Iceland has a colder climate than most of the United States—the mean temperature is just 11 degrees Centigrade in summer, which makes the heating season last all year long, and no cooling is required in summer [23:139-184].

Closer to home, many utilities located in large, urban areas have supplied district heating for many years. There are presently forty-four city systems, with the New York City system as large as the next eleven systems put together. The systems average about one-third of maximum capacity over the year, and only the New York City system supplies any appreciable amount of air conditioning. Consolidated Edison's steam system load curves reveal that the minimum usage occurs in fall and spring, amounting to about 27 percent of the year's peak demand [7:24-28].

District heating will make the most useful contribution to reduction of thermal pollution by providing an outlet for waste heat in those locations where the usage of heat energy is at a nearly constant level throughout the year. This condition is most likely to be realized where the climate requires both heating and cooling of homes and buildings. Thus, for example, the average three bedroom home in New York City or St. Louis might require a peak of 50,000 BTUs per hour in winter and 24,000 BTUs per hour in summer. The lithium bromide water refrigeration system which uses low temperature heat as a power source is about 50 percent efficient so the demand for heat is nearly constant summer and winter [17:13]. In spring and fall, there is an excess of heat which would have to be disposed of as waste heat.

A very complete study of the economic possibilities of an EUS in a new city has been made by the Oak Ridge National Laboratory under contract with the U.S. Department of Housing and Urban Development. The study developed data which could then be applied to a theoretical city with the climate of Philadelphia. The new city would have a population of 400,000 people and employ two base load nuclear reactors producing a total of 500 megawatts sited at the center with an industrial area of five miles in radius surrounding the energy center. In this area, process steam would be provided. District heating and cooling would be provided up to a radius of twelve miles. The energy center would supply all electric, space heating, and cooling needed in the area plus process steam to industry and waste heat to greenhouses adjacent to the energy center, which is connected to an electrical grid for power distribution so that at peak electrical loads, power can be transferred from plants outside the new city.

From an economic standpoint, EUS would be comparable in cost to other systems presently in use for the northern two thirds of the United States. South of this area, the demand for air conditioning is too great, and its lower efficiency would make EUS less attractive economically. The area of the new city would have to be densely populated (the study assumed a total area of 16 square miles) and of flat terrain. As one of the largest cost items is for underground piping, EUS would be most practical for new cities where streets would not have to be torn up [7].

EUS would have a very favorable environmental impact as far as thermal pollution is concerned. Thermal pollution would be reduced to about 63 percent of that from a present nuclear power plant and down to as little as 21 percent in mid-winter or summer. Two hundred acres of greenhouses would be sufficient to dissipate the remaining waste heat [24:75].

EUS could use enough heat if applied on a national basis to greatly reduce the extent of thermal pollution resulting from power production. As has been indicated, thermal pollution cannot be reduced below a certain level (21 percent in this study) by beneficial uses because some provision must be made for variation in load and outages. Application of EUS would present many problems, two of which will be mentioned. The basic technical problem presented by any complex system is maintaining the necessary balance among its counterparts. Changes in the economic life of a city cause an ebb and flow of industries in and out of the area. Population or lifestyles may also change. Thus, changes in economic geography may unbalance the energy system to the extent that what was once a miracle of efficiency may become monstrously inefficient. This is the main reason why utility supplied process steam has not become more popular with industry. If a utility changes its generating plant operation or industry its manufacturing process, the balance is upset to the detriment of both. The remaining problem is one of politics and economics. There is no economic unit large enough with the breadth of organizational abilities and the power to enforce decisions on third parties aside from government. In Iceland, Reykjavik was the first city to make extensive use of geothermal energy because the city government developed, owned, and operated the system from the beginning [23:143]. In a country with a market economy and a tradition of economic freedom, it would be politically difficult for government to organize and coordinate the various productive elements required.

Conclusion

This chapter has explored the creation of thermal pollution and the methods for dissipating it in ways less harmful to the environment or perhaps turning it into a benefit. The first section on the generation of electric power demonstrated that there is no method of generating power "just around the corner" which will reduce thermal wastes. In fact, thermal pollution from power plants will get rapidly greater in extent as the demand for electric power soars. The wider spread use of low cycle efficiency, nuclear fueled, power production will further increase the amount of thermal pollution in the short run. The methods of dissipation of waste heat to the environment discussed in the second section to alleviate thermal pollution in waterways are possible but will entail great capital investment and increased costs to all power users. There may be environmentally undesirable side effects such as local weather alteration, increased water usage, or visual pollution.

Waste heat can be capitalized and used to increase crop yields. The "new" industry of aquaculture can be created to supplement the dwindling supply of a national food resource. Through a systems approach thermal wastes could heat and cool buildings and supply heat to industry. In these uses thermal wastes would be substituted for the heat generated by vast amounts of oil which has been in such short supply at various times in recent years.

In the immediate future, it seems likely that nearly all thermal wastes which are treated will be dissipated through cooling towers. Low temperature heat in the form of process steam sold to industry will gain greater popularity due to new anti-pollution laws which make the alternative of in-house fuel combustion less attractive economically, but EUS as the norm is likely to remain a distant dream. Expansion of present district heating and cooling systems located in central cities is likely, however, as users seek to avoid the costs of pollution control equipment. Agricultural uses for waste heat will gradually grow and are the best way to use waste heat at present because they dissipate heat year round in a way which is least damaging from an environmental standpoint (e.g., the waste heat is dissipated over a large area) and should return at least their capital costs.

The Projected Cost of Thermal Pollution Abatement

There have been many studies made of the total cost of thermal pollution control. The results of these studies vary widely for a number of reasons: (1) The degree of application of expensive control methods, such as the use of cooling towers, is uncertain at this time as individual decisions are mainly the result of compliance with environmental protection laws. These laws are subject to unpredictable changes resulting from future political decisions. In the face of uncertainty, many studies simply assume the blanket utilization of cooling towers. Sight of the objective—the cost of thermal pollution abatement—tends to be lost and the cost of mass application of cooling towers substituted. Further, the mass application of cooling towers is not supported by a projection of statistics from a Federal Power Commission survey, which shows that only 40 percent of new plants over 300 megawatt capacity now under construction will utilize cooling towers to dissipate waste heat [18:32]. Since most of these plants will not be finished until the late 1970s, and as less than 18 percent of 1969 power plants utilized cooling towers, it is foolish to predict that all plants will use cooling towers by 1980 [27:XIX]. (2) The future cost factors associated with various types of thermal pollution control are not well established and subject to change with future technological improvements. (3) Other factors are also unknown, such as the future cost of capital.

Although these factors tend to make the results of present studies of low validity, a study by the Federal Power Commission was presented so as to give a general idea as to the costs involved (see Table 2-3). This study was chosen because of its realistic assumptions. On the basis of these results shown in Table 2-3 two conclusions are evident: First, the cost of thermal pollution control is not very great as a percentage of total electric power costs but it is large in gross figures: Second, there is a significant difference between the cost of reasonable and prudent control represented by Study "A" and rigid and dogmatic control represented by Study"C."

Factors Effecting Future Beneficial Utilization of Waste Heat

To the extent that it is economically practical, utilizing thermal wastes is preferable to spending the large amounts on cooling systems indicated as being necessary to meet standards according to the F.P.C. study referred to earlier. In order to fully develop the potential for waste heat use, two things are necessary: research and entrepreneurs. The need for research is obvious, but entrepreneurs are also sorely needed who will bear the risks to pioneer an industry in hopes of ultimate profit. Unfortunately, inflexible or outmoded governmental policies tend to discourage both.

Privately owned power companies are in the best position to provide the research and to set up and operate waste heat utilization projects. They have the ability to coordinate power plant requirements and waste heat usage requirements, and they should be given the economic motivation. Rates in the electric power industry are set so that a fixed percentage of utilized investment will be profit. With such a system, there is no incentive to innovate—any increase in gross income from beneficial uses of waste heat would be lumped with income from the sale of power, and the percentage profit would remain the same with electric customers of the utility paying a lower rate. Entrepreneurship is entered into with the hope of making a profit, and no well-managed business will enter into a new area where there is no hope of profit. Changes must be made which will allow the utility to keep most of the profits earned as a result of developing and operating beneficial uses for waste heat. These changes must be set with finality so that all concerned will *know the rules* before the project is begun.

Often, rigid anti-pollution regulations discourage projects which are just getting under way. While some of these projects may violate present regulations, their eventual success is the best way to control thermal pollution in the long run. Industries which are just starting up cannot afford the capital costs of meeting these regulations, nor is technology sufficiently developed to make it possible in the short run. Where such ventures are successful, anti-pollution regulations may be gradually applied to them.

Long Island oyster farms, located on Long Island Sound in New York State, were set up with the intention of oyster farming in thermally enriched water of 90 degrees Fahrenheit. Although the environmental effects of the project were approved by the state, a later change in the law so rigidly defined the maximum allowable increase in the water surface temperature of Long Island Sound, that oyster farming became impractical. Long Island oyster farms have reached an accommodation with the state which would permit continued operation, but future operation is in doubt as long as the law is not changed, and it is unlikely that any new such operations will be started in New York State [24:264-267]. Cases such as this must not be decided on blanket standards but on individual merits, and all potential entrants into areas of beneficially using waste heat must *know in advance* that the state will not make some future arbitrary law which will make their continued existence economically impossible.

In conclusion, with regard to public policy on the beneficial uses of waste heat, economic incentives must be encouraged or at least not penalized, and regulations must be flexible. The watch words in this area must be *incentive* and *flexibility*.

Glossary

Base Load Power Generation. The methods of power generation used to produce most of the electric power required on a continuous basis.

Cycle Efficiency. Pertains to the thermodynamic efficiency of the steam cycle to convert heat into mechanical energy.

Low Temperature Heat. Steam or hot water taken out of the steam-electric generation system before all of its usable potential to do work has been expended.

Peak Load Power Generation. The methods of power generation generally used to provide intermittent increases in response to demand changes.

Rejected Heat. The heat energy remaining in steam after its use to generate mechanical energy. When the heat energy is dissipated in the environment, it is also called waste heat.

System Efficiency. Pertains to the overall efficiency in the use of heat for one or several uses.

Thermal Pollution. Results when the acts of man increase the temperature of the air or water to the extent that some beneficial use of the air or water is proscribed.

Waste Heat. Heat which is dissipated to the environment at a sufficiently low temperature level that all useful work has been extracted from it given the limits of present technology.

Bibliography

Books and Pamphlets

1. Abrahamson, Dean E. (compiler). *Environmental Cost of Electric Power.* New York: Scientists' Institute for Public Information, 1970.
2. Edison Electric Institute. *Statistical Yearbook of the Electric Utility Industry for 1971.* Report of the Statistical Committee of the Edison Electric Institute. New York: Edison Electric Institute, 1972. Number 39. Publication No. 72-25.
3. Eisenbud, M. and Gleason, George (eds.). *Electric Power and Thermal Discharges.* New York: Gordon and Breach, Science Publishers, 1969.
4. Massachusetts Institute of Technology (ed.). *Energy Technology to the Year 2000.* Cambridge, Massachusetts: Technology Review, 1972. Pp. 34-51.
5. National Academy of Engineering. Committee on Power Plant Siting. *Engineering for Resolution of the Energy Environmental Dilemma.* National Academy of Engineering, 1972.
6. Saltonstall, Richard, Jr. and Page, J.K., Jr. *Brownout and Slowdown.* New York: Walker, 1972.

Government Publications (Except Symposiums)

7. Miller, A.J., et al. *Use of Steam-Electric Power Plants to Provide Thermal Energy to Urban Areas.* U.S. Department of Housing and Urban Development and Oak Ridge National Laboratory. Springfield, Virginia: U.S. Department of Commerce, January 1971. ORNL-HUD-14.
8. U.S. Congress. Joint Committee on Atomic Energy. *Selected Materials on Environmental Effects of Producing Electric Power.* Washington, D.C.: U.S. Government Printing Office, 1969.
9. U.S. Congress. Senate. Committee on Public Works. *Hearings Before Subcommittee on Air and Water Pollution: Thermal Pollution–1968.* 90th Congress, Second Session, 1968.

10. U.S. Department of Commerce. Bureau of the Census. *Statistical Abstract of the United States–1972.* Washington, D.C.: U.S. Government Printing Office, 1973.
11. U.S. Department of the Interior. Federal Water Pollution Control Administration. *The Cost of Clean Water, Volume II, Detailed Analysis.* Washington, D.C.: U.S. Government Printing Office, January 10, 1968.
12. U.S. Federal Power Commission. *The 1970 National Power Survey.* Washington, D.C.: U.S. Government Printing Office, December 1971.
13. Yarosh, M.M., et al. *Agricultural and Aquacultural Uses of Waste Heat.* Oak Ridge National Laboratory. Springfield, Virginia: U.S. Department of Commerce, July 1972. ORNL-4797.

Magazines

14. Aschoff, A.F. "Water Reuse in Industry." *Mechanical Engineering*, April 1973, Pp. 21-25.
15. Beall, S.E. "Uses of Waste Heat." *Oak Ridge National Laboratory Review* 3, 4 (Spring 1970): 9-14.
16. "Cooling Towers, A *Power* Special Report." *Power*, March 1973.
17. Lusby, W.S. and Somers, E.V. "Power Plant Effluent–Thermal Pollution or Energy at a Bargain Price?" *Mechanical Engineering*, June 1972, Pp. 12-15.
18. Olds, F.C. "Cooling Towers." *Power Engineering*, December 1972, Pp. 30-37.
19. Wneh, Walter. "Thermal Wastes: A Potential Resources." *Frontier* 30 (Winter 1970): 26-29.

Reports, Single Studies

20. Brady, D.K. and Geyer, John C. *Development of a General Computer Model for Simulating Thermal Discharges in Three Dimensions.* Edison Electric Institute Report No. 7. New York, New York: Edison Electric Institute, February 1972. EE-1. Publication No. 72-902.
21. Parker, Frank L. and Krenkel, Peter A. *Thermal Pollution: Status of the Art.* National Center for Research and Training in the Hydrologic and Hydraulic Aspects of Water Pollution Control. Report No. 3. Nashville, Tennessee: Vanderbilt University School of Engineering, December 1969. Chapter 5.

Symposiums, Conferences, and Reports

22. Krenkel, P.A. and Parker, F. (eds.). *Biological Aspects of Thermal Pollution.* National Symposium on Thermal Pollution, Portland, Oregon, 1968. Nashville, Tennessee: Vanderbilt University Press, 1969.

23. Mathur, S.P. and Stewart, R. (eds.). *Proceedings of the Conference on the Beneficial Uses of Thermal Discharges.* New York State Department of Environmental Conservation, September 17-18, 1970. Thruway Hyatt House, Albany, New York.
24. Yarosh, M.M. (ed.). *Waste Heat Utilization Proceedings of the National Conference, October 27-29, 1971, Gatlinburg, Tennessee.* Electric Power Council on the Environment and Oak Ridge National Laboratory. Springfield, Virginia: U.S. Department of Commerce, May 1972. CONF-711031.

Miscellaneous

25. Hauser, L.G., Manager. *Electric Light and Power.* Energy/Generation Edition. Boston, Massachusetts. Vol. 49, No. 1. Pp. 32-34.
26. National Economics Research Associations, Inc. *Possible Impact of Costs of Selected Pollution Control Equipment on the Electric Utility Industry and Certain Power Intensive Consumer Industries.* Springfield, Virginia: U.S. Department of Commerce, National Technical Information Service, January 8, 1972.
27. U.S. Federal Power Commission. *Steam-Electric Plant, Air and Water Quality Control Data for the Year Ending December 31, 1969.* (Based on FPC Form No. 67) Summary Report. Washington, D.C.: U.S. Government Printing Office, February 1973. FPC-S-229.

3 Opportunities for Savings from By-Products of Stack Gases of Electric Power Plants

A.A. Ferrara

The power plants of America consume about 25 percent of our energy input. Can they be a source of significant by-product recovery?

America's economy, more than any other nation's, relies heavily upon electrical power. Although containing less than 6 percent of the world's population, the United States consumes nearly 35 percent of the total power generated [53:3]. In the past, the electrical power industry has relied heavily upon fossil fuels—particularly coal and oil—as its primary energy source. As a result, electric power plants have been a major source of air pollution. In general, air pollution is caused by the injection of gas and particulate matter into the atmosphere. Based on statistics compiled by the Federal Power Commission (FPC), in the year 1968 electric power plants were responsible for over 50 percent of the sulfur oxides (SO_x) and nearly 20 percent of the particulate matter and nitrogen oxides (NO_x) absorbed by the atmosphere (Table 3A-1).

Coal burning power plants are primarily responsible for the atmospheric pollution caused by this industry. Incombustible substances in the coal produce a fine, powdery substance known as fly ash, which mixes with the plant's waste flue gas. Present technology is capable of removing up to 99 percent of this pollutant from stack gas emissions. However, once collected this ash becomes a solid waste that must be disposed of. Many utilities feel, therefore, that they are simply converting an air pollution problem to one of waste disposal. Sulfur oxides are caused by sulfur contained within both coal and oil, which is oxidized during the combustion process. At present, experts disagree as to whether or not an adequate technology exists to limit SO_x emissions from fossil-fueled power plant stacks. Nitrogen oxide emissions are caused by the need for air to support the combustion process. At present, the primary means of controlling this pollutant is by modifying existing equipment.

Despite the increased use of nuclear energy, which does not contribute to atmospheric pollution, Federal Power Commission (FPC) projections indicate that due to the constantly increasing demand for electrical power, more than twice as much energy will be generated in 1990 by fossil fuels, than was produced in 1970 (Table 3A-2). Consequently, although the Environmental Protection Agency (EPA) has imposed strict regulations on this industry, the EPA projects that even if met, SO_x emissions from fossil-fueled power plants in the year 2000 will not show any substantial decrease compared to present levels [35:15]. Hence, pollution from fossil-fueled power plants will continue to be a problem through the end of the century.

In the next section, the cost of controlling fly ash and SO_x emissions from fossil-fueled power plants is discussed. NO_x emissions will not be included since control techniques are as yet undeveloped, and no means of recovering nitrogen oxides has been advanced.

Since effective economical methods are available for limiting fly ash emissions, the emphasis for this pollutant will be on collection and disposal costs and how they can be offset in part by this waste product. It is the intention to examine specific cases for analysis, and accordingly the Detroit Edison Company was selected as a representative utility for evaluating the overall costs of fly ash disposal.

Regarding SO_x emissions, few control systems are presently in operation. The EPA, however, is presently conducting exhaustive studies of SO_x control costs. Using some of this information, the important parameters affecting this expense are discussed; e.g., plant size, load factor, and sulfur content of fuel. In addition, the EPA has entered into joint ventures with several utilities to evaluate the reliability and costs of owning and operating various control processes. Here again several of these plants will be examined as case studies. Finally, the utility is basically faced with two choices to limit SO_x emissions: that is, it may install one of the control processes being developed, or it may elect to burn a higher quality (i.e., lower sulfur content) fuel. Therefore, in order to be meaningful the costs of alternate SO_x control processes must be compared to the cost penalty incurred by burning lower sulfur content fuels. In performing this analysis, the potential benefits from the sale of sulfur and/or sulfuric acid by-products realized from these processes must be included. By comparing the "true" costs of these alternatives, therefore, their relative economic merits may be evaluated.

Methods of Controling Fly Ash

Fly ash is caused by incombustibles in the fossil fuels, particularly coal, and consists of those particles which are too small to settle out in the combustion chamber and become suspended in the high velocity flue gases. The primary means of removing these particles in power plants is with mechanical collectors and/or electrostatic precipitators. It is common to use these devices in series and the resulting system is typically capable of removing up to 99 percent of the fly ash initially present.

Mechanical collectors make use of gravitational, or centrifugal force to remove the fly ash. Collection efficiencies of these devices typically range from 40 to 80 percent, depending upon the size of the fly ash particles and the draft drop across the tubes [12:72].

In electrostatic precipitators, a high intensity electrical field is generated which induces an electrical charge in the fly ash particles. The charged particles are then attracted to collector surfaces which may either be dry or wet. This device is highly efficient and capable of removing better than 90 percent of the fly ash passed through it. However, the efficiency of these devices varies directly with the sulfur content of the coal, i.e., the higher the sulfur content, the greater the precipitator efficiency [8:65]. In order to meet present SO_x emission standards, some utilities are electing to use low sulfur content coals. Although this is an effective means of limiting SO_x emissions, it lowers the efficiency of fly ash collection.

As a further means of lessening the air pollution problem, many utilities build higher stacks than would otherwise be required. Although stack height does not reduce the total emission of pollutants into the atmosphere, it does disperse them over a larger area and may decrease concentrations to acceptable levels [77:11].

Methods of Controlling Sulfur Oxides, SO_x

A great deal of work is presently being done to investigate methods of reducing emissions of sulfur oxides from power plants. However, to date, experts disagree on whether or not an adequate control technology is available. Perhaps a hundred control techniques can be identified in various stages of development, ranging from methods that have been evaluated in full scale unit tests, to pilot plant facilities [6:10, 111].

In general, SO_x control systems may be classified as throwaway and recovery techniques. In throwaway systems, the sulfur oxides are converted into various sulfites and sulfates for disposal, which utilities claim converts the pollution problem from one of SO_x emission to solid waste disposal.

Recovery systems produce either sulfuric acid or elemental sulfur. Of the two, sulfur is easier to store and the more useful by-product. To be of commercial value, a high concentration of sulfuric acid is desirable. Of the systems presently under study, it is anticipated that three will be of significant benefit in the near future. They are: wet lime/limestone scrubbing, wet magnesium oxide (MgO) scrubbing, and catalytic oxidation (CAT-OX). Of these, the wet lime/limestone scrubbing system is considered to be the most fully developed. It is projected that each of these systems will remove approximately 90 percent of the SO_x contained in the waste flue gases, thereby meeting established standards. Each of these systems is briefly described below.

Lime/Limestone Process

Since they are relatively cheap elements, lime or limestone are the absorbents principally employed in the throwaway systems. The essential difference betweeen these elements is that lime is more reactive with SO_x than limestone and, therefore, less of it is needed for a given removal requirement. Limestone is available in most areas of the country and may be calcined to lime by heating. Since calcining prior to injection in the furnace adds significantly to the cost, various systems have been proposed using limestone as extracted from the earth. Dolomite is another substance which behaves similarly to limestone and has been used in throwaway systems. The following discussion, therefore, although described as limestone systems, may make use of either lime, limestone or dolomite.

The limestone-injection/wet scrubbing system uses an aqueous solution to absorb the SO_x in the waste flue gas (Figure 3-1). The limestone reacts with the sulfur oxides to form calcium sulfite. In addition to removing the calcium sulfite and unreacted limestone, the wet scrubbers may also remove the particulate fly ash matter. Basically, the aqueous solution is fed into the top of the scrubber and the flue gas carrying the particulates into the bottom of the unit. The resulting counterflow causes the liquid to separate the particulate matter from the gas. Subsequently, the solution is transported to a settling tank from which the reaction products are removed for disposal. It is anticipated that this system will exhibit SO_x and particulate removal efficiencies of better than 95 percent respectively [7:36].

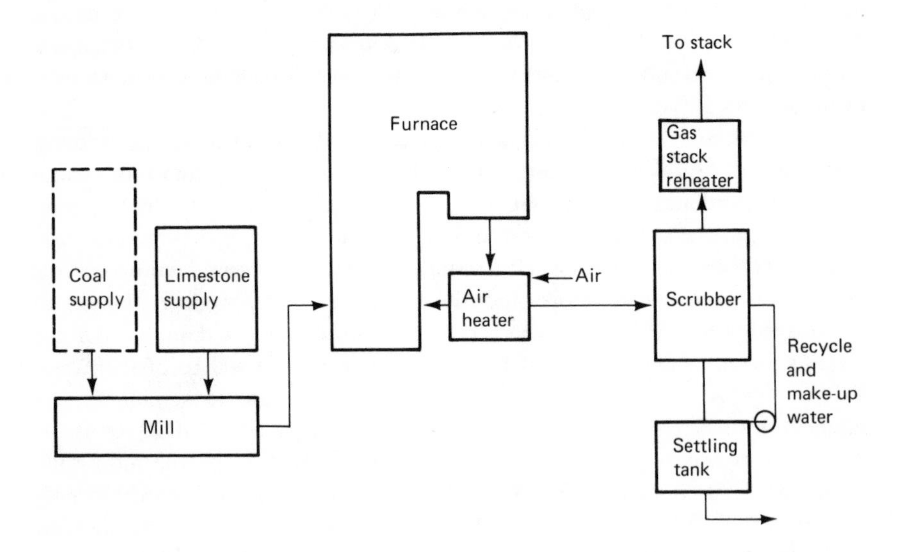

Figure 3-1. Limestone Injection—Wet Scrubbing Process. Source: U.S. Dept. of Health, Education and Welfare, *Control Techniques for Sulfur Oxide Air Pollutants* (Washington, D.C.: Government Printing Office, Jan. 1969), p. 53.

For older units requiring a retrofit, an add-on system may be employed. The add-on is similar to the "wet" system, except that the limestone is injected into the scrubber rather than in the boiler area. Since the add-on system does not make use of the boiler heat to calcine the limestone, however, the efficiency of this device per pound of limestone is less.

In addition to mechanical problems caused by platting, caking, and plugging of the limestone-based materials, the possibility of groundwater contamination must be considered in using wet limestone systems. Magnesium is contained in at least trace amounts in most of the available limestone. Unlike calcium sulfite, magnesium sulfite is soluble in water. Potentially, this sulfite could be leached from the settling tanks into the groundwater supply. Another disadvantage of these systems is the large amount of water vapor, absorbed by the flue gas in passing through the scrubber, which is subsequently emitted from the stack.

Wet MgO Scrubbing

Due to the problems mentioned in regard to limestone scrubbing systems, a considerable amount of work has been done to develop alternate cleaning systems.

The Chemico/Basic design is typical of MgO wet absorbent systems (Figure 3-2). This process is similar to the "add-on" wet limestone process described above, except that it makes use of magnesium oxide (MgO) rather than limestone. The MgO combines with the sulfur oxides to form magnesium sulfite which is subsequently converted to sulfuric acid. As shown in Figure 3-2, a portion of the solution feeding the scrubbers is tapped off to crystallizing equipment, where solid crystals of magnesium salts are formed. Subsequently, a centrifugal process is used to separate the liquor from the crystals and the crystals are dried to remove the water of crystallization. The anhydrous salts can then be shipped to a chemical processing plant where magnesium oxide is regenerated and gaseous sulfur dioxide produced. The ultimate by-product is sulfuric acid produced from the gaseous sulfur dioxide.

Catalytic Oxidation (CAT-OX)

As shown in Figure 3-3, in the CAT-OX process the flue gas must first be passed through an electrostatic precipitator to remove particulate matter. Subsequently, the gas flows through a catalytic bed, which is maintained at elevated temperatures of 800-900°F. The catalyst acts to convert the SO_2 into SO_3, which is then dissolved in water to form sulfuric acid. Although the process is old, in recent years Monsanto has done the only significant work and is presently marketing this technology.

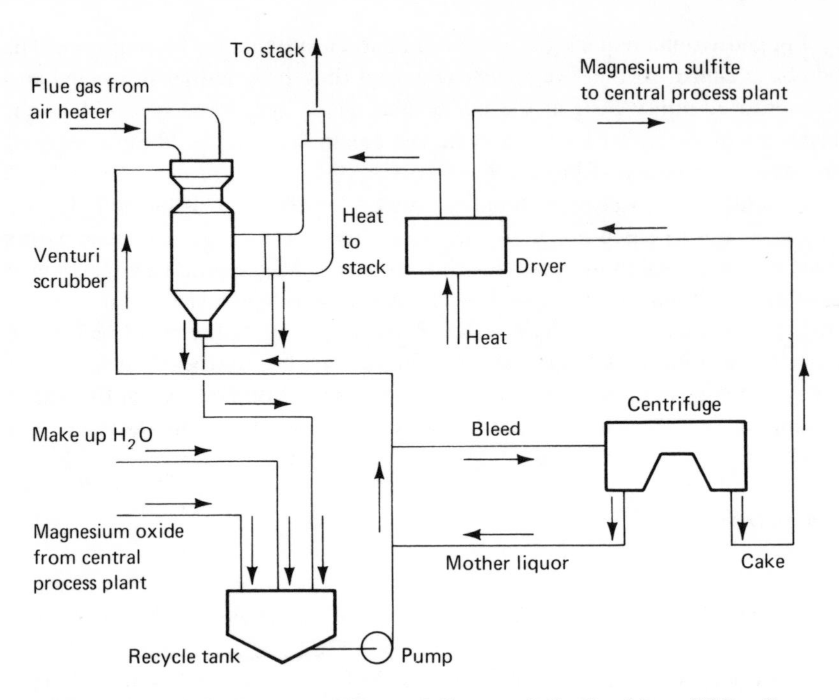

Figure 3-2. MgO Recovery System. Source: J.G. Farthing, "SO_2 Removal Systems: Development Proceeds at Quickening Pace." Reprinted from May 15, 1971 issue of *Electrical World* © Copyright 1971, McGraw-Hill, Inc. All rights reserved.

Characteristics of Electric Power Generation

For many years, fossil fuels—particularly coal—have been the primary source of energy for electric power plants in the United States. Over the period 1966 through 1971 fossil fuels accounted for more than 97 percent of the power generated (Table 3A-3). Although the consumption of coal decreased from 65 to 53 percent of the total power generated over this period, due to the limited supplies of natural gas and the increasing cost of fuel oil, it is projected that even greater quantities of coal will be consumed in the future. Since the FPC has established a strong correlation between the burning of coal to produce electrical energy and air pollution, the trend to increased consumption of coal will be significant to air quality levels (Table 3A-4). This is primarily because most of the available coal is high in both sulfur and ash content, whereas low sulfur and ash content oils are available. Although natural gas is a highly desirable fossil fuel since it contains negligible amounts of ash and sulfur, its use will be inhibited due to an insufficient supply.

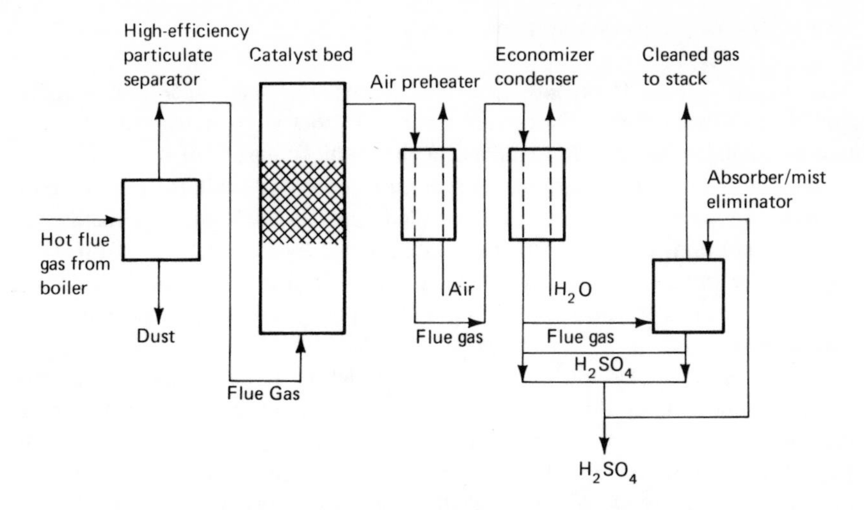

Figure 3-3. Catalytic Oxidation Process. Source: U.S. Department of Health, Education and Welfare, *Control Techniques for Sulfur Oxide Air Pollutants* (Washington, D.C.: Government Printing Office, Jan. 1969), p. 55.

The East North Central region—Illinois, Indiana, Michigan, Ohio, and Wisconsin—was the major source of air pollution in the United States for the year 1969, accounting for about 39 percent of the total pollutants emitted (Table 3A-5). This region also consumed about 38 percent of the coal used in the United States for electric power generation and produced more electrical energy than any other area. Moreover, based on information compiled by the Edison Electric Institute, in 1969 coal was the fuel utilized to produce nearly 90 percent of the power generated in this region [76:23].

Study Methodology

In keeping with the intent to examine specific cases as illustrations, the East North Central Region was selected for investigating the costs of disposing fly ash. Having thus narrowed the research, a major utility in the area, Detroit Edison, was then selected for detailed investigation. Detroit Edison services a fairly large population area—over four million people according to the 1970 census [25:24-291]—and generated about 60 percent of the power generated in the state of Michigan in 1969 [69].

Regarding SO_x removal, the study draws upon projections of the EPA and available case data. This approach was necessitated since few SO_x removal processes are presently in operation and Detroit Edison was unable to provide actual cost data for these installations.

Cost of Fly Ash Collection Systems

The Detroit Edison Company, as is true of many of the larger coal burning utilities in urban centers, uses mechanical collectors and electrostatic precipitators to collect the fly ash contained in the waste flue gases of its fossil-fueled power plants. Typically, data in the literature reports precipitator capital costs of from $6 to 40 per kilowatt of installed capacity, depending primarily on retrofit conditions [40:IV-5]. Using data compiled by the FPC and information provided by the Detroit Edison Company, however, a more thorough investigation of the annual owning, operating, and disposal expenses of these systems may be performed.

In the year 1969, Detroit Edison generated slightly over 30 million megawatt-hours of power. Over 90 percent of this energy was produced in coal-burning facilities with tested precipitator efficiencies of better than 97 percent [23: 143-145]. The total capital cost reported for precipitators was about $23 million, on an installed capacity of slightly more than 900 kw; i.e., a capital cost of $4.05/kw of installed capacity. Using a 15 percent interest factor indicates that it costs the firm about $3.5 million per year to own these facilities. In addition to this expense, the firm spent about $2.8 million to operate and maintain the equipment [69], and nearly $1.2 million dollars [23:143-145] to collect and dispose of the ash. Consequently, the total expense for Detroit Edison was about $7.5 million dollars for 1969 to own and operate its fly ash collection system. Normally, the utility industry expresses expenses in terms of total power generated. On this basis, fly ash pollution control cost Detroit Edison about .25 mils/kwh of power generated in coal burning facilities. In comparison, the firm spent between about 3.5 to 9.7 mils/kwh for fuel in 1971 at its various plants [24:70, 71]. Since fuel costs are by far the major part of total operational costs, fly ash pollution control represents a small operational expense for this utility.

Based on this evaluation, it appears that efficient, fairly economical methods exist for collecting the fly ash emitted from power plants. However, once collected the fly ash becomes a solid waste that must be disposed of. Although studied for over thirty years, this solid waste disposal problem continues to plague the utility industry and appears to be the main problem associated with fly ash pollution [47:2]. In the past, investigators have attempted to show that by marketing recovered fly ash, utilities could realize a profit. Some, however, argue that this is not practical and utilities must recognize that in today's environment, fly ash disposal can only be treated as an operating expense [47: 16]. A good part of this disagreement may involve the basic definition of what are the costs of fly ash disposal. Should the utilities, for example, include owning, operating, and maintenance expenses incurred in collecting the ash as a part of the total cost, or should they simply treat the cost of transporting it from the power plant to a land fill area? In this analyst's opinion, the question of profitability of fly ash collection is academic because the accumulation of fly ash is a natural by-product from burning coal. Therefore, it is felt that system owning and operational expenses should be treated as normal operating expenses and disregarded in evaluating the "profitability" of fly ash disposal. Consequently, in the following paragraphs a closer look at the costs and revenues of disposing fly ash will be presented.

Fly Ash Disposal Costs and Revenues

For the year 1969, the FPC reported a total ash collection and disposal expense of nearly $40 million was incurred by U.S. utilities to dispose of nearly 35 million tons of ash; i.e., a cost of about $1 per ton of ash collected [23:9, 13]. This expense represents the cost for the utility to transport the ash to a land fill area and grade the surface [69]. Offsetting this expense, revenues of approximately 1.9 million dollars was realized from the sale of about 2.5 million tons of ash, or about 7 percent of the total generated. The return was thus about $.76 per ton of ash sold. In this context, ash refers to both bottom ash, cinders, slag, and fly ash. Cinders and slag are easier to sell than fly ash, due to their mechanical properties. For example, since they are essentially sintered products, they are not powdery and hard to control, and are more easily used as aggregate material [48, 84]. Detroit Edison feels that it could market all of the bottom ash, cinders, and slag generated. At present the firm does not market all of this material as some is used for in-house purposes, e.g., to create road surfaces at dump sites.

The bulk of the ash marketed in the United States is being sold by a relatively few companies specializing in this area, or utility companies having their own sales organizations [50:11]. Based on information provided from various industry sources, fly ash marketing firms typically pay from $.25 to $1.00 for each ton of ash purchased from the utility. The exact payment depends primarily upon the quality of the ash, the amount available, and the proximity of markets. In general, the larger the amount to be marketed and the further it has to be transported, the less the utility will receive for the fly ash. Assuming that the utilities were to receive $.50/ton of ash, however, indicates that for the year 1969, fly ash collection and disposal expenses could have been cut to less than $20 million if all the ash collected were sold. Therefore, even if the utilities could not realize a net profit from the sale of ash, they could significantly reduce their collection and disposal expenses.

At present, the primary means of disposing fly ash is by carting it to a land fill or dump area. Since the majority of coal burning plants are located in densely packed urban centers (Figure 3-4) where land is at a premium, it is becoming increasingly expensive for plants to dispose of this waste. At the current production rate, fly ash disposal costs typically vary from $.50 to $2.00

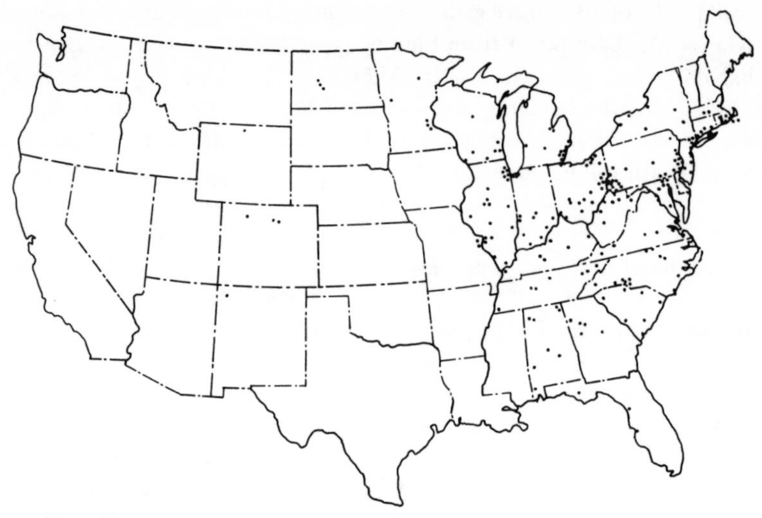

• Denotes power stations with annual coal burn exceeding 400,000 tons per year

Figure 3-4. Major U.S. Coal-Burning Power Plant Locations. Source: Environmental Protection Agency, *Technical & Economic Factors Associated with Fly Ash Utilization* (Springfield, Va.: National Technical Information Service #PB-208-480, July 1971), pp. 4-5.

per ton depending on plant location. This gross variation is primarily caused by the distance the ash must be carted.

Utilities expect even larger disposal costs in the future [30:4-6]. At a recent conference of the National Fly Ash Association, it was reported that fly ash disposal costs are continuing to rise and that a cost of $2.00 per ton is no longer exceptional, but becoming average [56.3]. Table 3-1 presents a breakdown of the ash disposal expenses and revenues for plants of (a) the East North Central region, (b) Michigan, and (c) Detroit, respectively, for the year 1969. As shown a fairly wide variation in expenses and revenues can be realized in a given region. Further, data compiled for each individual plant of the Detroit Edison Company for the year 1969 show that even one utility may realize a fairly wide variation in ash collection expense and revenue for its various power plants (Table 3-2).

At present, Detroit Edison estimates that the cost to load the ash into trucks, transport it to a land fill area, and bulldoze the ash amounts to approximately $1.00 per ton of ash. In 1972, the firm disposed of about 900 thousand tons of fly ash, for a total expense of nearly $1 million. Regarding ash revenues, this firm is one of the few that has set up its own ash sales organization and realizes a higher return than firms using an external agency. In 1972, the firm sold about

Table 3-1
Comparison of Ash[a] Collection Expenses and Revenues for Steam-Electric Plants of East North Central Region, Michigan and Detroit–1969

	Ash Collection & Disposal			By-Product (Ash[a]) Sale		
	Expenses (1,000)	Collected (1,000 tons)	Average Cost ($/ton)	Revenues ($1,000)	Sold (1,000 tons)	Average Revenue ($/ton)
East North Central	16,701	12,980	1.29	432	656	.66
Michigan	2,160	2,241	.97	261	229	.88
Detroit Edison	1,172	1,322	.89	155	185	.84

[a]Ash here refers to fly ash, bottom ash, cinders and slug.

Source: The Federal Power Commission, *Steam-Electric Plant Air and Water Quality Control Data, for the year ended Dec. 31, 1969* (Washington, D.C.: Government Printing Office, Feb. 1973), pp. 9, 13, and 143-145.

Table 3-2
Individual Ash[a] Collection and Disposal Expenses of Detroit Edison Plants for 1969

Plant Site	Ash Collection & Disposal Expense ($/ton)	Return from Sale of Ash ($/ton)
Corners Creek	1.40	1.98
Marysville	2.56	7.15[b]
Pennsalt	1.33	–
River Rouge	.64	.29
St. Clair	.67	.29
Trenton Channel	.67	1.50
Wyandotte	1.87	–

[a]Ash here refers to fly ash, bottom ash, cinders and slag.

[b]Ash sold at this facility was small, amounting to about 2 percent of that collected.

Source: The Federal Power Commission, *Steam-Electric Plant Air and Water Quality Control Data, for the Year Ended Dec. 31, 1969* (Washington, D.C.: Government Printing Office, Feb. 1973), pp. 143-145.

110 thousand tons of fly ash, for which it received about $1.60 per ton. It reports that the bulk of its ash sales is to large users for $1.25 per ton [84].

In the future, the constantly increasing demand for electrical energy will result in a greater consumption of coal, leading to increased production of fly ash. The EPA projects that fly ash production will increase from today's level of about 30 million tons to nearly 100 million tons by the year 2000 (Figure 3-5). Consequently, fly ash collection and disposal will become even more of a problem than it is today, particularly for utilities located near urban centers. It is likely that in the future these utilities will have to transport the ash further to land fill areas, incurring higher transportation costs. Therefore, it will be increasingly important for utilities to find additional usages for fly ash, to lessen the expenses of–if not profit from–ash collection and disposal.

At this point, in order to evaluate the potential return from the sale of fly ash, a discussion of its marketability is required.

Fly Ash Market Evaluation

A summary of the total usage of fly ash for the year 1970 is presented in Table 3-3. As shown, total utilization was less than 7 percent of the amount generated. The data have not changed significantly over the past several years, with total fly ash utilization presently estimated to be on the order of 10 percent of that generated [48]. Many reasons may be cited for this limited use of fly ash,

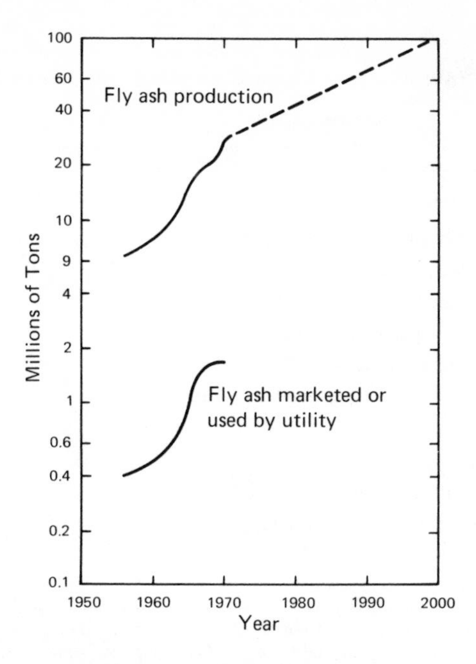

Figure 3-5. Fly Ash Production and Utilization. Source: Environmental Protection Agency, *Technical & Economic Factors Associated with Fly Ash Utilization* (Springfield, Va.: National Technical Information Service #PB-208-480, July 1971), pp. 4-14.

including buyer resistance, material inconsistency, competitive products, and existing specifications.

Buyer resistance may be attributed to bad past experiences and a disinterest in using an unfamiliar product. In the past, many utilities marketed poor quality fly ash which failed to meet buyer specifications. For example, during World War II Detroit experienced a cement shortage. Since fly ash may be used as a partial replacement for cement, many cement users became interested in using fly ash. Unfortunately, too much fly ash was used as a cement replacement and structural failures were experienced. Although Detroit Edison now monitors the quality of its fly ash and definite standards for replacing cement with fly ash have been established since this time, the firm is still having to overcome buyer resistance dating from nearly thirty years ago [84].

Another excellent use for fly ash is as a binder and filler in asphalt paving. As a filler, the fly ash may be used to replace aggregate material which represents about 80 percent (by volume) of the total asphalt. To be used as an aggregate, fly ash must be sintered. Sintering turns fly ash into a lightweight aggregate, weighing about one-half as much per cubic yard as natural material. Detroit Edison estimates that at the present price for lightweight aggregate of $8-

Table 3-3
1970 Fly Ash Utilization

	Tons (In Thousands)
Total Fly Ash Collected	26,538
Total Fly Ash Utilized	1,631
Partial replacement for cement in concrete	536
Fill material for roads	320
Lightweight aggregate	207
Raw material for Portland cement	158
Filler in bituminous products	131
Base stabilizer for roads, parking areas, etc.	111
Grouting	76
Control of mine fires	14
Oil and gas well conditioning agent	8
Foundries–manufactured products	4
Pipe coating	22
Miscellaneous	64

Source: Environmental Protection Agency, *Technical and Economic Factors Associated with Fly Ash Utilization* (Springfield, Va.: National Technical Information Service # PB-208-480, July 1971), pp. 4-13.

$10/ton, they can profit by $1.25/ton of fly ash. As a binder, between 8-10 percent of the total asphalt (by weight) may be fly ash, as a replacement for tar. Although it has been shown that fly ash asphalt has superior non-skid properties to normal asphalt, many states have yet to set up specifications and approve its use. Texas, for example, has approved its use and is building highways with fly ash asphalt, while Michigan has not. In the state of Michigan approximately 4 million tons of asphalt was used in 1972. It is estimated that about 2 million tons of fly ash could have been used in this market as an aggregate and binder material [84].

Federal Power Commission data show that for 1969 the ash collected by Detroit Edison represented about 59 percent of the state total [10:9, 13]. Assuming this ratio has remained about the same, based on the slightly more than 900 thousand tons of fly ash collected by Detroit Edison in 1972, much less than 2 million tons of fly ash was collected in Michigan. Hence, asphalt paving could have utilized all of the fly ash generated in the state. The economics of this situation indicates that an investment of $8 million to sinter the fly ash would enable Detroit Edison to sell all of the fly ash generated and make a "profit" from this market alone. If we assume a return of $1.25 per ton of fly ash and a savings of $1 per ton for not having to transport the ash to a land fill area, it appears that for every ton sold, Detroit Edison "profits" by

$2.25. Therefore, if the firm were able to sell all of the fly ash collected, it could realize a return of over $2 million.

Industry sources recognize that the aggregate market is an excellent means of utilizing fly ash [48]. However, aggregate companies are also aware of this application and its potential to curtail their business. As a result, they have established lobbies to protect their interests, and are major critics of its use. It is estimated that today's total U.S. aggregate (stone, sand, gravel, etc.) market of approximately 3 billion tons per year will grow to about 6 billion tons by the year 2000 [48]. Figure 3-5 indicates that total fly ash production is not expected to exceed one-sixtieth of this latter figure. Therefore, even if all the fly ash generated in the coming decades were used in this market, it would not significantly affect the aggregate industry.

At present, however, it is felt that the rate of return of sintering plants is not sufficient for normal commercial ventures. Therefore, in order to penetrate this market, the utilities may have to construct and operate these facilities. Since utilities are in the business of producing electrical power they may be understandably reluctant to enter other, unfamiliar markets. Also, they are presently experiencing difficulty in raising capital funds due to a "tight" money market and investors who are frequently unhappy with their rate of return. Furthermore, a Detroit Edison spokesman indicated that the firm would want to include this investment in its rate base. This would require that the firm submit detailed justification to the state utility commission for approval. Since the proceedings of these agencies are time-consuming, typically taking one to two years to grant a rate change, utilities are hesitant to appear before them. Therefore, in order for utilities to enter the lightweight aggregate market more aggressively, state utility commissions would have to express an interest in this venture and cooperate with the utilities.

As previously discussed, an excellent use for fly ash is as a replacement for cement. Since the bulk of the structural concrete used in the cities is ready-mixed concrete, it was decided to examine this industry. Figure 3-6 shows the cement used in ready-mixed concrete production in the various states for 1972. In preparing this figure it was assumed that the percentage of cement produced and used in the state for ready-mixed concrete production was the total percentage of cement produced in the state. It verifies that a significant market exists for cement in many of the states producing electrical power by burning coal. For example, in the state of Michigan slightly over 2 million tons of cement was used in the production of ready-mixed concrete. Of this total, Detroit accounted for about 40 percent [79]. It is generally accepted that approximately 20 percent of the cement in ready-mix concrete may be replaced by fly ash. Therefore, if this market were to accept this product, Detroit Edison could sell nearly 170,000 tons of fly ash to producers of ready-mixed concrete. In recent years this market has become more attractive due to a shortage of portland cement in the United States and its ever increasing price. Up to 1972,

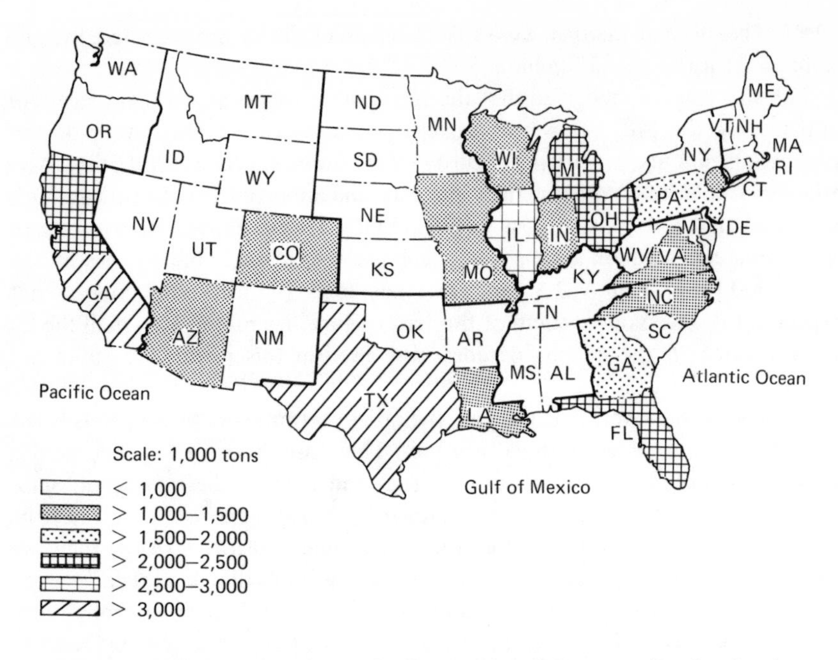

Figure 3-6. Cement Used in U.S.: Ready-Mixed Concrete Production by Region 1972. Source: U.S. Department of the Interior, Bureau of Mines, *Mineral Industry Surveys: Cement in 1972* (Washington, D.C.: Bureau of Mines, Oct. 1973), pp. 2, 4, and 9.

portland cement was plentiful and available for less than \$20 per ton. However, in some areas of the country portland cement is presently selling for \$28/ton. Detroit Edison estimates that ready-mixed concrete producers pay about \$5 per ton to transport the fly ash to their facility. Adding the cost of purchasing the fly ash itself, \$1.25 per ton, brings the total fly ash cost at the manufacturer to about \$6.25 per/ton. Since cement is currently selling for between \$22 and \$25 per ton in Detroit, the ready-mixed concrete producer can save between \$16 and \$19 for every ton of fly ash used. Obviously this is a significant impetus for its future application. Detroit Edison sold only slightly more than 100 thousand tons of the approximately 900 thousand tons of fly ash collected by the firm in 1972. They estimate that approximately 40,000 tons was marketed to ready-mixed concrete producers. Therefore, although this market could not totally solve the utility's fly ash disposal problem, it could absorb significantly more of this product. Moreover, it should be pointed out that Detroit Edison markets its fly ash more aggresively than many other utilities and it is conceivable that the increased usage of fly ash in the ready-mixed concrete market could have a larger impact on other utilities.

An additional promising use of fly ash is in place of cement clinker in the production of portland pozzolan cement. In this application, the fly ash is ground into the cement clinker. It is estimated that up to 40 percent of the cement normally used can be replaced by fly ash in producing this product. Presently, the Dundee Portland Cement Company is purchasing fly ash from Detroit Edison to manufacture portland pozzolan cement. Although they estimated a market of 100,000 tons for 1973, they are presently using about 25,000 tons. Since this operation has been underway for a relatively short time, it is anticipated that a greater utilization of fly ash will be made in the future [84].

Summary: Costs of Fly Ash Disposal

Mechanical collectors in combination with electrostatic precipitators are effective methods for collecting fly ash emissions.

It appears that the total owning and operational expenses for fly ash collection systems are a small fraction of overall operational expenses, which include the costs of fuel. In comparison, fuel costs at various plant sites varied from 3.5 to 9.7 mils/kwh in 1971 [24:70, 71]. Therefore, reliable and economic systems are available to collect fly ash emissions. The primary problem remaining appears to be disposing of the ash once it is collected. As the power generated by coal burning plants increases in the coming decades, the fly ash collected will also increase. The EPA projects that the total fly ash collected by the year 2000 will be more than triple the amount collected in 1970. The present trend toward increased consumption of coal due to the unavailability of fuel oil will result in an even greater production of fly ash than previously predicted. Therefore, fly ash collection and disposal will be an ever increasing burden to the utility industry.

Depending on the location, the cost to dispose of fly ash can vary from $.50 to $2.00 per ton. Detroit Edison estimates that it costs them $1.00 for every ton of fly ash removed. Since this firm collected slightly less than one million tons of fly ash, even if the utility were to give it away, a significant benefit would be realized.

Looking at the market for fly ash, it has been shown that although only 10 percent of the total fly ash produced is being sold, significant markets exist for its use. Buyer resistance, material inconsistency, competitive products, and existing specifications are significant impediments to an increased use of fly ash. The aggregate market could conceivably make use of all the fly ash generated in the United States. For example, all of the fly ash generated in Michigan in 1972 could have been used as a binder and filler material in the production of asphalt for highways. Moreover, the U.S. aggregate industry would not be significantly affected by an increased use of fly ash. It is estimated that total fly ash use in this market would decrease the total demand for aggregates by less than 2 percent through the end of the century. Since, in order to penetrate this market, utilities may have to construct and operate plants to sinter the fly ash, the cooperation of state utility commissions will be necessary for utilities to commit the required capital and enter this market. A significant market also exists for fly ash as a replacement for cement in ready-mixed concrete production and portland pozzolan cement. The economics of these applications suggest that increasing amounts of fly ash will be used in this market in the future. As shown for Detroit Edison, for every ton of fly ash sold, the firm benefits by better than $2. Hence, if this utility were able to sell all the fly ash generated, they would benefit by over $2 million per year. Therefore, whether or not this truly represents an accounting type of profit, it surely represents a cash flow which cannot be overlooked.

In the following paragraphs, attention is now turned to the costs of removing sulfur oxides from power plant stack gases and the potential benefits from the sale of sulfur by-products.

Projected Costs of SO_x Removal Systems

Typically, the capacity of power plants ranges from 200 to 1000 megawatts (mw), although units as small as 50 and larger than 2000 mw are in use. Industry sources generally agree, however, that a 500 mw plant may be considered normal.

With the cooperation of major contractors and utilities, the EPA has expended a considerable effort to develop evaluation techniques to predict the costs of alternate SO_x removal system [36, 39]. Using actual cost data, these techniques have been shown to be accurate to within 10 to 20 percent of capital costs. Based on these analytical procedures, cost projections have been prepared for alternate systems as a function of the following parameters: sulfur content of fuel, plant size, load factor, indirect costs, waste disposal costs, and retrofit factor. In the following paragraphs these data will be used to indicate the scope of the problem and the potential economies of scale that may be realized.

A cost comparison for a typical 500 mw plant burning 3.5 percent sulfur coal shows that although capital costs are higher, on an annualized basis recovery systems are less expensive than lime/limestone scrubbing systems (Table 3-4). Annualized cost figures for the MgO and CAT-OX systems include a revenue of $15 per ton of sulfur recovered while for the lime/limestone processes a waste disposal cost of $3 per ton of wet sludge was included. Since raising capital is a problem for utilities, it is likely that initial capital costs are weighted heavily in their economic decisions. However, financial experts generally agree that this is a short-sighted philosophy, and one must consider the time-value of money in reaching capital budgeting decisions.

Table 3-4
Comparison of Capital and Annualized Costs of Alternate SO_x Removal Processes

(Basis: 500 mw, 3.5 percent sulfur coal, retrofit, 60 percent load, waste disposal cost at $3/ton wet sludge, sulfur credit at $15/ton, particulate removal included.)

	Comparative Process Costs	
Process	Capital, $/kw	Annualized, mils/kwh
Throwaway		
Lime scrubbing	40[1]	2.40
Limestone scrubbing	41[1]	2.45
Regenerable		
CAT-OX	55	2.75
MgO	49	2.40

[1] Including disposal facility cost of $5/kw.
Source: G.T. Rochelle, *Economics of Flue Gas Desulfurization* (Research Triangle Park, North Carolina: EPA, May 1973), p. 13.

A sensitivity analysis performed by the Environmental Protection Agency for the wet lime/limestone scrubbing process has shown that the most significant parameters are plant size, fuel sulfur content, and load factor. Assuming that this analysis is equally valid for the other systems, some general conclusions may be reached from these data.

Regarding plant size, the analysis concludes that both initial capital and annualized cost economy of scale benefits may be realized; e.g., a capital cost of about $40/kw is projected for a 1000 mw power plant burning 3.5 percent sulfur coal, in comparison to approximately $70/kw for a 100 mw facility (Figure 3A-1). Similarly, annualized costs are shown to decrease as plant size and load factor increase. For example, for a 100 mw facility burning 3.5 percent sulfur coal, an annualized cost of about 4.6 mils/kwh versus 2.9 mils/kwh is projected as the load factor increases from 40 to 70 percent (Figure 3A-2). The sulfur content of the coal also has a significant effect on the annualized cost of limestone scrubbing systems. An approximately linear relationship exists between annualized cost and coal sulfur content for a process designed to remove 90 percent of the SO_x present in the flue gas. Annualized cost varies from about 1.9 to 2.7 mils/kwh as the coal sulfur content increases from 1 to 5 percent, for a 500 mw plant, operating at 60 percent load factor (Figure 3A-3).

This analysis is indicative of the results of varying plant size, load factor, fuel sulfur content, and emission standards, as it serves to scale the problem and indicate the relative sensitivity of cost to these parameters. Although the trends indicated by these curves are generally accepted—e.g., the industry accepts the notion that economy of scale benefits will be realized with increasing plant size—the validity of the absolute values projected is questioned. Far too often contractors and "paper" studies have projected lower system costs and greater system reliability than were ultimately realized.

Power companies have been reluctant to accept the validity of such analyses, however reasonable the assumptions and however sound the empirical data until verified by actual case data. Accordingly the EPA has entered into joint ventures with various utilities to evaluate and demonstrate the costs and performance of alternate SO_x removal systems.

In the following paragraphs, the cost data for selected individual plant sites are presented.

Case Studies: Cost of Alternate SO_x Removal Systems

At present, the available cost data are primarily limited to those SO_x control processes in operation which are joint ventures of the EPA and a utility, and have been operational for less than one year.

In order to evaluate the reliability and costs of wet limestone scrubbing, the Commonwealth Edison Company of Chicago, Illinois, has installed this system on its Will County plant. The unit selected, designated #1, is a 163 megawatt coal-fired power plant that was initially put in operation in 1955 [45:1]. The power required to operate the wet scrubber system has been estimated to be 7 megawatts, or slightly more than 4 percent of the plant rating. The utility estimates a capital cost for this facility of $85 per kilowatt neglecting the cost of sludge disposal equipment [45:23]. Including this latter expense raises the total cost to $96 per kilowatt. On an annual basis the company estimates that it costs them about $4.5 million to own and operate this facility, with a sludge disposal cost of $7/ton (Tables 3A-6 and 3A-7). For a load factor of 70 percent, therefore, and a plant rating of 153 mw, an annualized cost of 4.74 mils/kwh is calculated for this installation. It should be pointed out that these figures are estimates and based on a scrubbing system for which the longest single period of continuous operation of one unit was 270 hours, or slightly more than sixteen days (up to mid-1972, based on a 70 percent load factor). The overall system consists of two units in parallel, which up to late 1972 operated simultaneously for only 322 hours (i.e., slightly more than nineteen days).

Regarding sludge disposal, it is interesting to note that full load operation of the scrubber system requires 15 tons per hour of limestone and produces 19 tons per hour of waste sludge. Based on the utility's estimates, the cost to dispose of the sludge is between $7 and $10 per ton [45:24]. This cost takes the sludge from the plant to the waste disposal pond.

EPA representatives feel that the wet lime system installed at Louisville Gas & Electric's Paddy's Run facility is a more efficient and more representative system for analyzing lime/limestone scrubbing type system costs. At this coal burning plant, a wet lime scrubbing system was installed on a relatively old unit (twenty-five years), with a rating of 70 megawatts. However, due to the unit's age, it is inefficient and the scrubbing system is actually performing as though it were servicing a new facility of approximately 100 mw [80]. Louisville Gas and Electric estimates that the facility cost them $50 per kw to purchase, and that owning and operating expenses amount to approximately 2.0 mils/kwh, including sludge disposal. To date, this process has been operational for approximately three months, from September to December of 1973, and the firm is highly pleased with its performance. Based on this, they presently have plans to install this system on two 180 mw facilities and a 425 mw plant. They estimate capital costs of $45/kw and annual costs of between 1.5 and 2.0 mils/kwh for these projects.

Recently, the EPA has spent considerable funds to research and evaluate the Magnesium Oxide (MgO) scrubbing system. As part of this effort, they have entered into a joint venture with the Boston Edison Company to install an MgO scrubbing system on Boston's Mystic Unit #6. This plant is a 155 megawatt, oil-fired unit. Following a successful demonstration of the system on this unit, plans call for installing this design on units 4, 5, 7, for a total plant capacity of 1,050 megawatts.

Boston's Mystic Unit #6 was put in operation late in 1972, and is the only operational plant using MgO scrubbing. As of late 1973, this system is still undergoing de-bugging operations and cost projections are being revised. For the total project, 1050 mw, the utility is presently projecting a capital cost of about $42 million, or $40 per kw. On an annualized basis, they anticipate that it will cost about $12 million to own and operate; about $6 million dollars to own the system (including interest and depreciation), and a similar amount to operate it [81]. The total system will produce about 32,000 (long) tons of sulfur per year, which at present can be marketed for about $400,000 ($12.50 per ton of sulfur). Hence, accounting for this credit reduces the total annualized expenses to $11.6 million. Based on a load factor of 72 percent, therefore, it will cost the utility 1.75 mils per kilowatt-hour of energy produced to own and operate this system.

In another joint venture, the EPA and Illinois Power are presently evaluating the CAT-OX process. The first commercial sized installation of Monsanto's Catalytic Oxidation system was placed in operation on September 4, 1972 on the 100 megawatt Wood River #4 unit of the Illinois Power Company [42:1]. This power plant burns approximately 275,000 tons of coal per year, with a sulfur content of 3.1 percent. Using these figures, it is estimated that the CAT-OX system will produce about 19,000 tons per year of equivalent 100 percent sulfuric acid [42:2].

Although capital costs for this installation were initially estimated to be $6 million, the final cost amounted to approximately $8.5 million, or $85/kw [70].

Authorities estimate that operating expenses amount to between $500,000 and $600,000 a year. Here again, the final figure is higher than the initial projection of $400,000 per year. Moreover, Illinois Power estimates that the sale of sulfuric acid amounts to a return of approximately $100,000 per year. Industry sources [81] generally agree that a 15 percent factor applied to capital costs is a reasonable approximation for computing annualized costs; i.e., if one were to compute the present worth of the total expenses incurred over the operational life of the plant (approximately thirty-five years) and use a capital recovery factor to calculate equivalent annualized costs, the average of these figures would correlate with the simple 15 percent factor. Therefore, using this approach an annualized cost of approximately $1.8 million, or 2.96 mils/kwh based on a 70 percent load factor, was calculated to own and operate this facility.

Regarding the sale of sulfuric acid, the utility feels that a "good" market exists for it locally, which is able to absorb all that the firm presently produces. As previously mentioned, the CAT-OX process produces a relatively low concentration sulfuric acid. Illinois Power presently sells the 78 percent pure sulfuric acid which is produced to fertilizer manufacturers. They charge approximately $15 per ton of acid, of which $9 per ton is freight charges [70]. On the basis of a total annual cost of nearly $2 million therefore, the $100,000 realized from the sale of acid amounts to a savings of approximately 5 percent.

In summary, at this point, the EPA and various utilities are presently evaluating full-scale SO_x control systems to determine their reliability and economic feasibility. For comparative purposes, Table 3-5 has been prepared for the four systems described in the previous paragraphs. It shows that in these cases utilities pay a capital cost of from $40 to $85 per kw of installed capacity and an annualized cost of about 2 to 5 mils/kwh, depending on the system and its installation. As stated initially, the utilities are faced with the choice of installing SO_x removal systems, or paying a premium for lower sulfur content fuels if SO_x emission standards are met. In the next section, the costs of lower sulfur coal and oil to achieve comparable results are considered.

Higher Quality Fuel Costs

The present world shortage of oil caused by the Arab oil embargo has made it difficult, if not impossible, to project the future cost of this fuel. Although the United States possesses an abundant supply of low sulfur coal, limited supplies are presently available, causing its price to increase as utilities are forced to shift from oil to coal to meet energy demands.

In general, utilities enter long-term contracts with fuel suppliers and transporters to obtain their fuel. Since they purchase and transport such large quantities, they are normally given a discounted rate. Based on the best

Table 3-5
Comparative Costs of Actual SO_x Removal Systems

	Process	Plant Size mw	Capital Cost ($/kw)	Annualized Cost (Mils/kwh)
Will County Unit #1[a] (Chicago, Ill.)	wet limestone scrubbing	163	85[1] (96)[2]	4.74
Paddy's Run[b] (Louisville, Ky.)	wet lime scrubbing	70	50[2]	2.0
Mystic Unit # 6[c] (Boston, Mass.)	MgO	155	40	1.75
Wood River # 4[d] (Decatur, Illinois)	CAT-OX	100	85	3.10

[1] Neglecting sludge treatment equipment

[2] Including sludge treatment equipment

Sources:

[a]D.C. Gifford, "Will County Unit 1 Limestone Wet Scrubber," paper presented at the American Institute of Chemical Engineers 65th Annual Meeting, Nov. 1972, pp. 23, 24.

[b]R. Van Ness, Engineer, Louisville Gas and Electric (Louisville, Ky.), Personal Communication, Dec. 1973.

[c]C.P. Quigley, Head, Mechanical and Structural Division, Boston Edison Co. (Boston, Mass.), Personal Communication, Dec. 1973.

[d]E.A. Shultz and W.E. Miller, "The CAT-OX Project at Illinois Power," paper presented at the Electrical World Technical Conference (Chicago, Ill., Oct. 25-26, 1972), pp. 1-3.

estimates of industry sources, Table 3-6 was prepared, which shows the expected added costs of low sulfur fossil-fuel in several cities. Comparing these costs with the annualized costs of SO_x removal systems presented in Table 3-5, shows that in today's market control systems compete favorably with low sulfur content fuels as a means of meeting emission standards.

For example, Boston Edison estimated a 1973 cost of $8.00 per barrel of low sulfur ($<$.3 percent) fuel oil, in comparison to $5.50 for higher sulfur (2.8 percent) oil [81]. Since its 1050 mw Mystic power plant requires approximately 10 million barrels of fuel oil a year, the firm can save $25 million a year by burning a higher sulfur content fuel. Deducting the projected annualized cost of $11.6 million for the MgO scrubbing system, the utility may save better than $13 million a year, or about $1.30 per barrel of fuel oil, by cleaning its stack gas instead of using low sulfur oil. Since Boston Edison estimates that it costs oil processors about $1.50 per barrel to desulfurize oil, it is projected that this source will not be more economical than MgO scrubbing; i.e., even if the fuel oil processors wanted to respond to this competition by lowering their prices, the MgO scrubbing system would still be more economical.

As another illustration, in order to meet local emission standards, Detroit

Table 3-6
Typical Added Costs of Low-Sulfur Fossil-Fuels

	Low-Sulfur Fuel	Mils/kwh
Louisville, Ky.[a]	coal	8.34
Boston, Mass.[b]	oil	3.5
Detroit, Mich.[c]	coal	3.2

Sources: Computed from information provided by

[a]R. Van Ness, Engineer, Louisville Gas and Electric (Louisville, Ky.), Dec. 1973.

[b]C.P. Quigley, Head, Mechanical and Structural Division, Boston Edison Co. (Boston, Mass.), Dec. 1973.

[c]J. McCarthy, Fuel Buyer, Detroit Edison Co. (Detroit, Mich.), Dec. 1973.

Edison is presently experimenting with low sulfur coal (about .1 percent sulfur) from Montana and Wyoming. Since they are buying a relatively small amount of this coal, they anticipate that its cost may be less when they enter long-term agreements to buy this fuel in quantity. However, at present it will cost the firm an added expense of approximately 3.2 mil/kwh to purchase this fuel than the presently used high sulfur content coal (about 3 percent sulfur). Moreover, the Western coal has a lower heating value than coals presently being used, causing greater quantities of fuel to be consumed to meet energy demands. In addition, since it contains less sulfur, it lowers precipitator efficiencies and increases the expense to collect fly ash. Yet another problem with this fuel is that it is high in sodium content, which requires an added expense to modify existing equipment for its use.

Louisville Gas and Electric, estimates that due to the high freight costs to transport western coal to their facility, their fuel costs would double, from about $10/ton to $20/ton [80]. Since the heat content of this fuel is approximately 8,000 BTU per pound in comparison to the 11,500 BTU per pound of available local high sulfur content coals, an added cost of 8.34 mils/kwh would be incurred. These and other increases in costs due to low sulfur fuels are summarized in Table 3-6.

Therefore, on the basis of this evaluation it appears that SO_x control processes are competitive with the use of low sulfur content fuels. Although many argue that there are still engineering difficulties to be overcome with the SO_x removal processes, low sulfur coals also introduce boiler and ash collection uncertainties which are yet to be solved.

Regenerable systems such as MgO scrubbing and the CAT-OX process produce recoverable sulfur. Concluding this analysis, the impact of this by-product on existing markets and the effect of sulfur revenues on process costs will be examined.

Sulfur Market and Revenue Projections

Many argue that abatement sulfur extracted from power plant stack gases will glut the existing market and cause a significant decline in the value of sulfur. Some even feel that if regenerable processes are widely adopted, utilities will be forced to pay for sulfur removal as they are presently paying for ash removal. In the following paragraphs, the potential impact on the world market of sulfur recovered from U.S. power plants, herein designated as abatement sulfur, is examined.

Although prior to 1968 sulfur was in short supply and prices were rapidly rising, since that time an oversupply condition has existed [31:2]. The present world oversupply of sulfur is the result of by-product recovery from the refining of Canadian natural gas [31:67]. It is projected that this condition will exist for the next ten to fifteen years and will only shift to a situation in which demand exceeds supply when both the recovery of sulfur from Canadian natural gas and its stockpile has peaked. Over the past few years, the business slowdown that began in 1969 has decreased the industrial demand for sulfur. Thus, while the production of sulfur has been increasing, its demand has been level, or decreasing. Projections of the world demand for abatement sulfur are further complicated by the fact that sulfur demand is inelastic; i.e., as the price of sulfur decreases, the quantity demanded does not increase sufficiently to produce greater total revenue.

Transportation cost is a significant factor in evaluating the impact of abatement sulfur on existing markets. For example, the natural gas supplies in Alberta (Canada) and Louisiana are of the same order of magnitude, with the former being significantly higher in sulfur content. However, the increased transportation costs from the Alberta region to Louisiana prevents this source from significantly affecting the Louisiana sulfur market. If the opposite were true, the U.S. sulfur producers in the Louisiana area would have to shut down, due to the increased production of by-product sulfur. Transportation costs are particularly important for sulfuric acid users, where acid transportation costs are more than three times as much per ton of sulfur value than elemental sulfur.

Since, the fertilizer industry represents the predominate market for sulfur, accounting for about one-half of the worldwide demand, it would be desirable if fertilizer factories were located close to power plants. Unfortunately, as shown in Figure 3-7, most of the fertilizer production is in Florida and along the Gulf Coast, where sulfur emission is relatively small. In contrast, the Northeast is a potential major source of sulfur, but there is little fertilizer production. As a result, some utilities, the Florida Power and Light Company as one example, appear to be in a prime market for abatement sulfur (Figure 3-7).

A study performed by the Sulfur Institute indicated that if utilities widely adopted recovery stack gas scrubbing systems, about 15 million tons of sulfur would be recovered in 1985 [82:1]. Since sulfur supply from existing sources is

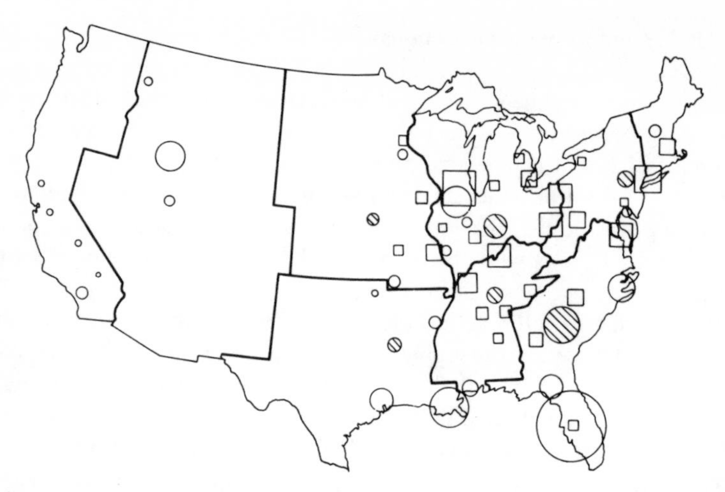

Size of circle indicates relative quantity of sulfur consumed in phosphate fertilizer production in adjacent area. Shaded circles indicate regional quantity of normal superphosphate. Blank circles indicate phosphoric acid plant complexes.

Size of square indicates relative quantity of sulfur emitted from power plants in adjacent area.

Figure 3-7. Sulfur Oxide Emission and Sulfur Consumption in the U.S. Source: Tennessee Valley Authority, *Sulfur Oxide Removal from Power Plant Stack Gas: Ammonia Scrubbing*, Prepared for the National Air Pollution Control Assoc. (Muscle Shoals, Ala.: TVA Bulletin Y-13, Oct. 1970), p. 53.

projected to exceed demand through this period, unless additional sulfur markets are developed utilities may be forced to pay to remove or stockpile sulfur. However, the Sulfur Institute has identified significant new markets which might absorb the increased supply, including sulfur-asphalt paving materials, sulfur concretes, sulfur coatings, and sulfur foams. These markets were selected since each might absorb large tonnages of sulfur by 1985 and are ecologically acceptable [85]. In addition, the time and money needed for commercial development appears reasonable. The study concludes that if these markets are developed, it will be possible for utilities to benefit from the sale of abatement sulfur [85].

The Illinois Power & Light Company and Boston Edison Company consider the sale of abatement sulfur an attractive market in their areas. However, Boston Edison is concerned that entry into this market by neighboring utilities will result in an over-saturation condition, causing price reductions.

In order to evaluate the potential impact of future sulfur price on projected annualized costs, a sensitivity analysis for Boston's MgO process was performed. As shown in Table 3-7, at the present recovery price ($12.50/ton), the firm

Table 3-7
Effect of Sulfur Revenue on Boston Edison's[1] MgO Scrubbing System Costs

Sulfur Revenue ($/ton)	Annual Owning and Operating Expenses (Mils/kwh)
−10.00[2]	1.86
0.00	1.82
12.50[3]	1.75
20.00	1.71

[1] Based on present cost estimates for Boston's Mystic Power Plant.
[2] That is, utility pays $10/ton to dispose of recovered sulfur.
[3] Present Boston Edison estimated return.
Source: Computed from Information Provided by C.P. Quigley, Head, Mechanical and Structural Division, Boston Edison Co. (Boston, Mass.), Dec. 1973.

estimates an annual owning and operating cost of 1.75 mils/kwh. Assuming the firm were forced to pay $10 per ton sulfur recovered, this expense would only increase to 1.86 mils/kwh. Further, if the sulfur revenue were to increase to $20 per ton, the annual owning and operating expenses would only drop to 1.71 mils/kwh. This analysis, therefore, indicates the relative insensitivity of this process to revenues from the sale of sulfur. In addition, since each of the projected costs is still below the present cost for low sulfur content fuels, 3.1 mils/kwh, it verifies the attractiveness of the MgO scrubbing process for this utility.

Summary: Costs of SO_X Removal Systems

At present the EPA and many utilities are spending considerable resources to investigate alternate SO_X removal systems. Industry sources still disagree as to whether or not these units will perform, reliably and economically. However, on the basis of EPA analysis and available plant data, it appears that SO_X control processes are at least competitive with burning a lower sulfur content fuel, the only other means of meeting present emission standards.

For a normal plant size of 500 megawatts, operating at a 60 percent load factor and burning 3.5 percent sulfur coal, the EPA estimates that the lime/limestone and MgO scrubbing processes will cost about 2.40 mils/kwh on an annual basis to own and operate (Table 3-4). Although the CAT-OX process is slightly more expensive, 2.75 mils/kwh, each of these control techniques appears cheaper than the projected costs of low sulfur content fuels at typical plants (Table 3-6). Many of the new plants being built are larger than 500 mw, indicating that economy of scale benefits may lower these costs further.

Selected case studies, have shown that on retrofitted power plants, the costs of SO_x removal varied from about 2 to 5 mils/kwh (Table 3-5). With the exclusion of Commonwealth Edison's unit, the costs average slightly more than 2 mils/kwh. Commonwealth Edison's plant is not considered representative of wet limestone scrubbing economics and that system costs are closer to 2 to 3 mils/kwh. Conversely low sulfur content fuels have been projected to vary from about 3 to 8 mils/kwh. Therefore, SO_x removal systems appear to offer a definite economic advantage over low sulfur content fuels, especially in certain locations. For example, Boston Edison estimates a savings of about 1.8 mils/kwh and Louisville Gas & Electric better than 6 mils/kwh will be realized by employing SO_x removal systems; i.e., Louisville Gas and Electric projects a premium fuel cost of 8.34 mils/kwh (Table 3-6) and a system cost of about 2.0 mils/kwh (Table 3-5).

Further, the sale of recovered sulfur does not significantly lower the cost of regenerable systems. Both Illinois Power and Light (CAT-OX), and Boston Edison (MgO) project small revenues from the sale of abatement sulfur. Consequently, although the future of this market is in doubt, Boston Edison's MgO process has been shown to be fairly insensitive to this return.

To achieve the advantages of lower annualized costs afforded by a process control system requires a significant capital investment and three years to become operational. In order to include this cost in the rate base, the utility would have to substantiate these expenses to the state utility commission. As the proceedings of these agencies are time-consuming and require considerable manpower, utilities are hesitant to request rate changes. The highly publicized proceedings and rate increases are obviously very unpopular to the end user. Consumer acceptance of rate increases is more apt to be obtained when the justification is readily understood.

Many state commissions have passed rulings which allow utilities to modify rates to reflect increasing fuel costs. Consequently, although the economics may indicate an advantage for SO_x control processes, utilities may still elect to purchase low sulfur oils to meet emission standards.

Study Summary and Conclusions

Fossil-fueled power plants are, and will continue to be a major source of air pollution in the United States. Although the overall percentage of power generated by fossil fuels will decrease, the FPC estimates that more than twice as much energy will be generated in 1990 than in 1970 by these fuels. Hence, although the EPA has established what many utilities feel are overly restrictive standards, it is projected that even if these regulations are satisfied, SO_x emissions will remain at present levels through the year 2000.

The three primary air pollutants emitted from fossil-fueled power plants are particulate matter, and sulfur and nitric oxides. At present, mechanical collectors and electrostatic precipitators are acceptable methods for limiting particulate matter (fly ash) emissions from power plants. The primary problem remaining with fly ash is how to dispose of this waste once it is collected. On the other hand, research is still underway to develop methods to control nitric oxide emissions and, to date, no method has been identified which will recover nitrogen oxides from the waste flue gas. Since fossil-fueled power plants are primarily responsible for the bulk of sulfur oxide emission into the atmosphere, a great deal of work has been done to develop methods of controlling SO_x emissions from these facilities. Although there is disagreement among the experts over the state-of-the-art of SO_x control processes, three appear to be gaining acceptance; viz., wet lime/limestone scrubbing, MgO scrubbing and the CAT-OX processes. Wet lime/limestone scrubbing systems produce a solid waste, calcium sulfite, that must be disposed of, whereas MgO scrubbing and the CAT-OX processes produce recoverable sulfur.

The purpose of this investigation has been to evaluate the costs of air pollution control for fossil-fueled power plants. Since the power generated by fossil fuels will continue to increase in the coming decades, the total costs of air pollution control will similarly increase, unless practical methods of offsetting these expenses are developed. Regarding fly ash pollution, it has been shown that economical and reliable methods exist to control fly ash emissions into the atmosphere. Using the Detroit Edison Company for this evaluation, it was concluded that fly ash control is only a small percentage of the cost of fuel. The primary problem remaining for the utilities is how to dispose of this waste once it is collected. Therefore, the study concentrated on the costs of fly ash disposal from the point that it is loaded on trucks for transportation to a landfill area. At present, utilities typically pay from $.50 to $2.00 per ton to dispose of fly ash. Over the past several years, Detroit Edison has been spending about $1.00 per ton of ash collected. In 1972, therefore, Detroit Edison spent about $900 thousand for fly ash disposal. Since the EPA projects that by the year 2000 the total fly ash collected will be more than triple the amount collected in 1970, the costs of fly ash disposal will increase in the coming decades. Additionally, since many utilities are located in urban centers where land is at a premium, it is projected that the disposal cost per ton of fly ash collected will also increase. In order to offset this expense, utilities are attempting to market fly ash more aggressively.

At present it is estimated that about 10 percent of the fly ash generated is sold. However, it has been shown that sizable, relatively untapped fly ash markets exist. At today's levels, if the total fly ash generated were used as a lightweight aggregate, approximately 2 percent of the aggregate market would be affected. However, in order to produce lightweight aggregate from fly ash, the ash would have to be sintered. Hence, if this market is to be explored, capital would have to be invested to construct sintering plants. Buyer resistance, existing specifications, and lobbying by the aggregate industry are significant impediments which must be overcome to gain acceptance in this market. For the Detroit Edison Company, it was shown that all of the fly ash generated could have been used as a binder and filler in asphalt paving of state highways in 1972. However, although fly ash asphalt possesses superior non-skid properties, Michigan has yet to approve it for highway usage. Another significant market for fly ash is as a replacement for cement in the production of ready-mix concrete and portland pozzolan cement. In the Detroit area, it is estimated that about 170,000 tons of fly ash could be used in ready-mixed concrete.

Therefore, it can be concluded that significant markets exist for fly ash, which could utilize this product more extensively. If Detroit Edison were able to receive a net-back of $1.25 per ton of fly ash collected, based on the present generation of ash, the firm would benefit by over $2 million annually. Based on this evaluation it appears that utilities can realize a significant cash flow from the sale of recovered fly ash and would benefit by entering these markets more aggressively.

Regarding SO_x emission control standards, the utilities are faced with the choice of burning lower sulfur content fuels, or installing what many feel are unproven control systems. If the utility elects to install a control process a fairly large capital investment is required and it will take approximately three years to build the system. On an annualized basis it has been demonstrated that SO_x control processes can be more economical than the burning of low sulfur content fuels. Typically the costs of these systems has been shown to be in the range of 2 to 4 mils/kwh in comparison to a premium, i.e., low sulfur content, fuel cost of 3 to 8 mils/kwh. Two of the processes discussed, MgO scrubbing and the CAT-OX, produce recoverable sulfur.

Examination of the market for abatement sulfur showed that some utilities, such as Florida Power and Light, are located close to fertilizer plants, the prime users of sulfur. The only two firms presently recovering sulfur from their stack gases, Illinois Power and Light and Boston Edison, each feel that excellent demand exists locally for their abatement sulfur. Many people are concerned that sulfur recovered from power plant stack gases will swamp the market, depressing the price of sulfur and increasing the cost of recoverable processes. The Sulfur Institute, however, feels that markets could be developed by 1985 to absorb the sulfur recovered from utility stacks.

Additionally, analysis of the effect of sulfur price on the annualized cost of MgO scrubbing at Boston Edison's Mystic plant showed an insensitivity to this variable. Although somewhat surprising to this analyst, it appears therefore, that sulfur control processes can be more economical than purchasing lower sulfur content fuels. In addition, the utilization of western low sulfur content coals also presents technical difficulties; i.e., they possess lower heat rates, decrease precipitator efficiencies, have a larger ash content than coal presently being used, and this ash contains large concentration of sodium salts resulting in lower boiler efficiencies. Utilities at present, therefore, also consider the utilization of these fuels to be in the development stage requiring that equipment be modified to burn this fuel.

At several points in the study it was pointed out that there are at least three factors affecting the utilities decisions regarding air pollution control measures: the price of premium fuel, the availability of capital, and the influence of regulatory agencies. Based on discussions with industry sources it appears that of the three, the most significant is the influence of regulatory agencies.

For example, although it has been shown that SO_x control systems can be less expensive on an annualized basis than purchasing lower sulfur content fuels, the attitude and procedures of state utility commission may induce the firm to choose the more expensive alternative. In many states, rulings have been passed which allow the firm to modify rate charges to reflect an increased fuel cost. Therefore, many utilities have chosen to take the easier path of burning a low sulfur content oil, than investing in a control process. Similarly, in the area of fly ash disposal, it was pointed out that in order to penetrate existing markets the utilities would have to invest capital to construct fly ash processing plants. Here again, the attitude of state utility commissions may be a deterrent. In order to include capital investments in the rate base, the utility must first make the investment and then substantiate the need for the expense to the commission. The required proceedings can take from two to three years before approval. Since the utilities are in the business of producing electrical power and not, for example, lightweight aggregate, the commission may elect not to approve this investment.

Moreover, local environmental regulations may require that emission control standards must be met by 1975. Since control techniques typically take two to three years to install and de-bug, the utility is again faced with the decision of going to a regulatory agency to ask for a variance if it decides to install an SO control process. The influence and interaction between federal and local regulatory agencies may also adversely impact economic decisions. At present, the EPA has yet to set up standards for the ponding of the wet sludge produced from the lime/limestone scrubbing processes [48].

It appears that before "pure" economics can be the primary motivator for these decisions by utilities, the regulatory agencies will have to more closely coordinate their activities. What is needed is an overall, or systems-minded review of the total problem. This direction would probably best originate with the FPC and work its way down to local agencies.

Finally, the evaluation presented has been prepared essentially on an accounting or financial basis. As such it has shown, for example, that although a significant market exists for fly ash as an aggregate or as pozzolanic material in the production of concrete, competition from existing products make it difficult for this by-product to enter these markets. One must ask, however, whether the United States as a nation can continue to strip mine aggregates, defacing the countryside, purely on the basis of economic considerations. Although the by-product ash and sulfur collected from power plants may alter present market conditions, and limit the return realized by existing firms, the benefits of cleaner and healthier air would appear to far outweigh this factor.

Appendix 3A

Projected Costs of Limestone Scrubbing SO_x Removal Systems

The capital cost, economy-of-scale benefits to be realized for a limestone scrubbing system is presented in Figure 3A-1. As shown, capital cost constantly decreases as plant size increases; e.g., a 50 megawatt plant would cost approximately $85 per kilowatt in comparison to $40 per kilowatt for a 1000 megawatt plant.

In Figure 3A-2, projected annualized cost, economy-of-scale benefits as a function of plant size and load factor are presented. Here again, it is seen that as plant size increases the annualized cost is predicted to constantly decrease. Moreover, the higher the load factor, the lower the annualized cost for any given plant size. For a 500 megawatt plant the annualized cost is anticipated to decrease by approximately 33 percent as the load factor increases from 40 to 70 percent; i.e., a decrease in annualized cost of from approximately 3.3 mils/kwh to 2.2 mils/kwh.

Figure 3A-3 presents projected annualized cost data as a function of fuel (coal) sulfur content for the base case. The uppermost curve of this figure assumes that all of the gas is cleaned (i.e., scrubbed) to remove 90 percent of the SO_2 present. The parametric curves below this case are based on cleaning a sufficient quantity of gas to meet various emission standards. A 0.5 percent S standard means that the process will emit no more SO_2 than would be emitted by burning a 0.5 percent sulfur content coal. Hence for this standard, the annualized cost for burning coal with sulfur content of 0.5 percent would be zero. However, as shown, as the sulfur content of the fuel increased, the annualized cost would also increase; e.g., for 3.5 percent sulfur coal, an annualized cost of approximately 2.4 mils/kwh is projected. For this illustration, then, it would not be worthwhile for a utility to pay a premium of more than 2.4 mils/kwh for a lower sulfur content fuel.

Table 3A-1
Estimated Nationwide Discharges of Airborne Pollutants for 1968

[Million tons]

Source	Carbon Monoxide	Particulate Matter	Sulfur Oxides[1]	Hydro-Carbons	Nitrogen Oxides[1]	Total
Transportation	63.8	1.2	0.8	16.6	8.1	90.5
Power plants	0.1	5.6	16.8	Neg.	4.0	26.5
Other fuel combustion in stationary sources	1.8	3.3	7.6	0.7	6.0	19.4
Industrial processes	9.7	7.5	7.3	4.6	0.2	29.3
Solid waste disposal	7.8	1.1	0.1	1.6	0.6	11.2
Miscellaneous	16.9	9.6	0.6	8.5	1.7	37.3
Total	100.1	28.3	33.2	32.0	20.6	214.2

[1] Sulfur oxides expressed as tons of sulfur dioxide and nitrogen oxides as tons of nitrogen dioxide.

Source: Federal Power Commission, *The 1970 National Power Survey* (Washington, D.C.: Government Printing Office, December 1971), p. I-11-2.

Table 3A-2
Estimated Electric Utility Production of Electric Power by Primary Energy Source

Fuel or Energy Source	1970 Million mwh[1]	1970 (%)	1980 Million mwh[1]	1980 (%)	1990 Million mwh[1]	1990 (%)
Fossil fuel	1,262	81.8	1,922	61.8	2,628	44.3
Nuclear	22	1.4	874	28.1	2,913	49.2
Hydro	257	16.8	317	10.1	381	6.5
Total	1,541	100.0	3,113	100.0	5,922	100.0

[1] Includes energy for pumped-storage pumping: 6, 38, and 94 million mwh (megawatt hours) respectively, for 1970, 1980, and 1990.

Source: Federal Power Commission, *The 1970 National Power Survey* (Washington, D.C.: Government Printing Office, December 1971), p. I-11-6.

Table 3A-3
Electric Power Generation by Fossil Fuels (1966 to 1971)

Year	Total Power Generated[1] (mwh)	Power Generated by Fossil Fuels[2] (percent): Coal	Oil	Gas	Total
1971	1,347	53	16	28	97
1970	1,283	55	14	29	98
1969	1,191	59	12	28	99
1968	1,106	62	9	28	99
1967	992	64	9	27	100
1966	949	65	8	27	100

[1] Values have been rounded to nearest mwh.

[2] Values have been rounded to nearest percentage.

Source: *Statistical Year Book of the Electric Utility Industry for 1971* (N.Y., N.Y.: Edison Electric Institute, October 1972), p. 22.

Table 3A-4
Percentage of Pollutants Emitted by U.S. Power Plants as a Function of Fuel Used–1969

Fuel	Particulates	SO_2	NO_X	Total
Coal	99.3	91.7	75.1	89.8
Oil	0.7	8.3	11.5	7.7
Gas	–	–	13.4	2.5

Source: The Federal Power Commission, *Steam-Electric Plant Air and Water Quality Control Data, for the year ended Dec. 31, 1969* (Washington, D.C.: Government Printing Office, February 1973), p. XVI.

Table 3A-5
Correlation of Coal Usage to Regional Distribution of Air Pollutant Emissions –1969

Geographic Region	Power Generated (percent)[1]	Pollutant Emitted[2] (percent)	Coal Burned[3] (percent)
New England	4	3	2
Middle Atlantic	14	14	14
E. North Central	23	39	38
W. North Central	6	7	7
South Atlantic	19	18	21
E. South Central	10	15	15
West South Central	13	1	–
Mountain	3	2	3
Pacific	7	1	–
Non-Contiguous U.S.	1	–	–
Total	100	100	100

[1] Numbers have been rounded to nearest whole percentage.

[2] Based on total emissions of particulates, SO_X and NO_X.

[3] (Coal Consumption in Region/Total Coal Consumed in U.S.) × 100%.

Source: The Federal Power Commission, *Steam-Electric Plant Air and Water Quality Control Data, for the year ended Dec. 31, 1969* (Washington, D.C.: Government Printing Office, February 1973), pp. 1, 5, and XVIII.

Table 3A-6
Estimated Investment Costs for Commonwealth Edison's Will County Unit 1 Limestone Scrubbing System

	Capital Lost ($1,000)
B & W Wet Scrubber	$ 2,928
Equipment Erection	5,556
Electrical Equipment and Erection	1,210
Foundations	923
Limestone Handling System	204
Professional Engineering	965
Mill and SO_2 Buildings	193
Structural Steel	375
Miscellaneous Equipment	946
	$13,300
Sludge Disposal System	1,700
	$15,000
Cost per kilowatt, with sludge treatment	$96
Cost per kilowatt, without sludge treatment	$85

Source: D.C. Gifford, "Will County Unit 1 Limestone Wet Scrubber," paper presented at the American Institute of Chemical Engineers 65th Annual Meeting, November 28, 1972, p. 23.

Table 3A-7
Estimated Annual Costs for Commonwealth Edison's Will County Unit 1 Limestone Scrubbing System

	Annual Cost[1] ($1,000)
Scrubber System	
Carrying charges on 13,300,000	1,890
Property tax on 13,300,000	266
Limestone @ $5/ton	459
One Shift Position	70
Auxiliary Power	412
Reheater Steam	57
Maintenance	266
Subtotal	$3,420
Sludge Disposal System[2]	
Carrying Charges on 1,700,000	241
Property tax on 1,700,000	34
Sludge disposal @ $7/ton	823
Subtotal	$1,098
Grand Total	$4,518

[1] Costs are based on 70% capacity factor, 15-year life.

[2] Sludge disposal includes sludge disposal plant operating, maintenance, hauling and land fill disposal costs.

Source: D.C. Gifford, "Will County Unit 1 Limestone Wet Scrubber," paper presented at the American Institute of Chemical Engineers 65th Annual Meeting, Nov. 28, 1972, p. 24.

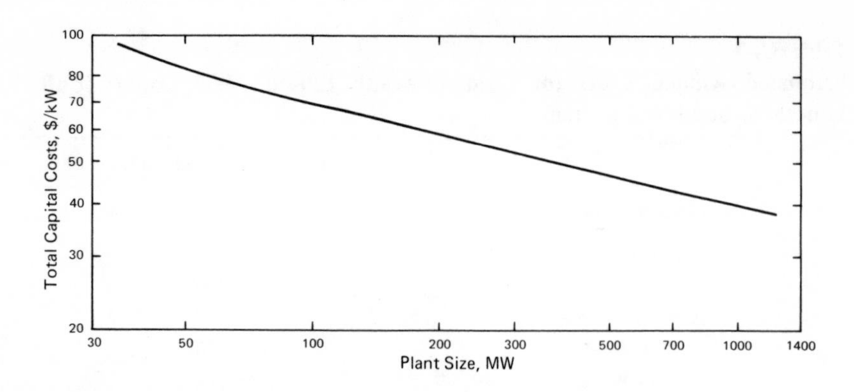

Figure 3A-1. Effect of Plant Size on Total Capital Costs for Limestone Scrubbing System. Source: G.T. Rochelle, "Economics of Flue Gas Desulfurization," Research Triangle Park, N.C. EPA, May 1973, p. 19.

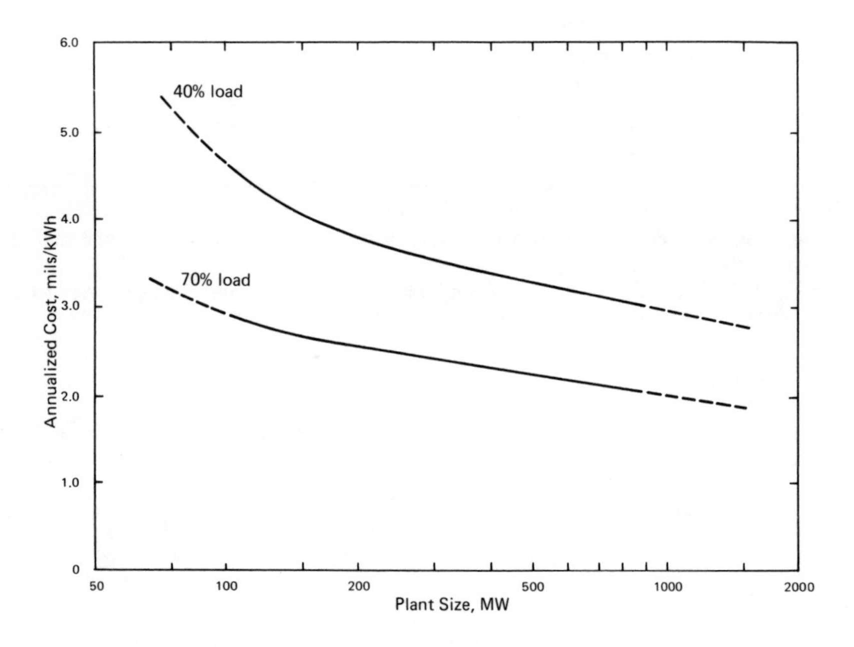

Figure 3A-2. Effect of Plant Size and Load Factor on Annualized Cost for Limestone Scrubbing System. Source: G.T. Rochelle, "Economics of Flue Gas Desulfurization," Research Triangle Park, N.C. EPA, May 1973, p. 21.

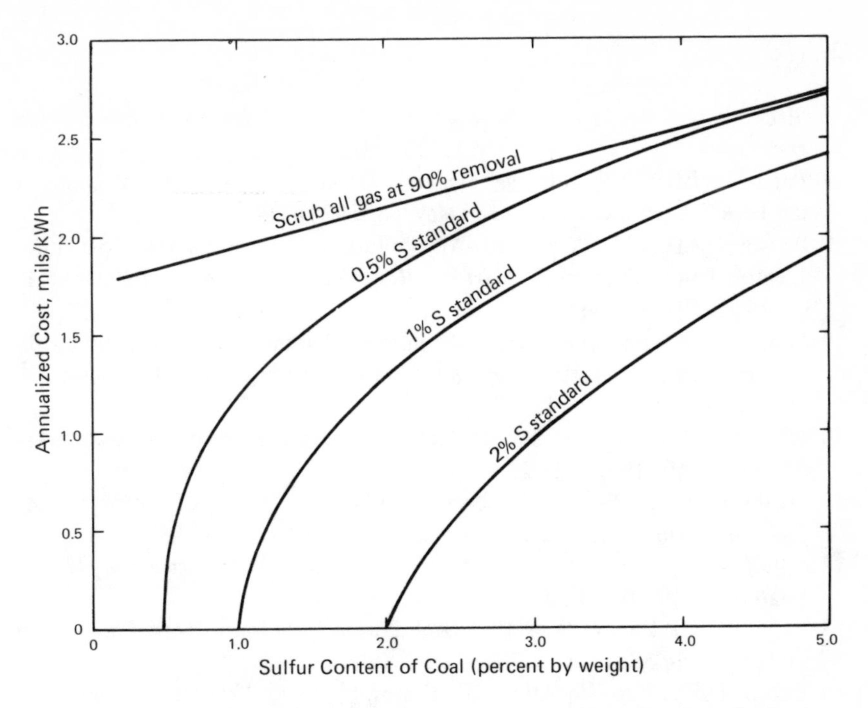

Figure 3A-3. Effect of Coal Sulfur Content and Emission Standard on Annualized Cost of Limestone Scrubbing System. Source: G.T. Rochelle, "Economics of Flue Gas Desulfurization," Research Triangle Park, N.C. EPA, May 1973, p. 22.

Bibliography

Books

1. Strauss, Werner (ed.). *Air Pollution Control, Part I.* New York: Wiley–Interscience, 1971.
2. Stern, Arthur C. (ed.). *Air Pollution, Vol. III: Sources of Air Pollution and Their Control.* New York: Academic Press, 1968.
3. Hottel, H.C. and Howard, J.B. *New Energy Technology, Some Facts and Assessments.* Massachusetts: MIT Press, 1971.
4. *Ashrae Guide and Data Book–Equipment.* New York, New York: American Society of Heating, Refrigerating and Air-Conditioning Engineers, Inc., 1969.
5. Garvey, Gerald. *Energy, Ecology, Economy.* New York: W.W. Norton & Co., Inc., 1972.

Articles

6. Slack, A.V. "Removing SO_2 From Stack Gases." *Environmental Science and Technology* 7, 2 (February 1973): 110-119.
7. Farthing, J.G. "SO_2 Removal Systems: Development Proceeds at Quickening Pace." *Electrical World* 175 (May 15, 1973): 34-39.
8. Watson, N.D., Jr. "Costs of Air Pollution Control in the Coal Fired Electric Power Industry." *Quarterly Review of Economics and Business* 12 (Autumn 1972): 63-85.
9. Pfeiffer, B.A. and Gilbert, R.D. "Pollution Abatement Expenditures by the Electric Power Industry." *Public Utilities Fortnightly* 90 (August 31, 1972): 21-28.
10. "Pollution Cost Studies Should Include Benefits We Buy." *Industry Week* 173 (April 10, 1972): 21-22.
11. Hardison, L.C. "Techniques for Controlling the Oxides of Nitrogen." *Air Pollution Control Association Journal* 20 (June 1970): 377-382.
12. "Budgets of U.S. Utilities Assure Fast System." *Electrical World* 177 (March 15, 1972): 44-50.
13. "The High Price of More Electrical Power." *Business Week* (August 19, 1972), pp. 54-58.
14. Booth, D.W. "Can We Afford Tomorrow?" *Public Utilities Fortnightly* 90 (September 14, 1972): 25-31.
15. "Generation in Focus." *Power Engineering* 76, 6 (June 1972): 20.
16. Roe, K.A. and Young, W.H. "Trends in Capital Costs of Generating Plants." *Power Engineering* 76, 6 (June 1972): 40-44.
17. Frankenberg, T.T. "What Sulfur Dioxide Problem." *Combustion* (September 1971), pp. 6-12.
18. "The High Price of More Electric Power." *Business Week* (August 19, 1972), pp. 54-58.

Public Documents

19. U.S. Department of Health, Education and Welfare. *Control Techniques for Sulfur Oxide Air Pollutants.* January 1969.
20. The Federal Power Commission. *The 1970 National Power Survey*. Washington, D.C.: December 1971, Parts 1-4.
21. U.S. Department of the Interior, Bureau of Mines Information Circular/1970 (IC-8488). *Ash Utilization,* 1970.
22. U.S. Environmental Protection Agency. *Conceptual Design and Cost Study–Sulfur Oxide Removal from Power Plant Stack Gas, Magnesia Scrubbing–Regeneration: Production of Concentrated Sulfuric Acid.* EPA-R2-73-244, May 1973.

23. The Federal Power Commission. *Steam-Electric Plant Air and Water Quality Control Data, for the year ended Dec. 31, 1969.* FPC-S-229, February 1973.
24. The Federal Power Commission. *Steam-Electric Plant Construction Cost and Annual Production Expenses–24th Annual Supplement, 1971.* February 1973.
25. U.S. Department of Commerce, Social and Economics Statistical Information. *General Social and Economic Characteristics–Mich., 1970 Census of Population.* April 1972.
26. U.S. Department of the Interior, Bureau of Mines. *Mineral Industry Surveys: Cement in 1972.* October 24, 1973.

Reports

27. Tennessee Valley Authority. *Sulfur Oxide Removal From Power Plant Stack Gas: Ammonia Scrubbing.* Prepared for the National Air Pollution Control Association, Muscle Shoals, Alabama: Bulletin Y-3, October 1970.
28. *Cleaning Our Environment, The Chemical Basis for Action.* A report by the Subcommittee on Environmental Improvement, Committee on Chemistry and Public Affairs, American Chemical Society. Washington, D.C., 1969.
29. *Engineers Orientation Course.* Con Ed Document.
30. *Technical and Economic Factors Associated with Fly Ash Utilization.* NTIS Document #PB-208-480. Springfield, Virginia. July 1971.
31. Farmer, M.H. *Long Range Sulfur Supply and Demand Model* NTIS Document #PB-208-993. Springfield, Va.: ESSO Research and Engineering Company. November 1971.
32. M.W. Kellogg Company. *Detailed Cost Breakdown for Selected Sulfur Oxide Control Processes.* NTIS Document #PB-209-024. Prepared for Environmental Protection Agency. Springfield, Va.: March 1972.
33. *Some General Economic Considerations of Flue Gas Scrubbing for Utilities.* Research Triangle Park, North Carolina: Control Systems Division, Environmental Protection Agency, National Environmental Research Center. October 1972.
34. McSorley, J.A., M.W. Kellogg Company. *Applicability of SO_2–Control Processes To Power Plants.* NTIS Document #PB-213-421. Prepared for Environmental Protection Agency: Springfield, Virginia: November 1971.
35. Princiotta, F.T. and Kaplan, N. *Control of Sulfur Oxide Pollution From Power Plants.* Research Triangle Park, North Carolina: Environmental Protection Agency, Control Systems Division, October 1972.
36. Burchard, J.K.; Rochelle, G.T.; Scholfield, W.R.; and Smith, T.O. *Some General Economic Considerations of Flue Gas Scrubbing for Utilities.*

Research Triangle Park, North Carolina: Environmental Protection Agency, Control Systems Division, October 1972.

37. Jones, J.W.; Stern, R.D.; and Princiotta, F.T. *Waste Products From Throwaway Flue Gas Cleaning Processes–Ecologically Sound Treatment and Disposal.* Research Triangle Park, North Carolina: Environmental Protection Agency, Development Engineering Branch, January 1973.
38. Waitkus, J.; Keppel, R.A.; Littlejohn, C.E.; and Lyerly, R.L. "An Evaluation of Sulfur Dioxide Removal Methods for Oil Fired Plants." A report prepared for the Florida Power & Light Company. Dunedin, Florida: Ray L. Liverly and Assoc., April 1972.
39. Rochelle, G.T. *Economics of Flue Gas Desulfurization.* Research Triangle Park, North Carolina: Environmental Protection Agency, Control Systems Laboratory, May 1973.
40. "Assessment of the State-of-Technology of Air Pollution Control Equipment and of the Impact of Clean Air Regulations on the Adequacy of Electric Power Supply of North American Bulk Power Systems: A Report by Clean Air Subcommittee of the Technical Advisory Committee." Princeton, N.J.: National Electric Reliability Council, October 1972.
41. "Load Growth and Environmental Protection, 1972-1982." A Report prepared by the Detroit Edison Company for the Michigan Public Service Commission, Detroit, Mich.: April 24, 1972.

Other

42. Shultz, E.A., and Miller, W.E. "The CAT-OX Project at Illinois Power." Paper presented at the Electrical World Technical Conference, Chicago, Illinois, October 25-26, 1972.
43. Hollinden, G.A., and Kaplan, N. "Status of Application of Lime/Limestone Wet Scrubbing Processes to Power Plants." Paper prepared for presentation at American Institute of Chemical Engineers 65th Annual Meeting. New York, November 26-30, 1972.
44. Sherwin, R.M.; Robin, I.A.; and Anas, P.P. "Economics of Limestone Wet Scrubbing System." Paper presented at International Symposium for Wet Limestone Scrubbing, New Orleans, Louisiana, November 8-12, 1971.
45. Gifford, D.C. "Will County Unit 1–Limestone Wet Scrubber." Paper presented at American Institute of Chemical Engineers 65th Annual Meeting, New York, November 28, 1972.
46. "Testimony on Sulfur Oxide Removal Systems–A.V. Slack." Lawrence County, Pennsylvania: Court of Common Pleas, January 1973. Available from Edison Electric Institute as enclosure of letter dated March 19, 1973 to chief executives of member firms.
47. Minnick, C.J. "Multiple By-Product Utilization." Paper presented at the Third International Ash Symposium, Pittsburgh, Pa., March 13-14, 1973.

48. Faber, John H. Executive Director, National Ash Association Washington, D.C. Personal Communication on Fly Ash Utilization, May 1, 1973.
49. Hill, Gladwin. "Sulphate Peril May Lead to Cleaner Air Law." *New York Times*, March 12, 1973, p. 62.
50. Brackett, C.E. "Production and Utilization of Ash in the United States." Paper presented at the Third International Ash Utilization Symposium, Pittsburgh, Pa., March 13-14, 1973.
51. Condry, L.Z.; Muter, R.B.; and Lawrence, W.F. "Potential Utilization of Solid Waste from Lime/Limestone Wet Scrubbing of Flue Gases." Paper presented at the Second International Lime/Limestone Wet Scrubbing Symposium, New Orleans, Louisiana, November 1971.
52. Blocker, W.V.; Morrison, R.E.; Morten, W.E.; and Babcock, A.W. "Marketing Power Plant Aggregate as a Road Base Material." Paper presented at the Third International Ash Symposium, Pittsburgh, Pa., March 13-14, 1973.
53. "Questions and Answers about the Electric Utility Industry." Edison Electric Institute, 1973 Edition, New York.
54. McGlamery, G.G. "Magnesia Scrubbing." Paper presented at the Flue Gas Desulfurization Symposium. New Orleans, Louisiana, May 14-17, 1973.
55. Brink, R.H. "Use of Waste Sulfate on Transpo '72 Parking Lot." Paper presented at the Third Ash Utilization Symposium, Pittsburgh, Pa., March 13-14, 1973.
56. Hyland, E.J. "Factors Affecting Pozzolan Marketing." Paper presented at the Third Ash Utilization Symposium, Pittsburgh, Pa., March 13-14, 1973.
57. McFalls, H., President, Western Fly Ash, Inc. Personal Letter on Fly Ash Utilization, September 5, 1973.
58. Haavik, D.J. Bechtel Power Corporation. Personal Letter on Fly Ash Utilization, September 11, 1973.
59. Martin, J.P., V.P. Corporate Planning, the General Crushed Stone Company. Personal Letter on Fly Ash Utilization, September 17, 1973.
60. Wechselblatt, P.M. and Quig, R.H. "Chemico-Basic Magnesium Based SO_2 Recovery Scrubbing Systems." Paper presented at the American Institute of Chemical Engineers 71st National Meeting, Dallas, Texas, February 20-23, 1972.
61. Koehler, G.R. "Operational Performance of the Chemico Basic Magnesium Oxide System at the Boston Edison Company, Part I." Paper presented at the EPA Flue Gas Desulfurization Symposium, New Orleans, Louisiana: May 14-17, 1973.
62. Sakanishi, J. and Quig, R. "One Years Performance and Operability of the Chemico/Mitsui Carbine Sludge (Lime) Additive SO_2 Scrubbing System." Paper presented at the EPA Flue Gas Desulfurization Symposium, New Orleans, Lousiana, May 14-17, 1973.
63. Prial, F.J. "Con Ed Seeks Suspension of High-Sulphur-Oil Bans." *New York Times*, October 27, 1973, pp. 1, 9.

64. Dennis, D.W. Representative of the Michigan Ash Sales Company. Personal Letter on Fly Ash Utilization, September 18, 1973.
65. Smith, C. Representative of G & W.H. Lorson Company. Personal Communication on Fly Ash Utilization, October 10, 1973.
66. Morton, W.E. President of Highway Materials Company (Brideport, W. Va.). Personal Communication on Fly Ash Utilization, August 27, 1973.
67. Rainey, Official of U.S. Census Bureau. Personal Communication on Ready-Mix Concrete Production, October 10, 1973.
68. McCormick, J.P., Vice President–Civic and Education Affairs, The Detroit Edison Co. Personal Communications on Fly Ash and SO_x Removal Costs, October 3 and 5, 1973.
69. Benson, G. Principal Engineer, The Detroit Edison Co. Personal Communications on Fly Ash and SO_x Removal Costs, October 23 and 24, 1973.
70. Miller, W.E. Director of Environmental Affairs, Illinois Power Company, Decatur, Illinois. Personal Communication on CAT-OX operation, August 28, 1973.
71. Quigley, C.P. Head, Mechanical & Structural Division, Boston Edison Company (Boston, Mass.). Personal Communication on MgO Process, August 23, 1973.
72. Quigley, C.P. "Progress Report, Magnesium Oxide Scrubbing System at Boston Edison Company's Mystic Station." Paper presented at *Electrical World*'s Technical Conference, "Sulfur in Utility Fuels: The Growing Dilemma," Chicago, Ill., October 25-26, 1972.
73. Quigley, C.P. "Operational Performance of the Chemico Magnesium Oxide System at the Boston Edison Company, Part II." Paper presented at the Flue Gas Desulfurization Symposium, New Orleans, Louisiana, May 14-17, 1973.
74. Maxwell, M. Environmental Protection Agency Engineer. Personal Communications on SO_x Removal Costs, April 4 and August 27, 1973.
75. Quig, R., V.P., Marketing–Pollution Control Division of Chemical Construction Corporation (New York, New York). Personal Communication on MgO Scrubbing System, September 25, 1973.
76. "Statistical Year Book of the Electrical Utility Industry for 1971." New York: Edison Electric Institute, October 1973, Pub. No. 72-25.
77. "The Tall Stack for Air Pollution Control on Large Fossil Fueled Power Plants." A compilation of papers provided by Congressman L. Wolff, N.Y.
78. Rochelle, G.T. Environmental Protection Agency Engineer. Personal Communications on SO_x Removal Costs, April 2 and May 17, 1973.
79. Glombowski, A., Director: Market and Economic Research Department, Portland Cement Association (Skokie, Ill.). Personal Communication on Cement Market, December 11, 1973.
80. Van Ness, R., Engineer, Louisville Gas and Electric (Louisville, Ky.). Personal Communication on Costs of Limestone Scrubbing, December 12, 1973.

81. Quigley, C.P., Head, Mechanical and Structural Division, Boston Edison Company (Boston, Mass.). Personal Communication on MgO Scrubbing, December 13, 1973.
82. Fike, H.L. and Platou, J.S. "New Uses for Sulfur–Their Status and Prospects." Paper presented at the Flue Gas Desulfurization Symposium, New Orleans, Louisiana, May 14-17, 1973.
83. McCarthy, J., Fuel Buyer, Detroit Edison Company (Detroit, Mich.). Personal Communication on Fuel Costs, Dec. 12, 1973.
84. Zimmer, F., Supervisor of Salvage Sales, Detroit Edison Company (Detroit, Mich.). Personal Communication on Fly Ash Disposal Costs, December 11, 1973.
85. Platou, J.S. "Elemental Sulfur–Problems and Opportunities." Paper presented at *Electrical World*'s Waste Disposal Conference, Chicago, Illinois, October 30-31, 1973.

4 Desulfurization of Fuel Oils

Gerald L. Neway

By-product recovery from the stack gases of the millions of homes and small businesses which burn fuel oil for space heating is not feasible. Can anything be done with recovery from central refineries?

Sulfur oxides are generally regarded as one of the most harmful forms of air pollution, especially in large cities. Chapter 3 examined the techniques for reducing the emissions of sulfur oxides from the stack gases of electric power plants employing sulfur bearing fuels. These techniques involve rather elaborate installations for removing the sulfur oxides in stack gases and are not economically feasible for the millions of residential and commercial consumers of fuel oil which may contain sulfur. For the latter, the only practical means of controlling sulfur oxide emissions is through the imposition of input standards that set the maximum permissible percentage of sulfur in fuel oils consumed.

Sulfur appears in fuel oil combined chemically with hydrocarbons and is converted to sulfur oxides upon combustion. The sulfur content of fuel oils is measured by the percentage by weight which the chemically combined sulfur represents of the total weight of the oil.

The Environmental Protection Agency, acting under the 1970 Amendment to the Clean Air Act of 1970, has been required to establish national ambient air quality standards and national emission standards for significant pollution sources. In this connection, the EPA has set the maximum permissible sulfur content of fuel oils used for space heating at 0.7 percent by weight. The problem of sulfur oxide pollution is most severe in the large cities of the northeast and several of these states or cities have established standards for the sulfur content of fuel oils more restrictive than the national level of 0.7 percent by weight set by the EPA. These standards for recent years for several northeastern states are set forth in Table 4-1.

Untreated crude oils contain varying amounts of sulfur. The imposition of upper limits on sulfur content by federal, state, and local environmental regulations has stimulated demand for low sulfur fuel oils which, accordingly, can command a premium price. This price differential makes it economically advantageous to treat fuel oils at the refinery so as to reduce sulfur content. This chapter investigates the cost of such treatment, the circumstances under which it is worthwhile, and the market for the sulfur recovered as a by-product of desulfurization.

Sulfur content is, of course, not the only parameter determining the relative price of fuel oils at any given state of the market. The specific gravity of the oil is one such parameter since it is directly proportional to the hydrocarbon concentration and hence to the heating value per unit of volume. The specific gravity is also related to the viscosity of the oil at a given temperature and hence to its handling qualities. In general, the price of oil per unit of volume increases, other things being equal, with an increase in specific gravity.[a] Transportation cost is another significant factor affecting the cost of oil and the differences in prices shown in Table 4A-1 reflect this factor as well.

Grades of Fuel Oil

Fuel oils are divided into grades reflecting source and the uses for which they are suitable. There are two broad classifications of fuel oils: distillates, which consist of the lower boiling fractions, and residuals, which are the bottoms remaining from the distillation or blends of these bottoms with distillates. The following are the grades of fuel oil as defined by the American Society for Testing and Materials (ASTM):

Grade No. 1 is a light distillate used in burners of the vaporizing type. In such burners, oil is converted to a vapor by contact with a heated surface or by radiant heat. Volatility must be high enough to insure complete vaporization of the fuel.

Grade No. 2, a heavier distillate than No. 1, is used in atomizing types of burners which spray the oil into a combustion chamber where droplets burn while in suspension. This grade of oil is used mostly for domestic burners, but also in many medium capacity commercial-industrial burners where its ease of handling and ready availability may sometimes justify its higher cost relative to residuals.[b]

[a]The Chevron Oil Trading Company, for example, in *Posted Export Prices* (Bulletin No. 20, 6/29/73), states the F.O.B. (free on board) price for Arabian Light Crude Oil, Ras Tanura Saudi Arabia, as $2.898 U.S. dollars per barrel, effective June 1, 1973. The same bulletin states specific gravity range for this oil as 34.00 to 34.09 degrees API (i.e., according to the American Petroleum Institute scale of specific gravities). The same bulletin also provides a formula for price adjustment if the specific gravity of the delivered oil falls outside the above range. Specifically, the Chevron bulletin provides that "for each 0.1 degree API above the indicated minimum gravity 0.15¢ per barrel will be added, and for each 0.1 degree API below the indicated maximum gravity 0.15¢ per barrel will be deducted." In accordance with this formula, if the Arabia Light Oil under discussion, when tested, showed a specific gravity of 34.1 degrees API the price would be adjusted upward to $3.048 per barrel. Conversely, if the test showed a specific gravity of 33.99 degrees API the price would be adjusted downward to $2.748 per barrel. Chevron bulletins 11, 12, 13, and 19, issued in the first six months of 1973, specify similar mechanisms for oil price adjustments in accordance with specific gravity variations (see also Table 4A-1).

[b]The ASTM specifications do not include any standard for No. 3 oil, presumably because it is unimportant. If it existed it would presumably be intermediate in density between No. 2 and No. 4 oil, and might be produced by mixing the two grades of oil.

Grade No. 4 is customarily light residual, but is sometimes a heavy distillate. It is employed in burners equipped with devices that atomize oil of higher viscosity than domestic burners can handle. Its viscosity range is such that it may be pumped and atomized at relatively low storage temperatures; accordingly, it requires no preheating for easy handling in all but extremely cold temperatures. This product is used in large apartment buildings, light industry, and small businesses.

Grade No. 5 fuel oil is a residual classified in two sub-types, Light and Heavy. The No. 5 (Light) is of intermediate viscosity for burners capable of handling fuel more viscous than No. 4 without preheating. No. 5 (Heavy) is more viscous than Light, but used for similar purposes. In using either type of No. 5 oil, preheating may be necessary in some types of equipment for burning and in colder climates for handling.[c] No. 5 is used for space heating of such medium sized facilities as schools and hospitals.

Grade No. 6 is a highly viscous residual oil used mostly in commercial and industrial heating. Sometimes referred to as "Bunker C," it requires preheating in the storage tank to permit pumping and additional preheating at the burner to permit atomizing. The extra equipment and maintenance required to handle this fuel generally precludes its use in small installations.

All of the above grades of fuel oil are used for space heating. The ASTM 1971 specifications give only casual attention to sulfur content, noting only that Nos. 1 and 2 are limited to not more than 0.5 percent or the legal limit, and that the sulfur content of the remaining grade is limited to the legal limit.

Prior to desulfurization, the sulfur content tends to increase as the grade number increases. This can be explained by reference to Table 4-1, which shows the fractions and their sulfur contents resulting from the vacuum distillation of a barrel of U.S. crude oil of 1.7 percent sulfur by weight, characterized by Dr. L.E. Swabb as typical [3:68]. It will be noted the most volatile fraction–naphtha and light distillate–has the lowest sulfur content and that the latter increases to the highest value for the residual. This corresponds roughly to a change in grade number from No. 1 to No. 6.

Sulfur Content and Sources of Crude Oil

With 6.4 percent of the world's population, the United States accounts for 36.2 percent of the total global energy consumption [4:180]. Moreover, oil is the source of 43 percent of this enormous consumption of energy [4:187] and the United States holds only about 7 percent of the crude oil reserves of the non-Communist nations of the world [3:77]. It is not surprising that the United

[c]Sulfur-bearing oils of the same grade will handle better than oils of lower sulfur content. Installations in New York City using oils of the higher grade numbers which had not required preheating equipment when burning high sulfur fuels found such equipment necessary when city authorities lowered permissible sulfur content [2:289].

Table 4-1
Fractions from Vacuum Distillation of a Typical[a] Barrel of Crude

Fraction	Percent of Barrel	Sulfur Content by Weight Percent
Naptha and Light Distillate	40	0.4
Light Gas Oil	20	1.4
Vacuum Gas Oil	27	2.3
Vacuum Residual	13	4.3

[a]Average sulfur content 1.7 weight percent regarded as typical.
Source: *Electrical World* Symposium, *Sulfur in Utility Fuels: The Growing Dilemma*, A Technical Conference sponsored by *Electrical World.* Data from presentation by Dr. L.E. Swabb, McGraw-Hill, 1972, slide 1.

States has had to turn increasingly to imports in recent years to meet its petroleum needs. In doing so, it had to draw on crudes from the Middle East with 69 percent of the non-Communist reserves and Venezuela, which accounts for 3 percent; the latter, however, is a more important source than its small percentage of world reserves would suggest because of its relative proximity to the United States. The oils of Africa, principally North Africa, are low in sulfur but account for only 10 percent of world non-Communist reserves [3:77]. Table 4-2, which shows the sulfur content of selected crude, residual and distillate oils by source, indicates the general nature of this relationship. The sulfur content of the crudes and residuals included in Table 4-2 is markedly higher for those from the Mideast than those from Africa and the Far East, with Domestics somewhere in between.

Desulfurization as the Answer

The federal, state, and local governments have imposed limitations on the sulfur content of fuel oils consumed within their jurisdictions. For example, Table 4-3 shows the progressively more rigorous restrictions on sulfur content imposed in the northeast where the sulfur oxide pollution problem has been most severe. Since high sulfur fuel oil has been increasingly barred, the demand for naturally occurring low sulfur fuels has been raised above the available supply. The result has been premium prices for low sulfur fuels. From the standpoint of economy, desulfurization at the refinery will pay so long as the added costs involved do not exceed the added price paid for low sulfur fuel oils obtained from low sulfur crudes. Put differently, "desulfurization will set a limit on the premium low-sulfur fuels can carry over high-sulfur fuels" [3:150].

As Dr. Bruce Netschert put it, desulfurization becomes the logical alternative when four facts are recognized:

Table 4-2
An Evaluation of Sulfur Weight Percentage Selected Crude, Residual and Distillate Oils, by Source

Source	Crude	Residuals	Distillates
DOMESTIC:			
Denver-Julesburg	0.12	0.30	0.02
Wyoming Sour	2.4	4.10	0.63
North Dakota Northern	0.5	1.90	0.07
Rocky Mountain Sweet	0.14	0.70	0.001
FOREIGN:			
Africa			
Algeria-Ohanet	0.06	0.13	0.015
Algeria-Zarzaitine	0.07	0.07	0.006
Algeria-Hassi Messaoud	0.14	0.37	0.009
Angola-Tobias	1.51	2.56	0.347
Libya-El Hofra	0.31	0.59	0.13
Libya-Dahra	0.36	0.82	0.25
Nigeria-Bomu	0.14	0.29	0.04
Far East			
Australia-Moonie	0.04	*	*
British Borneo Seria	0.1	0.20	0.02
Sumatra-Minas	0.08	0.40	0.05
Sumatra-Duri	0.18	0.2	0.07
Mid East			
Iran-Agha Jari	1.34	3.04	0.33
Iran Gach Saran	1.60	2.46	0.82
Qatar	0.96	2.3	0.18
Kuwait	2.50	4.06	0.70
Saudi Arabian mixture	1.63	3.58	0.85
Saudi Arabia-Khursaniyah	2.67	4.55	1.00

*Not listed in analysis
Source: *The Oil and Gas Journal*, April 15, 1963, pp. 116-146.

Item One —You are not permitted to burn high-sulfur fuels.

Item Two —There are supply and price problems involved in the use of all the low-sulfur fuels, not to mention the serious difficulties and high costs of converting existing plants to such fuels.

Item Three —Nuclear energy is at best a medium-term solution, given the delays.

Item Four —Although there are some problems with high-sulfur fuels there is no fundamental scarcity. [3:143]

Table 4-3
Eastern Seaboard Specifications For No. 6 Fuel Oil Maximum Sulfur Content Allowed (Percentage Sulfur by Weight)

	Effective Date			
	10-1-69	10-1-70	10-1-71	10-1-72
New Jersey	1.0	0.5	0.3	0.3
New York State	1.0	1.0	0.37	0.37
New York City	–	–	0.30	0.30
Massachusetts	2.2	2.2	2.2	2.2
Boston Area	1.0	1.0	0.5	0.5
Philadelphia	–	1.0	1.0	0.5

Source: Charles H. Watkins and Frank Stolfa, Universal Oil Products Company, for presentation at the API Division of Refining, 36th Midyear Meeting San Francisco, Calif., May 10-14, 1971.

Moreover, Dr. Netschert warns against expecting that the problem will go away for fuel oil users, discerning no "grounds for believing that the ban would be repealed or even significantly relaxed during the foreseeable future" [3:143]. The importance of desulfurization is that it increases the options available to the industry to make transitions and adjustments, depending upon varying conditions, be they economic, political or social.

As Dr. L.E. Swabb describes it: "The basic reaction in the desulfurization of fuel oil involves hydrogen reacting with sulfur in the presence of specially designed catalysts which is converted in auxiliary processes to elemental sulfur. . . . The reaction is carried out at high pressures (400 to 2000 psig) and at relatively high temperatures (650 to 800°F.) . . ." [3:71].

Catalytic hydrogenation is limited in some instances because some crude oil constituents–notably compounds with metals–may render catalysts ineffective. This is true of Venezuelan crudes, which have a high metal content. In such cases, "the most feasible method is to vacuum-distill the residuum and desulfurize only the distillate vacuum gas oil. The desulfurized vacuum gas oil usually contains less than 0.1 percent sulfur. This oil can be blended back with a portion of the high-sulfur residue from the vacuum distillation to give a product of the desired sulfur content" [6:72].

The foregoing describes in qualitative terms the basic processes for the removal of sulfur from crude oils. In the section that follows, quantitative operating data are presented for several actual cases, indicating by yield data the extent of success in removing sulfur from crudes, supplemented by Appendix tables on some of the cases reflecting operating costs and capital investments required.

Case I: Hydrodesulfurization of Residuals with H-Oil–Cities Service Oil Company, Lake Charles Refinery [7]

In 1963, a hydrogen cracking unit was constructed at the Lake Charles Refinery to convert approximately 2,500 barrels per day (BPD) of West Texas high sulfur content vacuum bottoms into lighter products. Initially no effort was made to remove sulfur, but around 1966-67 an attempt was made to use this facility to desulfurize atmospheric residuum from the same crude and this was successful in reducing sulfur from 2.5-3.0 percent by weight in the feedstock to 0.3-0.5 percent in the product.

The source on which this section is based presented data on results employing twelve different feedstocks, and the results for two of them are shown in Table 4-4 and will be discussed here. The two feedstocks selected for illustration are: (a) Kuwait atmospheric residual which is typical of feedstocks with *high* sulfur and *low* metal content; and (b) Venezuelan vacuum bottoms which exemplify a *low* sulfur and *high* metal feedstock. Both the Kuwait and Venezuelan feedstocks employed were of moderate specific gravity and high asphaltene content. The desired product in each case was a fuel oil product containing 1.0 percent sulfur by weight with a 400°F.± (material boiling) point. Table 4-4 illustrates two points: (1) In general, feedstock characteristics were found to be

Table 4-4
Selected Attributes and Results of Hydrodesulfurization with Two Feedstocks

Feed Stock	Kuwait Atmospheric Residual	Venezuelan Vacuum Residual
Sulfur weight percentage of feedstock	3.9	3.2
Sulfur weight percentage of product	1.0	1.0
Percentage yield of materials boiling 400°F.±	97.2	90.6
Percentage desulfurization	76.0	71.0
Hydrogen consumption, standard cubic feet per barrel	475	950
Feed rate, barrels per day	20,000	15,000
Capital investment (million dollars)	$7.3	$10.1

Source: Table 4A-4.

important for the operating results and the outcome of "H-Oil" desulfurization; and (2) In particular, better results were achieved using the Kuwait feedstock than the Venezuelan. Summarizing this differently, it will be noted that although the sulfur content of Kuwait feedstock was higher, the percentage yield of desired product was also higher as was percentage desulfurization and feed rate. In spite of these favorable comparisons, the hydrogen consumption (a major factor in operating costs) and the required capital investment were both lower using the Kuwait feedstock. The sulfur content for the product was the same regardless of the feedstock.

Case II: Universal Oil Products Company–RCD Isomax [8]

In recent years UOP has devoted much time and energy to the investigation and evaluation of two types of desulfurization techniques; they are (1) the indirect approach, and (2) the direct approach. The "indirect approach" involves recovery of distillate gas oils through vacuum fractionation for subsequent hydrotreatment and blending of the desulfurized gas oil with unprocessed vacuum residual. This approach, however, is restricted generally to crudes of lower sulfur content and higher API gravity. The second method, the "direct approach," hydrotreats the entire residuum. This technique is currently employed in the processing of higher sulfur Middle East crudes which contain low to moderate amounts of metals. UOP expects that this approach or some modification of it will be necessary to meet the increasingly stringent regulations concerning maximum sulfur content.

As of early 1972, UOP had three RCD Isomax units in operation, two in Japan and the other in the Middle East. Table 4-5 presents yield data for the second RCD Isomax Unit in Japan, which went on stream in mid-1970. It was designed to handle about 45,000 barrels per day of Kuwait reduced crude with its primary product being heavy fuel oil containing 1.0 percent sulfur by weight.

Data on other applications of UOP's RCD Isomax process are given in Table 4-6, which shows operating costs for desulfurization of two reduced crudes from Kuwait by two slight variations of the RCD Isomax process. In Case C, the sulfur content was reduced 0.5 weight percent by removal of 89 percent of the sulfur in the feedstock. The cost for this was 75.4 cents per barrel, which would have increased the cost in Case C from $2.373 per barrel (the December 15, 1972 posted price for Kuwait crude) to $3.127 per barrel. This gives some idea of the order of magnitude of the increase in oil price occasioned by reduction of sulfur content to 0.5 weight percent.

If the sulfur content is reduced even further as in Case D to sulfur weight percent of 0.32 requiring a 92.5 percent sulfur removal, the cost of desulfurization is raised to 84.4 cents, thereby increasing the $2.373 cost per barrel to

Table 4-5
Commercial Desulfurization of Reduced Crudes; RCD-Isomax Japan

	Volume %			
Flow Rates Charge				
Reduced Crude	100.0			
Products				
Unstabilized Gasoline	0.6			
Kerosine	4.0			
Heavy Fuel Oil	97.7			
Total	102.3[a]			
	Reduced Crude	Gasoline	Kerosine	Heavy Fuel Oil
Inspections				
Gravity, API°	15.9	58.2	40.0	22.2
Distillation, °F.				
IBP (Initial Boiling Point)	475	120	345	520
10%	621	–	–	605
EP (End Point)	–	330	470	–
Sulfur Wt–%	3.91	0.01	0.08	1.03

[a]Some of the difference between 100 and 102.3 percent volume is attributable to expansion from heat, but, in general, refining processes are not necessarily strictly additive so far as volume is concerned.

Source: *Technology for Utilization of High Sulfur Fuels*, Part 1, 51a "RCD Isomax–Production Route for Today's and Tomorrow's Low Sulfur Residual Fuels," by A.P. Krueding, Universal Oil Products Company, Des Plaines, Illinois, for Presentation at the 71st National Meeting of The American Institute of Chemical Engineers, February 20-23, 1972.

$3.217 per barrel. Cases C and D employed the same feedstock, so the increased cost is attributable solely to the expense of sulfur removal. These costs in both cases C and D only reflect production costs and do not allow for the refiner's profit margin or transportation cost in moving the products to the user.

The additional investment cost to produce 0.5 percent sulfur content fuel oil rather than 1.0 percent was estimated to be $6 million, adding an operating cost of 2 cents per barrel to achieve Case C. Appendix Table 4A-3 shows that the further reduction from 0.5 percent sulfur by weight (Case C) to 0.32 percent involved further investment of capital to the tune of $4.7 million, i.e., $27.7 million minus $23 million, as seen in the bottom line of Table 4A-3.

It is evident that desulfurization involves large-scale investment as well as substantial operating costs. A theoretical alternative, which has had only

Table 4-6
Estimated Operating Costs for the Desulfurization of Kuwait Reduced Crude by RCD Isomax

(Basis: 8,000 operating hours per year (333 operating day) utilities based on electric drive, and maximum air cooling.)

Case	C	D
Capacity, Barrels per standard day (BPSD)	40,000	40,000
Desulfurization Mode	Direct	Modified Direct
Heavy Fuel Oil Product Sulfur Wt–%	0.5	0.32
Operating Costs	¢/Barrel	¢/Barrel
Labor	1.0	1.6
Utilities	8.3	9.4
Catalyst	12.5	8.9
Maintenance, Taxes and Insurance at 6% of Estimated Plant Cost	9.6	11.9
Total Direct Operating Costs	31.4	31.8
Overhead at 100% of Direct Labor	0.7	1.3
Interest at 5% of Estimated Plant Cost	8.0	10.0
Depreciation–10% of Estimated Plant Cost	16.0	20.0
Total Indirect Operating Cost	24.7	31.3
Total Direct and Indirect Operating Costs	56.1	63.1
Hydrogen at $0.25/1000 Standard Cubic Feet	19.3	21.3
Total	75.4	84.4

Source: *Technology for Utilization of High Sulfur Fuels*, Part 1, 51a "RCD Isomax–Production Route for Today's and Tomorrow's Low Sulfur Residual Fuels," by A.P. Krueding, Universal Oil Products Company, Des Plaines, Illinois, for presentation at the 71st National Meeting of The American Institute of Chemical Engineers, February 20-23, 1972.

secondary interest except for power generating stations, is desulfurization at the point of use.

Desulfurization at the Point of Combustion

We have seen that a reduction in the sulfur content of fuel oils costs money to accomplish, but low sulfur fuels are nevertheless the only ones acceptable in many cities seeking to reduce the concentration of sulfur oxides in their atmospheres. There is an inverse correlation between sulfur content and oil prices and a theoretical alternative to the payment of premium prices for low sulfur fuel oil is to burn cheaper high sulfur oils and capture the resulting sulfur oxides before they are emitted to the atmosphere. This is done by removing sulfur oxides through use of stack gas "scrubbers."

Some large-scale consumers of fuel oil, notably certain electric utilities, have opted for the stack gas scrubber route, and Chapter 3 recounts their trials and tribulations as well as some pioneering successes. But stack gas scrubbing requires major capital investment and is completely infeasible for the overwhelming majority of fuel oil users. Fortunately, sulfur content is always low in the grades of fuel oil used in the burners of individual homes. For the apartment houses, commercial users, and small industrial consumers, who use fuel oils of significantly higher sulfur content, desulfurization at time of use is economically and practically entirely infeasible.

By-Product Sulfur

It was noted in Chapter 3 that the short supply of sulfur that had prevailed prior to 1968 was ended through by-product recovery of sulfur, principally from the desulfurization of Canadian natural gas. Since 1972 world oversupply of sulfur has been the norm. Nevertheless, the Sulfur Institute, a U.S. trade association, has been actively exploring possible new uses for sulfur and its spokesmen see reason for optimism that new uses for sulfur may be found sufficient to absorb the surplus [9]. In addition to sulfur from desulfurization of natural gas, sulfur from refineries desulfurizing fuel oil may make a substantial and fairly concentrated contribution to the sulfur supply, and public utilities using stack gas scrubbers may make a smaller and more scattered one. Accordingly, it seems likely that if new markets can be found for sulfur, the refineries desulfurizing fuel oils will stand to benefit more than will the electric utilities.

Conclusion

Desulfurization is a necessity if society wishes to achieve an environment where the harmful effects of sulfur emissions can be minimized. However, the cost to achieve this level, and maintain it, is substantial. Those who control the production of heating oils, in order to meet the legal requirements set forth by the EPA, will sustain these costs, both for raw material and for desulfurization. The cost of desulfurization is partially offset by the sale of the extracted sulfur. However, ultimately the entire cost of raw materials and desulfurization must be absorbed by the consumer. If we want a clean and healthy environment we must be aware the financial burden will rest upon the citizens.

Appendix 4A

Table 4A-1
Posted Export Prices–Middle East Crudes

Crude Oil	Gravity Range Degrees API[b]	Loading Port	F.O.B. Price U.S.$ per Barrel[a]	
			*3/1/73	**6/1/73
Arabian Light	34.00/34.09	Ras Tanura, Saudi Arabia	2.742	2.898
Arabian Light	34.00/34.09	Sidon, Lebanon	3.677	3.884
Arabian Medium	31.00/31.09	Ras Tanura, Zuluf, Saudi Arabia	2.626	2.776
Arabian Heavy	27.00/27.09	Ras Tanura, Saudi Arabia	2.481	2.623
Iranian Light	34.00/34.09	Abadan, Iran	2.657	2.808
Iranian Light	34.00/34.09	Kharg Island, Iran	2.729	2.884
Iranian Heavy	31.00/31.09	Abadan, Iran	2.614	2.764
Iranian Heavy	31.00/31.09	Kharg Island, Iran	2.674	2.826

[a]Comparative prices are presented to show to the reader the relative cost of crude oils. Prices listed are per barrel of 42 U.S. gallons at 60° F.O.B., port of loading indicated, in bulk cargo lots, for crude oil of gravity within the range indicated, and subject to availability.

For each 0.1 degree API below or above the indicated minimum or maximum gravity; 0.15¢ per barrel will be subtracted or added, for example, if a contract called for Arabian Light from Ras Tanura, with API as presented above but when the crude was tested the API was actually 33.7°, the adjusted price would be ($2.7420–$.0045) $2.7375 per barrel.

[b]Gravity Range Degrees API, is the American Petroleum Institutes expression of the density or weight of a unit volume of material [1:166].

Sources: Chevron Oil Trading Company, *Posted Export Prices*.

*Bulletin No. 19–Issued 3/29/73.

**Bulletin No. 20–Issued 6/29/73.

Table 4A-2
Yields From the Hydrodesulfurization of Residuals by the H-Oil Process

	A	B
Feed Stock	Kuwait Atmospheric Residual	Venezuelan Vacuum Residual
Type of Catalyst-Extrudate		
Feedstock Data		
API Gravity	16.5	7.5
Sulfur weight %	3.9	3.2
Carbon Residuum weight %	9.5	21.4
Vanadium and nickel, ppm	60	760

Product Cuts	A Weight %	A Volume %	A Sulfur %	B Weight %	B Volume %	B Sulfur %
1. Methane–Ethane	3.2			2.5		
2. Propane–Butane	0.3			2.3		
3. Butane thru Material boiling at 400°F.		4.6	0.1		13.2	0.1
4. 400°-650°F. (Material boiling range)		17.4	0.3		26.8	0.3
5. 650°-975°F.		46.3	0.7		27.7	0.9
6. 975°F.± (Material boiling at more than 975°F.)		33.5	1.7		36.1	1.6
Total yield of materials boiling above propane		101.8	0.9		103.8	0.9
400°F.±		97.2	1.0		90.6	1.0

Table 4A-2 (cont.)

Product Cuts	Weight %	Volume %	Sulfur %	Weight %	Volume %	Sulfur %
650°F.±		79.8	1.1		63.8	1.3
Conv. 975°± Vol.% (High material to low material Feed stock converted to product)		22			64	
% of Desulfurization		76			71	
Hydrogen Consumption SCF/Bbl (Standard Cubic Feet per barrel)		475			950	
Feedrate, Barrels per day		20,000			15,000	
Capital Investment (millions)[a]		$7.3			10.1	

[a]Installed cost on a 1971 U.S. Gulf Coast basis. Includes the H-Oil Unit, process royalty and initial catalyst inventory. Does not include provisions for product fractionation, hydrogen recovery or sulfur recovery facilities.

Source: *Hydrodesulfurization of Residuals with H-Oil*, A.A. Gregoli and G.R. Hartos, Cranbury, New Jersey: Cities Service Research and Development Company, p. 27.

Table 4A-3
Estimated: Yields and Investment Costs by the RCD Isomax Process For the Desulfurization of Kuwait Reduced Crude With Unit Capacity of 40,000 BCD and Feed Gravity of 16.3 API

Case		C			D	
Heavy Fuel Oil Product Sulfur Wt–%		0.5			0.32	
Desulfurization Mode		Direct			Modified Direct	
Yields	BPSD*	Vol–%	Sulfur Wt–%	BPSD*	Vol.%	Sulfur Wt–%
Hydrogen Sulfide	–	–	3.7	–	–	3.85
Methane, Ethane, Propane & Butane	–	–	1.15	–	–	1.00
Light Gasoline	300	0.75	0.55	400	1.00	0.70
Naptha	1,180	2.95	3.35	780	1.95	1.60
375°-650° Distillate	2,700	6.75	6.00	3,220	8.05	7.15
650°F. + Heavy Fuel Oil	36,200	90.50	87.45	35,940	89.85	87.05
Total	40,380	100.95	101.20	40,340	100.85	101.35
Total Fuel Oil Product Boiling above 375°F.	38,900	97.25	93.45	39,160	97.90	94.20
Hydrogen Chemical Consumption, Standard Cubic Feet/Barrel		770			850	

Table 4A-3 (cont.)

Product Properties	Gravity Range API Degree	Sulfur Wt. %	Viscosity CS at 122°F.	° API	Sulfur Wt. %	Viscosity CS at 122°F.
Light Gasoline	82.2	Nil	–	80.6	Nil	–
Naptha	53.5	0.05	–	53.0	0.05	–
Distillate	35.5	0.0	3.0	35.2	0.1	2.6
Heavy Fuel Oil	21.5	0.5	150	21.1	0.32	150
Total Fuel Oil	22.4	0.47	105	22.1	0.30	95
Plant Investment						
Material & Labor RCD Isomax			$17,600,000			$21,900,000
Design Engineering & Contractor Expense			3,700,000			4,600,000
Total Estimated Erected Cost			$21,300,000			$26,500,000
Initial Catalyst Inventory			1,700,000			1,200,000
Total Initial Cost			$23,000,000			$27,700,000

*BPSD Barrels Per Standard Day

Source: *Technology for Utilization of High Sulfur Fuels*, Part 1, 51a "RCD Isomax–Production Route to Today's and Tomorrow's Low Sulfur Residual Fuels." A.P. Krueding, Universal Oil Products Company, Des Plaines, Illinois, paper presented at 71st National Meeting of The American Institute of Chemical Engineers, February 20-23, 1972.

Notes

1. American Society for Testing and Materials, *The 1971 Annual Book of the American Society for Testing and Materials*, Part 17, Easton, Md., 1971, pp. 162-167.
2. Alfred J. Van Tassel (ed.), *Environmental Side Effects of Rising Industrial Output* (Lexington, Mass.: Lexington Books, D.C. Heath and Company, 1970).
3. *Electrical World* Symposium, *Sulfur in Utility Fuels: The Growing Dilemma*, A Technical Conference sponsored by *Electrical World* (New York: McGraw-Hill, 1972).
4. Sam H. Schurr (ed.), *Energy, Economic Growth, and the Environment* (Baltimore, Md.: The Johns Hopkins University Press, 1972).
5. *Oil and Gas Journal*, April 15, 1963, pp. 116-146.
6. R.L. Duprey, "The Status of SO_x Emission Limitations," *Chemical Engineering Progress* 68, 2 (February 1972): 70-76.
7. A.A. Gregoli and G.R. Hartos, *Hydrodesulfurization of Residual With H-Oil*, Cities Service Research and Development Company.
8. A.P. Krueding, "RCD Isomax Production Route to Today's and Tomorrow's Low Sulfur Residual Fuels," *Technology for Utilization of High Sulfur Fuels*, 71st National Meeting of American Institute of Chemical Engineers, New York, February 20-23, 1972, Part 1, 51a.
9. H.L. Fike and J.S. Platou, "New Uses for Sulfur–Their Status and Prospects," Paper presented at Flue Gas Desulfurization Symposium, New Orleans, La., May 14-17, 1973.

Glossary (Terms and Abbreviations)

BPCD. Barrels per calendar day.

BPSD. Barrels per standard day.

Catalyst. A substance which changes the rate of a chemical reaction, but is present in its original concentration at the end of the reaction.

Degrees°Fahrenheit. Units of measure of temperature.

EP. End point, the final boiling point.

Gravity API°. A unit of measure of specific gravity standardized by the American Petroleum Institute.

High or Low Metal Content (as applied to oil). A term used to indicate amount of metal contained in a feedstock either as metal or a compound of one, frequently vanadium or nickel.

H-Oil Process. A conversion and hydrodesulfurization process to produce low sulfur distillates and fuel oils patented by Hydrocarbon Research.

Hydro-catalyst. A catalyst used to hasten the hydrogenation reaction in hydrocarbons.

Hydrogen Partial Pressure. The portion of the total system pressure attributable to the presence of hydrogen, expressed in pressure units.

Hydrotreatment. Treatment of feedstock with hydrogen at high temperatures.

IBP. Initial boiling point.

RCD Isomax. a process for conversion and desulfurization of high boiling feedstocks to low sulfur distillates and fuel oils patented by Universal Oil Products Company.

Sulfur Weight Percent. Percentage of sulfur in a substance by weight.

Vacuum Distillation. Distillation under a pressure less than the external pressure in the vicinity.

Volume Percent. Percentage by volume of the component referred to.

5 Savings of Energy and Recovery of By-Products from Solid Waste

Stewart Winnick

America is known as the affluent society. Does recovery from its waste offer a partial solution to some of our problems?

Introduction

Today, in urban centers in particular, refuse collection is a vital necessity from the standpoint of public health. The volume of refuse collection can best be illustrated by the fact that in the fiscal year ending June 1971, the New York City Department of Sanitation collected approximately 3.8 million tons of solid waste [3:112]. This operation cost $114 million, or $30.00 per ton collected.

Refuse collection comprises about 80 percent of the total cost of collection and disposal. Yet collection is only an intermediary stage in the process of disposal. For this reason, collection costs are not *directly* related to the final disposal problem.

Collection of municipal solid wastes usually involves combined classes of refuse [12:34]. It also employs many varied techniques for transporting the refuse to collection vehicles. Other aspects concern where the refuse is deposited for collection, and the use of satellite vehicles to reduce collection costs.

However, it must be noted at this time that the collection of solid wastes, though more costly than their disposal, does not create a pollution problem. It is the disposal technique employed which creates an environmental problem.

Disposal Methods

Solid waste disposal involves "the orderly process of discarding useless or unwanted material" [19:7]. It is this aspect in the life cycle of municipal refuse which must be analyzed in the light of environmental protection.

The public awareness of environmental pollution has done much to point out the contaminants of the atmosphere and waterways. However, there has been a less dramatic reaction to pollution caused by solid wastes.

Total solid wastes have been accumulating in the nation at the astounding rate of 4.3 billion tons per year [3:23]. In 1969, about $3.5 billion was spent to handle 190 million tons of collected solid wastes, at an average cost of $18 per ton [3:23]. Of this collected solid waste expense, 80 percent of the cost ($14 per ton) was for collection; the remainder was for disposal costs.

The problem of solid waste pollution is even more dramatic if it is noted that most open dumps are inadequate, most incinerators are air polluters, and costs of collection and disposal are on the rise.

In today's environment, the United States occupies a precarious position. Accounting for only 6 percent of the world's population, the United States uses 40 percent of all non-replaceable resources, and 40 percent of the world's energy. Since continued growth of United States consumption is predicted, the implications for the world supply of resources is tremendous. There seems to be only one possible solution to this world economic dilemma–reclamation and reuse.

Reclamation is defined as the obtaining of useful materials from solid waste [19:14]. It is this technique which may become important as a method for rechanneling waste into useful resources.

There are no available nationwide data on the relative popularity of various methods of disposal. However, some indication can be obtained from a summary of the disposal practices employed by fourteen cities studied earlier at Hofstra University [26:chaps. 22-26]. There were four eastern, five midwestern, and five western cities studied. All but one of the cities studied in each region were larger than average population for the region, though not necessarily largest in the region. Sanitary landfilling was employed in all fourteen cities; was the exclusive method in six; and accounted for the bulk of disposal in all but three. Incineration was an important method of disposal in three of the cities. One of the cities had operated a composting plant for a portion of its refuse, but was forced to abandon the experiment as too expensive.

Sanitary Landfilling

As noted above, the most prevalent solid waste disposal technique is sanitary landfilling. Sanitary landfilling is "an engineered method of disposing of solid waste on land in a manner that protects the environment, by spreading the waste in thin layers, compacting it to the smallest practical volume, and covering it with soil by the end of each working day" [19:15]. There are various approaches to this technique, such as area, quarry, ramp, trench, and wet area. All these approaches, however, employ the same basic idea–cover up the solid waste.

This chapter will not attempt to define the requirements for a true sanitary landfill site. However, it may be noted that theoretical requirements are never truly achieved, and that sanitary landfilling is the direct descendant of open dumping.

Open dumping is the use of available land upon which to dump solid wastes. These wastes are left untreated and uncovered, to decompose in the atmosphere. As a result, open dumping creates scenic blights, health hazards (attracting nuisances and rodents), and depressed land values in land areas adjacent to the open dump. Yet, with 90 percent of the earth's solid waste disposed on land, open and burning dumps are all too prevalent [15:1].

Sanitary landfilling is the logical solution to open dumping. Some of the requirements for sanitary landfill are established by the American Public Works Association as follows:

1. Vector breeding and sustenance must be prevented.
2. Air pollution by dust, smoke, odor must be controlled.
3. Fire hazards must be avoided.
4. Pollution of surface and ground water must be precluded.
5. All nuisances must be controlled [18:82].

The basic operation of a sanitary landfill involves, as a first step, the selection and surveying of a site. Geological and hydrological factors must be evaluated; the site must be prepared; access roads, drainage, and control grades established; and equipment provided. Other provisions may be necessary depending on climate and location.

The basic advantage of sanitary landfills that are properly established are numerous. Little capital investment is required. Unused land may be reclaimed. Costly segregation of solid wastes may usually be avoided. Air pollution created by this process is minimal, and disposal is complete.

Several disadvantages, however, cause sanitary landfilling to be somewhat of a problem. Since large areas of land are required, communities have been running out of available sites. Utilization of available land in rural areas may require high transportation costs from urban centers.

Sanitary landfilling costs have been estimated at $1.13 per ton of solid waste, as opposed to $5.25 per ton for incineration, and $3.50 per ton for composting [5:328]. Moreover, it should be noted that sanitary landfilling provides reclaimed land which serves to provide more space for living. Nevertheless, most operations which are designated as sanitary landfills do not meet the accepted standards. If the requirements for sanitary landfills were completely adhered to, the costs would be much higher than they are now [26:462].

What is the nature of the costs of a sanitary landfill operation? The total cost of a landfill site involves land cost, plus site development, operating costs, and equipment, landscaping, and maintenance costs.

In the past, land costs have been a small part of the total cost. Metropolitan landfill sites increase in value after they are filled, and rural landfill sites are inexpensive initially. Today, however, there isn't much available land for landfill sites near most metropolitan areas. This has resulted in high transportation costs to distant sites and increased purchase costs for those sites still available near the cities.

Land for disposal in the Los Angeles metropolitan area has been purchased for $2,000 to $20,000 per acre since 1965, while leased land costs there have varied from 6 cents to 50 cents per ton of refuse, depending on the degree of compaction [7:123].

The site location is extremely important, since collection including cost of transport to the disposal site costs are the major portion of total refuse removal costs. In many cases, large communities suffer little disadvantage due to economies of equipment.

Experience shows, however, that sanitary landfilling costs vary from $0.75 to $4.00 per ton of collected refuse [7:124]. This includes land, labor, equipment, and contingencies. The wide cost variations are caused by different requirements, as well as size of the fill. Small operations (under 50,000 tons per year) cost from $1.25 to $5.00 per ton, while fills over 50,000 tons per year cost between $0.75 and $2.00 per ton [7:124].

The variations in sanitary landfill operating costs are shown in Table 5-1, which reflects the influence of operation size on unit cost [7:514]. Note that unit costs are very much higher for populations under 122,000 (based on an average of 4.5 pounds of refuse produced per person per day). However, economies of scale are almost totally absent in the range of populations from 122,000 to 610,000. This seems to illustrate that scale economies associated with operational costs are overwhelmed in larger communities by the higher costs of land. The cost of transportation to the disposal site, which is not included in Table 5-1, may also be expected to increase as the population served increases.

At this point, an analysis of the costs of sanitary landfilling, and possible recovery values, in several specific cases might yield some justifiable conclusions.

Table 5-1
Sanitary Landfill Operating Costs

Population (in thousands)	Range of Costs per Ton[a] (in dollars)
122	1.05 - 1.75
244	.90 - 1.40
366	.85 - 1.30
488	.80 - 1.20
610	.70 - 1.15

[a]Includes land, labor, equipment and contingencies, but not transportation to the disposal site.

Source: Institute for Solid Wastes of the American Public Works Association, *Municipal Refuse Disposal* (Chicago, Illinois: Public Administration Service, 1970), Figure 144, p. 374, as taken from "Sanitary Landfill Facts," U.S. Department of Health, Education, and Welfare, Public Health Service, Solid Waste Program, 1968.

Beaumont, Texas, is one region where sanitary landfilling is extensively employed. Beaumont is an area which has much swampy terrain. Open dumping had been used to the point where it created a health hazard and scenic blight. As a result, recent state landfill regulations were designed to create industrial parks and recreational areas, instead of unsightly open dumps [27:54].

The Neches River swamp, within a mile and a half of downtown Beaumont, had been an open dump since early 1900 logging cleared away the trees. With the exception of garbage deposited in a small landfill, all combustibles were permitted to be burned in the open dump. Furthermore, the nearby landfill's only means of compaction was by the action of the dump trucks. Finally, the state's ban on burning, and restrictions on sanitary landfilling required that changes be instituted.

Total daily volume of refuse in Beaumont is about 290 tons (based on 1971 statistics), comprised of refuse and appliances [27:54]. Of this daily volume, 170 tons is collected by the city, and 120 tons by private collectors paying annual dump fees, based on the cubic yard capacity of their equipment.

The program for improving solid waste disposal management in Beaumont has been of a twofold nature. First, the elimination of open dumps has improved the smoke pollution, and health and scenic blight. Second, the state restrictions on landfilling have resulted in the purchase of modern compaction and collection equipment. It has also resulted in the use of efficient shredding equipment, and a 24-inch sand cover between compacted layers, instead of an 18-inch sand cover.

The changes in the solid waste disposal techniques employed in Beaumont undoubtedly provide for more efficient and safer solid waste disposal. The only question remaining concerns the cost factors involved.

Operating at a daily volume of 290 tons, Beaumont's costs in 1971 ran at $1.85 per ton [27:56]. This included the sand cover material of 25,000 yards purchased in 1970 at a cost of $13,000.

More importantly, however, the new sanitary landfilling techniques employed in Beaumont have reclaimed 75 acres of the original 218 acre dump owned by the city [27:56]. Of this reclaimed land, 30 acres were sold to Bethlehem Steel, and 25 acres were sold to a boat club.[a] The balance of the reclaimed land is used as a training ground for the local fire department.

Chicago, Illinois, is another area where sanitary landfill has been successful, though on a smaller scale since the bulk of Chicago's solid waste is incinerated. Here, the emphasis has been on efficient, economical operation of a 600-acre landfill site by a private entrepreneur working for the city [28:92].

The landfill site servicing the Chicago area is located in Des Plaines, Illinois. Every effort is made to provide an attractive site, appealing to a public weary of unsightly dumps, and simultaneously to provide an economical disposal facility. To accomplish this, many artificial and natural techniques are employed.

An 800-foot-long road leads to the facility's office building and garage. The

[a]The sale prices were not available.

entrance is guarded by two attractive concrete walls advertising a sand and gravel company. The road is well constructed to sustain easily the weight of the heavy laden disposal vehicles. Top quality equipment is employed to provide proper processing of refuse, and the natural landscape is kept intact and attractive. These techniques, claim management, result not only in the elimination of public animosity, but in a low price for a quality service.

Operational facilities at the installation cost about $250,000. However, the cost of disposal to the city of Chicago is only about $1.60 per ton. This is based on a daily volume of six to seven tons [28:93].

Recovery values have not been broken down based on total volumes of refuse processed, nor in terms of value per ton of refuse processed. However, the reclaimed land has commercial as well as social value. In Chicago, a $35 million project on a 77-acre landfill site completed in 1962, includes golf courses, drive-in theatres, parks, community centers, and apartment buildings.

Fort Worth, Texas, is the third experience with sanitary landfills to be analyzed. Fort Worth, with a population of 365,000, is about three times the size of Beaumont (pop. 116,000) [29:102].

The solid waste management program in Fort Worth came under public attack in the mid-1960s due to the blanketing of the city's skies with fly ash and incinerator soot. An extensive study resulted in the adoption of strictly controlled sanitary landfilling operations.

After a thorough analysis of the city's solid waste disposal requirements through 1980, staff engineers recommended the elimination of incineration facilities, and conversion of burning dumps to sanitary landfills. This was easily accomplished by closing one old landfill, and purchasing two tracts of strategically located low-value land.

Sanitary landfilling was advisable basically due to the favorable projected cost estimates. These estimates placed the cost of sanitary landfills at $1.13 per ton through 1980. This compared with cost estimates of $5.25 per ton for properly controlled incinerators, and $3.50 per ton for composting facilities [29:103].

Based on an average of 4.5 pounds of refuse per person per day [7:514], the population of Fort Worth produces about 820 tons of refuse per day.

The $5.25 per ton cost for incineration covered the costs for operating two proposed new incinerators, as well as operating costs for the old incinerators with new proposed fly ash controls. The cost for incineration at the old facilities equipped with new fly ash controls alone came to $4.21 per ton [29:103]. Furthermore, estimates showed that conversion to incineration instead of sanitary landfilling would require $10 million more in investment and debt service costs than that for conversion to sanitary landfilling.

Finally, composting was discounted as unfavorable because of its high cost. Estimates put the cost of composting at $3.50 per ton for complete conversion [29:103]. Also noted was the fact that $3.25 per ton is the lowest known composting contract cost in the United States [29:103]. Another factor in rejecting composting was the fact that no ready market was available to purchase the compost soil conditioner.

As a result of the survey made, the city of Fort Worth adopted the proposal for complete conversion to sanitary landfilling at the end of 1966. There are now five such sites, covering 350 acres. They are attractive in appearance, and economically successful. Because of low operating costs due to the employment of modern equipment, sanitary landfilling costs in Fort Worth ran at $.51 per ton after conversion to sanitary landfilling. At the same time, future parks and salable land were being created.

Summary: Beaumont, Chicago, Fort Worth

What are the comparative statistics for the three cases discussed above for sanitary landfilling costs? Also, what conclusions can be drawn from the cases studied?

At this juncture, some observations can be made on the possible economies of scale implied by the results of the three cases studied. This concept is very important because the contention that solid waste management planning can decrease unit costs by the more efficient performance of larger facilities offers a possible solution to the problem of future solid waste disposal needs.

The comparative statistics for the three cases studied, covering relative unit costs per operation size for sanitary landfilling, are given in Table 5-2. The results do not necessarily yield irrefutable conclusions. However, the comparison between Fort Worth and Beaumont data indicates that there is a considerable cost savings in sanitary landfilling operations where the size of the operation is increased. However, though economies of scale appear to exist in the exercise of sanitary landfilling operations, there are many independent regional, and specifically local, factors, i.e., availability of cheap land and proximity of land to city, which may influence the cost factors in sanitary landfilling situations.

Incineration

While sanitary landfilling is presently the predominant method of disposal, incineration (particularly in its more sophisticated versions) gives promise of being the most important in the major metropolitan areas, where the shortage of suitable landfill sites is most acute. Possibilities for using garbage to offset energy shsrtages, and as a source of by-product recovery add interest to some forms of incineration which will be discussed presently.

It may be useful to point out some advantages and disadvantages of incineration. The advantages of incineration can be summed up as follows:

1. less land is needed than for sanitary landfilling;
2. a centrally located site is easier to obtain;
3. incinerator residue is inorganic;
4. volume of residue is greatly reduced;
5. incineration is unaffected by climate;
6. incineration is flexible;
7. cost offsets are available through recovery of heat and metal residues. [7:151]

There are, however, several disadvantages that must be considered in the use of incineration:

1. capital costs are very high;
2. operating costs are high;
3. incineration is uneconomical where sanitary landfilling sites are readily available;
4. incineration is not a method of complete disposal—ash and fly ash require further disposal. [7:151-152]

The cost of operating municipal incineration is much higher than that of sanitary landfilling. For example, in New York City, in late 1971, Mayor Lindsay cancelled plans for a super-incinerator in the Booklyn Navy Yard. That incinerator would have handled 6,000 tons of refuse per day, cost over $200 million to build, and $10 per ton to operate [3:113]. At the same time, it would have dumped 3,000 tons of particulates into New York City's air, despite advanced pollution control mechanisms.

Operating costs for incinerators are higher than those of sanitary landfills.

Table 5-2
Landfilling Costs in Three Cases Studied

City Facility Observed	Size of Operation (In Tons Of Refuse Processed Daily)	Unit Cost (Per Ton)
[a]Ft. Worth, Texas	820+	$ .51
[b]Chicago, Illinois	7	1.60
[c]Beaumont, Texas	290	1.85

+Based on 1960 census population figure of 365,000 times estimate of 4.5 pounds of refuse produced per person per day from *Municipal Refuse Disposal*, Institute for Solid Wastes of the American Public Works Assoc. (Chicago, Ill.: Public Admin. Service, 1970), p. 514.

Sources:

[a]Keith Davis, "Planned Landfills Cut Costs and Complaints," *American City* 3 (Dec. 1968): 102-104.

[b]John Sexton, "How to Make a Landfill Attractive," *American City* 81 (Nov. 1966): 92-93.

[c]Clarence R. Cowant, Jr., "Dumps are Signs of Poor Administration," *American City* 86 (Dec. 1971): 54-56.

Incineration operations costs vary from $3 per ton to $7 per ton of refuse burned [7:154]. The large differential is caused by area requirements as to pollution levels permitted, degree of burning, nature and type of incinerator used, profitability of residue recovery, and the general economic conditions in the region.

Unfortunately, at present, incineration has acquired a bad reputation with the general public. People object to the fly ash and the sometimes released in quantity, incinerator residue, as well as the excessive noise created by incinerator operations. However, recent local campaigns to inform the public of the new incinerator devices for controlling residue dispersion and noise have somewhat lessened the public criticism of incineration facilities.

Some typical unit operating costs for modern refractory incineration facilities are given in Table 5-3 [7:154]. These costs are seen to be much higher than those for sanitary landfilling.

Despite the lower operating costs of sanitary landfilling, projected space available to New York City will be exhausted by 1975 [3:112-113]. Furthermore, New York City's refuse volume is predicted to grow to 36,000 tons per day by 1985 [3:112]. Other large cities such as San Francisco face a similar lack of space for landfill disposal [26:515]. With the threat of inundation by our own wastes becoming more of a reality, and in view of the mounting costs of waste collection and disposal, plus the eventual exhaustion of natural resources, reclamation and reuse appear ever more logical.

There are several innovations in the area of refuse incineration which offer two major improvements. These techniques provide for a greater elimination of residues and pollutants, while at the same time reducing costs through high volume operations and by product recovery. One such technique is the use of water walls in place of refractory linings in incinerators. Although this increases operating costs, recovery of the energy of hot gases may more than offset the added operating costs [7:149].

Table 5-3
Operating Costs for Municipal Incinerators in Three American Cities (per ton of refuse handled)

Cost	City
$6.16	Philadelphia (N.W.)
5.87	Chicago
6.04	Milwaukee

Source: Institute For Solid Wastes of the American Public Works Association, *Municipal Refuse Disposal* (Chicago, Illinois. Public Administration Service, 1970), Table 19, p. 154.

Note: Costs are for 1968 operations. Chicago cost is average of three plants' figures. Costs include amortization and residue disposal.

Another development to improve incineration and lower costs is the Mono-Hearth type furnace. Refuse is spread out more evenly by distribution plates, forced through combustion chambers by vibrating grates, and burned at 2,800 degrees F. This results in a slag residue which is crushed and combined with hot gases at the bottom of the furnace. The combination is then diverted to a steam-generating boiler.

A technique of incineration which has received a good deal of attention is pyrolysis, which involves the destructive distillation of refuse in the absence of oxygen. The products are carbon and gases and the latter would be adaptable for use as turbine, diesel or boiler fuels. The gaseous output might also be scrubbed and refined for use in internal combustion engines. Additional work is needed to determine whether or not the products of pyrolysis of refuse will be competitive with fuels from other sources.

A promising experiment in use of garbage as fuel has been conducted by the city of St. Louis [20:10]. Here, shredded solid waste has been employed in conjunction with coal, as a fuel source for generation of electricity. The process works as follows: raw municipal solid waste is taken to a processing plant and ground into small particles. Next, iron particles are removed. Trucks then haul the ground waste to the power plant, where it is air-fired into a 140-megawatt boiler in the ratio (by heat value) of one part to nine parts of coal.

Problems with the St. Louis project have come from three areas. Glass and metal particles have caused damage to the pipelines. Air feeders have been jammed by large pieces of plastics, metal, or rubber. Finally, the sale of ash residue for highway construction has been precluded by the deposits of glass and metals. These problems are presently under investigation, and the Environmental Protection Agency has approved a $570,000 expenditure for an air classification system to remove heavy materials [20:10].

One attempt at innovation in the area of incineration is taking place in Chicago. Chicago employs the use of refuse incineration with an important innovation–production of recoverable steam. In 1971, Chicago's Public Works Department completed one of the western hemisphere's largest incinerators [31:38]. The incinerator is designed to burn 1,600 tons per day. A by-product of this incinerator's operation is the production of 440,000 pounds of steam per hour. Of this total, 200,000 pounds of steam per hour will be sold, and the rest will be used to operate plant facilities [31:38].

To gain acceptance by the community, the Chicago Department of Public Works invited the local press to observe its presentation to the Zoning Commission. The incinerator is located in the heart of an area zoned for light industry and residences. The Northwest Incinerator, as it is called, is guaranteed free from dust, odor, and vermin. Except for its twin 250-foot high brick smokestacks, it resembles modern office buildings.

The operation of the incinerator is based on a battery of four 400 ton per day furnaces. These are connected by a 10,000 cubic yard storage bin, holding 1.2

days' refuse. Trucks unload refuse into the bin, where cranes pick it up for deposit into charging bins.

Bulky wastes are shredded in a machine driven by the plant's steam output. Ferrous metals are separated by magnetic means, and the balance of refuse goes back into the bins.

The furnace has inclined, reverse-action grates, onto which hydraulic rams push the waste. The grate has teeth which move back and forth, pushing burning waste under the newly added waste. As a result, the three stages of incineration (drying, ignition, and combustion) occur almost simultaneously. This is responsible for an approximately 96 percent burnout on the grate. (Burning is also facilitated by two forced-draft fans which are powered by the facilities steam production.)

At this point, residue is cleaned in a water bath and filtered through a rotary screen. It is then loaded on trucks for deposit in landfills.

The key feature of the Northwest Incinerator, designed by I.B.W. Martin Company of Munich, West Germany [31:38], is a water wall circulating cooling water through tubes which form the firebox wall. The wall prevents slag buildup found on refractory furnaces. As a result, the incinerator need only be shut down for one weekend each year for cleaning.

The water wall also operates as a feed-water heater for a steam boiler. Chicago's Commissioner of Public Works, Milton Pikarsky, notes that the exact income value of the steam generated by the Northwest Incinerator is difficult to estimate [31:38]. However, he expects that it will yield about $2.10 per ton of solid waste burned. According to Mr. Pikarsky, the plant will cost $5.00 per ton to operate. This includes operating costs of $3.95 per ton, and twenty-year capital amortization of $3.40 per ton.

Another advantage of the water wall is that temperature and volume of gases is lower than in refractory furnaces. This permits the use of electrostatic precipitators to remove fly ash, instead of more costly mechanical devices.

The Chicago officials note that the incinerator facility is the least expensive for Chicago, where the problems of rail haul and landfill are overwhelming. Incineration is also the most acceptable method to the public.

Three older incinerators are used in conjunction with the newest to burn all household wastes collected by the city, and to generate steam. Other wastes collected by private carters are disposed of by landfill [26:499-500].

Another area where incineration is very important is the vicinity of metropolitan New York City. Here, land is very costly, and becoming increasingly unavailable for sanitary landfilling. As a result, incineration is being employed more extensively in the New York City area. Incineration, however, results in high operating and investment costs, and pollution control problems. One attempt to offset the high costs of incineration will begin operations in the Town of Hempstead, just east of New York City, on Long Island. Hempstead is a suburban area with a population of over 800,000.

On December 11, 1974, the Town of Hempstead signed a contract with Black Clawson Fibreclaim, Inc. to construct and operate a Resource Recovery System in Hempstead. Black Clawson will construct the $45 million facility within three years at no cost to the Town of Hempstead, and thereafter will have a 17 year operating contract to process a guaranteed minimum of 2,000 tons of refuse daily. The Town has the right to expand the volume to 3,000 tons per day [38:40].

Both Black Clawson and Hempstead will share in sales of recycled metals to scrap dealers and of steam to Long Island Lighting Company to be used in the generation of electricity. Accordingly, Hempstead is more interested in processing costs and recovery offsets than in construction costs [37:26]. Final figures estimate resource recovery of 140 tons of ferrous metals, 20 tons of aluminum, 2 tons of brass, and 80 tons of glass daily, yielding up to $3.5 million yearly [38:1]. Annual yields are estimated at 41,000 tons of recycled metals and 20,800 tons of recycled glass [40]. Also, non-recoverable materials will be converted to fuel for generators to produce 225 million kilowatts of electricity yearly. This equals about one-fourth of the Town's needs. Initially, Black Clawson had proposed a sliding scale based on weekly garbage tonnage, ranging from a fee of $9.68 to $12.67 per ton of refuse processed, with the Town getting 0 percent to 15 percent of recovery values and 10 percent to 40 percent of steam sales to LILCO. The Town presently pays $11.40 per ton of refuse processed, compared with an estimated *net* cost of the Black Clawson system of $6.68 per ton [38:2]. The new facility will be operational some time in 1977.

It is interesting to note how this unique resource recovery system will operate. The Black Clawson Disposal System is a Hydrasposal concept, using water and mechanical separators. Trash is conveyed into a Hydrapulper, where materials are ground and removed in water waste. This is done by high-speed cutting rotors.

Metal is removed by a junk remover, and washed. It is then conveyed to a magnetic separator, where ferrous metals are separated. Bits of glass are then removed from the water slurry by centrifugal force. Inorganic materials are dried and used as a source of fuel to generate electric power.

A comparison of the cost of incineration in the Chicago and Hempstead facilities based on preliminary estimates of returns from the Hempstead facility are shown in Table 5-4.

From the results, comparing the two cases studied, it is apparent that economies of scale are not necessarily present in the incinerator operations observed. This is probably explained by the fact that the two facilities examined are very close in actual size. Also, the cost offsets available to the Chicago facility are greater than those available to the Hempstead facility. This may be the result of better planning, or better *regional* sales opportunities.

Nevertheless, it should be noted that the net cost of incineration in the Chicago and Hempstead facilities comes to $2.90 per ton and $4.77 per ton,

Table 5-4
Cost of Incineration at Chicago's Northwest Facility and Hempstead's Proposed Westbury Facility

Facility Observed	Size of Operation	Unit Operating Cost	Unit Cost Offset
Northwest Chicago	1600 tons daily	$5.00 per ton	$2.10 per ton
Town of Hempstead (Westbury)	2000 tons daily	$5.75 per ton	$ .70 per ton[a]

[a]Based on estimate of 25 percent of $3,510,000 annual sale of reclaimed materials plus 35 percent of $4,581,738 annual sale of steam to Long Island Lighting Company, going to the Town of Hempstead. Hempstead returns are preliminary estimate. Unit operating cost does not include debt service.

Sources: 1. "Chicago and London Take the Financial Sting Out of Garbage," *Engineering-News Record* 183 (Dec. 4, 1969): 38-39; 2. Ken Cynar, "Black Clawson: Resource Recovery System," *Environs*, published by the Nassau County Environmental Management Council, Sept. 1973, p. 8; 3. Personal letters to the editor from Robert C. Williams, Counsel, Town of Hempstead, December 18, 1974, and from H.R. Creelman, Commercial Manager, Black Clawson Fibreclaim, Inc., 200 Park Avenue, New York, N.Y. 10017, January 23, 1975.

respectively. Both of these figures are clearly below those shown in Table 5-3. This is a clear indication that the adoption of methods to provide recoverable by-products during refuse disposal is a workable plan to offset the high costs of incineration. This may provide a positive alternative for the large urban centers, where rail haul and high land costs have made sanitary landfilling very expensive.

Technological developments in the techniques of incineration and by-product recovery constitute the principal bases of the Chicago and Hempstead systems. A major advance in the technique of government is the essential factor in a very much larger project for recovery of energy and by-products planned to begin operations in 1976 in Connecticut. It has been noted that successful programs in this field demand the availability of very large quantities of household and commercial garbage. Since collection and disposal is traditionally a function of local governments, the requisite volume of solid waste is usually found only in large cities such as Chicago or major suburban concentrations such as Hempstead. Connecticut proposes to overcome this obstacle by conducting the refuse disposal operations of its 190 towns at ten regional treatment centers operated by the Connecticut Resources Recovery Authority. The plan is to develop a system to convert all of the state's urban household and commercial garbage to low-sulfur fuel for electric generator plants and commercially salable scrap iron, aluminum and glass [36:11]. Connecticut's Resources Recovery Authority will begin operations at the first two plants by 1976, with all ten operating by 1980, and processing 10,000 tons daily [36:11].

The first plan to be functional, in Bridgeport, along with a second in

Hartford, will process 3,600 tons of refuse per day, supply 10 percent of electric fuel in the area, recycle 80,000 tons of iron scrap, 4,000 tons of scrap aluminum, and 40,000 tons of glass per year.

The Authority will charge $3.00 to $5.00 per ton for disposal, less recovery values [36:11]. The program is expected to save taxpayers $100 million by 1985, reduce air pollution from dumps and incineration, and reduce the need for new landfill sites.

Other Methods and Problems

Another major solid waste disposal technique is composting. Composting is "a controlled process of degrading organic matter by microorganisms" [19:6].

Properly managed composting at a high rate of digestion can produce a product which is suitable for garden and agricultural use. Although it is often billed as a fertilizer, compost is merely a useful organic soil conditioner which provides soil with better water and nutrient retention capacity. It does not, however, increase the nutrient content of the soil in itself [1:1].

The most important aspect of composting as a means of solid waste disposal involves the ability to recycle organic waste back to the soil with no significant pollution. Because composting appears to typify the ideal of recycling, it has received attention from environmentalists and government agencies responsible for environmental questions, out of all proportion to its economic or even scientific importance. Thus, the Environmental Protection Agency has commissioned studies on composting [1], and popular literature is full of references to composting as a means for achieving recycling. However, economically, composting does not compete favorably with incineration and sanitary landfilling.

Moreover, the composting process is seriously hampered by many constituents of today's solid waste. Plastics have come into increasing use and the volume of their output has gone up enormously [35:25]. But most plastic materials resist biodegradation, and accordingly interfere with composting [5: 325].

What exactly are the established estimated costs for composting and what are the offsets to disposal costs in the way of compost sales values? Capital costs for a range of capacities in windrow composting plants based on actual costs for such a plant in Johnson City, range from $16,560 per ton of capacity for a 50-ton-per-day plant to $5,460 per ton of daily capacity for a 200-ton-per-day plant (see Table 5-5) [1:58]. These estimates include equipment for sewage processing and, therefore, preclude comparisons with incineration and landfilling projects which do not include such equipment.

Operating costs for various windrow composting plants, again based on the Johnson City plant's actual costs, ranged from $13.65 per ton for a 50-ton-per-day plant to $8.70 per ton for a 200-ton-per-day plant on a two shift operation (see Table 5-6) [1:64].

Table 5-5
Estimated Capital Costs for Windrow Composting Plants

Cost Item	50 Tons/Day[1] (1 Shift)	52 Tons/ Day Johnson City Plant–1 Shift[2]	200 Tons/Day[3] (100 Tons/Day– 2 Shifts)
Buildings	$210,000	$368,338	$ 251,000
Equipment	482,700	463,251	607,100
Site Improvement	126,800	126,786	152,000
Land Cost	8,400	7,600	21,200
Total Cost	$827,900	$965,980	$1,031,300
Total Cost Per Ton Daily Capacity	$ 16,560	$ 18,580	$ 5,156

[1] Based on Johnson City cost data adjusted for building and equipment modifications.

[2] Actual cost of the research and development PHS-TVA composting plant at Johnson City, Tennessee.

[3] Estimates based on actual Johnson City cost data projected to the larger daily capacity plant.

Source: Federal Solid Waste Management Research Staff under the direction of Andrew W. Breidenbach, *Composting Of Municipal Solid Wastes in the United States* (Washington, D.C.: Environmental Protection Agency, 1971), Table 6, p. 59.

The total cost of composting ranged from $3.85 to $20.65 per ton (see Table 5-7) [1:68]. However, the windrow plants ranged in total cost from $11.23 to $20.65 per ton.

Against these figures, however, the recovery of costs through sale of compost must be considered. The value of compost for sale is based upon how much it adds to the soil for a particular crop. With the case of corn, for example, a benefit of $4.00 per ton has been estimated as the value of the first year of application of compost [1:69]. This value would just about cover the cost of transportation. With other more expensive crops, compost may be more valuable.

The University of California estimated in 1953 that farmers would pay from $10 to $15 per ton for compost. In fact, the farmers proved to show little interest in the purchase of compost, although in 1964, a plant in San Fernando sold compost for $10 per ton. Although other experiences of compost sales throughout the United States have ranged from $6.00 per ton in bulk sales to $42.50 per ton in bagged sales, composting does not offer a viable source of cost recovery through sale of compost [1:70].

A good example of the failure of composting sales to offer cost offsets to solid waste disposal costs is the case of Mobile, Alabama [26:505]. In 1965, Mobile established a plant to produce and sell a soil conditioner under the trade name "Mobile Aid." The initial problem was that freight costs to buyers exceeded the cost of the product. Furthermore, the bagged conditioner tended

Table 5-6
Estimated Yearly Operating Costs for Various Capacity Windrow Composting Plants

Plant Capacity Tons Refuse Processed Per Day	Number of Shifts	Plant Operating Costs			Operating Costs Per Ton Refuse Processed
		Operations	Maintenance	Total	
1968 Johnson City-52 Tons/Day (7,164)[1]	1	$ 99,575 (139,817)[2]	$32,590 (41,200)[2]	$132,165 (181,017)[2]	$18.45 (13.40)[2]
50 Tons/Day (13,000)[3]	1	133,950	43,700	177,650	13.65
200 Tons/Day (52,000)[3]	2	357,015	95,400	452,415	8.70

[1] Figure in parentheses is total tons of raw refuse actually processed in 260-day work year.
[2] Costs projected for operating USPHS-TVA composting plant at design capacity of 52 tons per day (13,520 tons/year) in 1969.
[3] Estimated costs based on USPHS-TVA composting project operating cost data.

Source: Federal Solid Waste Management Research Staff under the direction of Andrew W. Breidenbach, *Composting Of Municipal Solid Wastes In The United States* (Washington, D.C.: Environmental Protection Agency, 1971), Table 9, p. 65.

Table 5-7
Summary of Total Costs for Composting Plants[1]

Capacity Tons/Day	Number of Shifts	Plant Type*	Capital Cost (Per Ton/Day)	Cost Per Ton Refuse Processed			Notes
				Capital	Operating	Total	
50	1	W	16,560	6.12	14.53	20.65	2
100	1	W	10,000	3.68	10.62	14.30	2
157	1	HR	8,830	2.97	7.56	10.53	3
200	2	W	5,460	2.01	9.22	11.23	2
300	NA	W	5,000	1.53	5.00	6.53	4
300	1	HR	5,000	1.45	2.40	3.85	
346	2	HR	4,420	1.64	6.94	8.58	5

*W, Windrowing; HR, enclosed high-rate digestion

1. Cost data provided for plants other than Johnson City and Gainesville, were used without adjusting to current economic conditions.
2. Projected from Johnson City project data, at 100 tons per day for 260 days per year; straight-line depreciation of building and equipment over 20 years. Bank financing at 7½ percent over 20 years. Includes disposal of rejects into landfill.
3. Actual data from Gainesville plant with interest at 7½ percent over 20 years, at 45,000 tons per year (286 workdays). Includes sludge handling equipment and disposal of non-compostables remaining after paper salvage.
4. Actual data from Mobile, Alabama, composting plant. Components of costs not known. (G.J. Kupchick, "Economics of Composting Municipal Refuse in Europe and Israel, with special reference to possibilities in the U.S.A." *Bulletin of The World Health Organization* 34: (1966) 798-809.
5. Gainsville plant at 90,000 tons processed per year.

Source: Federal Solid Waste Management Research Staff under the direction of Andrew W. Breidenbach, *Composting of Municipal Solid Wastes In The United States* (Washington, D.C.: Environmental Protection Agency, 1971), Table 11, p. 68.

NA = Not Available

to deteriorate rapidly. Also, about half of the $40,000 yearly income from the sale of the compost product was spent on advertising. The result was that the net cost of compost disposal of solid waste to the city of Mobile was $11.00 per ton, compared to a cost of $2.00 per ton at Mobile's landfills. Finally, the composting project was abandoned in 1971 as a failure.

It may be noted that the situation in Mobile may well be typical of many smaller cities, perhaps most with populations less than 500,000. In such situations, recovery of energy through combustion of solid waste may not be feasible because of the absence of scale economies. In such cases, landfill costs are likely to be low as well because of the availability of relatively inexpensive land within easy hauling distance. In such areas where no socially useful application of solid waste can be found, it may be erroneous to insist that composting operations should yield a profit, or even pay for themselves. In such instances, it might be advisable from an environmental protection standpoint to pursue composting, even if its return did no more than cover the additional costs for composting and marketing the product. Subsidization of composting projects might even be advisable to recover some social value from an otherwise useless discard. Local governmental units, hard pressed to meet rising costs, are unlikely to undertake composting in the absence of state or federal subsidies.

Before any general discussion of recovery values as offsets to disposal costs is made, one other disposal problem should be mentioned.

Plastic Waste. Although it lends itself to a separate investigation, plastic waste must be considered in any discussion of solid waste disposal.

In 1970, the total United States production of plastic materials reached 9.35 million tons [10:7]. Of this total, about 25 percent was used in packaging. This volume of plastic packaging waste creates an enormous solid waste disposal problem. Incineration of plastics creates air pollution. Since they are inorganic, they can not be used in composting. Therefore, plastic wastes are best disposed of in sanitary landfills.

Incineration of refuse plastics can create several problems in obsolete incinerators. Incomplete combustion in inefficient incinerators can cause particulate pollution. Moreover, gases may be released which decompose metal parts of the incinerator when only organic materials are burned.

Plastics help to reduce pollutants, where incinerators are properly designed and operated. Further, because of their high heat content, plastics are helpful in the incineration of wet garbage.

European experiences provide data to suggest that refuse containing larger proportions of plastics do *not* create new problems. However, most European incinerators are also power plants utilizing the BTU content of the refuse, especially plastics [10:12].

Recycling, Recovery, and Reuse

Though the terms recycling, recovery, and reuse are often used interchangeably, they are different phenomena. For the purposes of this chapter, recycling will refer to the refluxing of materials back into the production stream; recovery will denote the salvage of materials or energy from municipal refuse otherwise discarded, after it has entered the usual refuse disposal channels; and reuse will encompass the concept of commodities re-entering the production stream in quantities representing a sizable portion of total consumption, whether they re-enter in original form, or otherwise.

This chapter will be limited to recycling through recovery and reuse from the consumption sector. Recycling within the production and commercial sectors will not be considered here. It should be noted that recovery doesn't become recycling until the materials enter the commercial salvage channels. Thus, recycling takes place mostly in the production sector. Also, it should be pointed out that this chapter will deal only with municipal solid waste recovery, completely omitting the large area of industrial waste recovery, where many highly efficient recycling procedures are employed.

The realization by the federal government that policies for recycling and recovery must be encouraged is evidenced by the advice given to Congress in 1970 by the Council on Environmental Quality. The council advised Congress to meet head-on the problems of reducing the mounting solid waste volume, and pursuing every conceivable method to encourage increased applications of recycling and recovery [3:23].

Presently, however, there are economic barriers and disincentives to the use of recycling and recovery approaches. Proof of this phenomenon is the fact that reclaimed materials make up a smaller portion of industrial raw materials than in the past. For example, the use of scrap metals in the aluminum, zinc, and copper industries from 1963 to 1965 was 40.9 percent of total materials used, but declined to 38.6 percent of total materials used in 1967 [3:24]. It should be noted, however, that this phenomenon may simply reflect the fact that there is a time lag for recovered materials to find their way back into the system, and that during a period of rising industrial output the stock of materials is not wearing out as fast as it is increasing.

With the growth of consumption greater than that of the population, some means must be established to provide for economical reuse of waste materials. General requirements of recycling involve new methods to recover materials from solid wastes, development of new uses and markets for recovered resources, and improved methods for collection and separation to encourage recovery.

Specific techniques for stimulating the use of recycling and recovery would be: 1. federal procurement to encourage recycling and 2. the use of subsidies, incentives, depletion allowances, capital gains treatment, and other tax policies to encourage recycling and reuse of solid wastes.

The potential impact of recycling on our urban economy is most impressive. Jeffrey S. Padros makes the point well in the following statement before the Subcommittee on Fiscal Policy:

> The potential impact of recycling on the urban economy should be evident. As we run out of landfill space, we will have to build expensive new processing plants, or, if regional waste handling plans can be worked out, pay heavily to export our refuse. In contrast each ton of what is now called solid waste that could be recycled back into the industrial process would mean a $10 to $15 or more toll charge that the City (New York City) would not have to pay. For every 1,000 tons of secondary materials we can remove from our daily solid waste load, the City can avoid spending $10 to $15 million or more to build a processing plant that could cost up to $5 million per year to operate [3:113].

What are the specific recovery values which are available to the major solid waste disposal methods as cost-offsets?

The cost of establishing an incineration facility is higher than that for a sanitary landfill site. For example, a sanitary landfill site established in Glastonbury, Connecticut (population 19,500), cost $55,000 for the landfill site, plus $34,000 in development costs [18-2:69]. This capital cost of $89,000 for a facility built in 1967 to handle 40 to 80 tons of refuse daily compares favorably against the cost of $320,000 to $640,000 for a comparable incineration facility, based on a cost of construction of $8,000 per ton of rated capacity [7:152].

However, there may be offsets to the costs of incineration which are not available to sanitary landfilling. Incinerator residue processing provides an area for resource recovery as a part of a community solid waste program.

It should be noted, however, that there is a basic need for research and development to provide new uses and products from metal and glass residues. Most important is the expansion of existing markets, and the creation of new applications for recoverable materials.

What is the nature and value of incinerator residues? Some pertinent data are provided by a report by the American Public Works Association on "Resource Recovery From Incinerator Residue" [14]. Six studies provide data on salvage of incinerator residues in Atlanta, Baltimore, Cincinnati, Los Angeles, Milwaukee, and New Orleans. The only city studied which profitably salvaged from incinerator residue was Atlanta, although the cost versus return of the operation was only marginal (see Table 5-8) [14:20]. It should be noted, however, that recovery from incinerator residues may very well be sub-marginal, if substantial selling costs are involved.

Local market outlets for non-ferrous metal residues were observed to be unlimited and lucrative in the above studies. The quantities of such non-ferrous metals in incinerator residues are shown in Table 5-9 [14:22].

Table 5-8
Atlanta, Georgia Hartsfield Incinerator

Cost to Process 1 Ton of Salvage Metal for Shipment	
Labor	$ 5.40
Power	.89
Maintenance	.91
Parts Replacement	.99
Original Equipment Pro-Rated on 20 Year Life	3.02
Total Cost	$11.21
Selling Price	$11.50
Profit	$ .29

Source: American Public Works Association Research Foundation in cooperation with American Public Works Association Institute For Solid Wastes, *Resource Recovery From Incinerator Residue, Volume 1–Findings And Conclusions* (Chicago, Illinois: American Public Works Association, 1969), Table VI, p. 20.

Ferrous metals are also being salvaged from incinerator residues. Their importance in incinerator residues is tremendous. Combined domestic refuse processed by incinerators in the United States contains 28 percent ferrous residue by dry weight [14:23]. Much ferrous scrap is being shipped to Europe from Milwaukee, where steel mills reported selling tens of thousands of tons per year, of late.

Glass is another solid waste residue in demand, today. Glass cullet ("clear color-sorted, crushed glass that is used in glass making to speed up the melting of

Table 5-9
Non-Ferrous Metals in Incinerator Residues

Constituent	Percentage of Sample
Aluminum	61.0
Zinc	19.7
Copper	1.8
Lead	1.4
Tin	1.0
Balance (Mostly Aluminum oxides and impurities including 1.4 percent of total as iron or its compounds.)	15.2

Source: American Public Works Association Research Foundation in cooperation with American Public Works Association Institute For Solid Wastes, *Resource Recovery From Incinerator Residue, Volume 1–Findings And Conclusions* (Chicago, Illinois: American Public Works Association, 1969), Table VII, p. 22.

the silica sand" [19:6]) comprises about 44 percent by weight of incinerator residues. In Baltimore, a glass company had a sign offering $20 per ton for clean flint cullet, and $18 per ton for dirty flint cullet [14:24].

At the Black-Clawson plant for processing the solid waste from the Town of Hempstead it is planned to crush all glass residue and then separate it precisely by color using a highly sophisticated optical separator. It is expected that this process will greatly increase the value of glass recovered from the town's refuse.

Presently, the cullet dealers get their supply from industrial breakage. The market for incinerator residue cullet of the 1950s declined in the 1960s due to the increased labor and transportation costs [14:24]. This is a good example of how goods, having entered the consumption stream, are difficult to recover.

Municipal solid waste residues are not being exploited at present for several reasons. This is basically due to the lack of adequate separation processes at the municipal incinerator source. The processing of tin cans from incinerator residue is not taking place on a large scale. Also, fly ash has become a disposal problem, rather than a source for land or construction filler (see Chapter 3).

The major reason that ferrous metal residues have not been reclaimed is the fact that tin is present in metal cans. If this tin were removed from can construction, and the yearly production of 7.3 million tons of cans were incinerated, the total recoverable residue would be 2.2 million tons [14:31].

There are many outlets for non-ferrous metal residues, but ferrous metal and glass residues are not in sufficient demand to provide recovery values. Thus, though the recovery of resources from incinerator residues may serve to alleviate a pollution problem, there exists no economic incentive for their recovery. The solution lies in either government subsidies, or special tax treatment to make recycling profitable. The ultimate key to the success of recycling, therefore, involves the creation of new sources of demand for recovered materials.

Another source of cost offsets to incineration involves heat recovery. The heat which results from the process of incineration is useful as a guarantee of continued combustion. However, the value of surplus heat requires circumstances in which the amount of heat generated makes its use worthwhile.

Waste can be burned without additives at 1500 to 1800 BTU/pound [11:29]. However, higher caloric values require higher combustion temperatures. The higher combustion temperatures increase corrosion and incrustations.

What impact does recycling have in various industries? This question is best answered by Table 5-10 [24:xvii]. From this table, it can be seen that the metals industries provide the best record of reused materials as a percentage of total raw materials consumed. These data are based on yearly averages of raw materials consumed and materials recycled or recovered during a two year period from January 1967 to December 1968. The paper industry utilized less than one-fifth reused material as a percentage of total raw material, while paper waste amounted to more than half of all household garbage [21:20]. The problem with reuse in the paper industry involves the high cost of reprocessed waste paper compared to virgin fiber.

Table 5-10
Impact of Recycled or Recovered Materials on Various Industries Raw Material Base

Industry	Total Recycled[1]	Total Consumed[1]	Percentage of Consumption by Recycling[2]
Paper	10.124	53.110	19.0
Iron and Steel	33.100	105.900	31.2
Aluminum	.733	4.009	18.3
Copper	1.447	2.913	49.7
Lead	.625	1.261	49.6
Zinc	.201	1.592	12.6
Glass	.600	12.820	4.2
Textiles	.246	5.672	4.3
Rubber	1.032	3.943	26.2
Total	48.108	191.220	25.2

[1] Million tons

[2] Study period was 1967-1968. During this period, 190 million tons of major manufactured materials were consumed yearly. During the same two-year period, 48 million tons of the same materials were recycled through the market yearly. The rate of recycling during the period was 25.2 percent of consumption. No data were available to indicate what proportion of materials recovered came from consumer versus production sector. Recovered materials were either fabrication wastes or discarded products returned to industry for reprocessing. Data were not available to indicate whether production sector inputs were from past or present production wastes.

Source: Arsen Darnay and William E. Franklin, *Salvage Markets For Materials in Solid Wastes* (Washington, D.C.: Environmental Protection Agency, 1972), p. 17.

Most of the difficulty in processing municipal wastes for salvage is caused by the need to separate refuse into basic material classes. This requires separation into metals, glass, paper, etc. Having been accomplished to date by manual labor, this techniqne costs about $16 per ton [24:xvii]. However, revenues have only amounted to between $4 and $9 per ton. This diseconomy involved in sorting municipal wastes has been the major obstacle to municipal waste recovery to date.

Now, however, two new approaches show signs of yielding economies in sorting out municipal solid wastes. The Black Clawson Company has developed a new process whereby municipal wastes are accepted and automatically separated into fibers, metals, and glass. This system is presently in the demonstration stage, and involves a derivation of paper pulping technology [24:xviii]. Another technique, developed by the Bureau of Mines, involves separation of metals and glass from incinerator residues using materials handling techniques developed in the mining industry [24:xviii].

Other technology employing magnetic and mechanical conveyor systems are also being tested. Two points, however, must be kept in mind:

1. New technological advances in recovery of solid wastes will require large scale operations, suitable only to densely populated areas, and
2. The problem of markets for these commodities will continue to exist as long as virgin materials are plentiful and market prejudices against reused materials persist.

Importance of Cost of Transportation to Landfill Sites

When does the cost of transportation to distant landfill sites makė incineration more economical? This is a question that is very important in solid waste disposal management today.

The phenomenon of increasing transportation costs involves two major cost areas. First, transportation costs are going up, especially due to gasoline price increases. Also, land values are going up, and land is becoming more scarce in areas near the central cities. Therefore, cheap land for landfill sites is available only in areas more distant from the cities than in the past. As a result, transportation costs to landfill sites have risen so sharply as to make incineration in the central city more economical in many areas.

Examples of this shift to incineration due to higher transportation costs are prevalent in big cities, such as Chicago and New York. Though no specific data are available concerning the lowest available landfill site cost (transportation cost to site, plus land cost of site), there are some cost data concerning current net landfill costs versus incineration costs in these two big cities.

In New York City, landfill disposal costs less than $5 per ton [39:47]. This disposal takes place in areas of New Jersey, where N.Y.C. garbage is deposited. Restrictions contemplated by the state of New Jersey on the dumping of N.Y.C. garbage could result in N.Y.C. having to send its garbage to parts of Ohio, at a cost of about $15 per ton [39:47]. As a result, incineration costs, estimated at $7-$8 per ton, would be highly favored over landfill costs of $15 per ton, if shipment of refuse to Ohio were required [3:113].

In the case of New York City, therefore, the impact of changing transportation costs on the decision of landfill versus incineration is obvious.

Another example of the above-described phenomenon is the city of Chicago. Chicago's new Northwest incinerator will run at an approximate net cost of $5 per ton [31:38]. However, local private landfill facilities can handle a daily volume of 7 tons at a cost of only $1.60 per ton [28:93]. These local facilities are in Des Plaines, Illinois, less than 10 miles from the Chicago city limits.

Clearly, the contrast in landfill versus incineration costs in these two big cities yields a pattern. Where available landfill sites are distant from the city, transportation costs are so great as to make incineration in the central city more economical in comparison. On the other hand, where landfill sites are available close to the city, landfill requires little transportation expense, and is more economical than incineration.

Presumably as the size of a city increases it will be necessary to haul refuse further to find suitable landfill disposal sites at reasonable cost. With modern methods of incineration, suitable sites for the latter may be located closer to the points at which refuse is generated. When does incineration become more economical, what is the critical population? With the information presently available, no such determination is practicable. This is true because the factor of land values differs greatly in many parts of the country. Also, transportation costs will differ throughout the United States. Thus, each locality must evaluate its own land values, distance for refuse transport to site, and alternative incineration costs to come up with an answer to the question of whether to use landfill or incineration techniques in disposing of solid wastes.

Break-even Volume for Resource Recovery from Incineration

Another important question concerns the volume required to operate a resource recovery system. Municipalities must consider at what volume of refuse is resource recovery economically feasible.

Some data on resource recovery systems are available for analysis. Connecticut's proposed incineration system will convert all of the state's refuse into low-sulfur fuels, and recyclable aluminum and glass, by 1980 [36:11]. This system will process 10,000 tons of refuse daily. The cost will average $3 to $5 per ton, after recovery offsets are considered [36:11].

Another new resource recovery incineration system is that of the Black Clawson Company, to be built for the Town of Hempstead in New York. This system will operate on the town's approximately 2,000 tons of daily waste, at an average net cost of $5.05 per ton [38:2, 40]. This cost takes into account cost offsets from recovered iron, aluminum, brass, and glass yielding $3,510,000 yearly, along with fuel sold for electric generators of Long Island Lighting Company [38:1,40].

Table 5-11 contrasts the data for Connecticut and Hempstead resource recovery systems. This suggests that the higher volume of operation in Connecticut yields a lower net cost. It must be underscored that neither set of data is based on actual operating experience at this point.

It is difficult to determine at what minimum level of refuse volume is a resource recovery system economically practical. It appears increasingly likely, however, that resource recovery in incineration facilities is a viable cost offset. Wherever sanitary landfill is not economically practical, and incineration is required, resource recovery should be evaluated. If volumes of refuse required for economic feasibility of resource recovery systems is not present, a system

Table 5-11
Comparison of Volumes of Operation to Net Costs of Incineration in Connecticut and New York Proposed Incineration Resource Recovery Systems

Location	Daily Volume in Tons	Net Cost in $ Per Ton
Connecticut	10,000	3 to 5
Hempstead, N.Y.	2,000	5.05

Note: Connecticut's proposed incineration resource recovery system will process the entire refuse volume of the state by 1980. The system proposed for Hempstead, New York, involves a resource recovery system of Black Clawson Company for the Town of Hempstead.

Sources: Michael Knight, "Plants to Make Fuel of Garbage," *New York Times* (May 17, 1974), p. 11; Art Thompson, "Hempstead Town Accepts Resource Recovery Bid," *Town of Hempstead News* (May 21, 1974); personal letters to the editor from Robert C. Williams, Counsel, Town of Hempstead, December 18, 1974, and from H.R. Creelman, Commercial Manager Black Clawson Fibreclaim, Inc., January 23, 1975.

combining several localities should be considered. Clearly, the future alternative to the increasing land and transportation costs involved in sanitary landfill will be incineration employing resource recovery to reduce the net costs.

Conclusion

It must be noted in conclusion that the two major methods of solid waste disposal are sanitary landfilling and incineration. The lower cost of sanitary landfilling is presently an attraction toward its use. However, the decreasing supply of landfill sites within economic transportation range of urban centers is creating a shift towards incineration.

The high cost of incineration construction and operation can be offset by a well-designed system of refuse sorting. New techniques can make recycling a viable cost offset to incineration. Preliminary studies, however, indicate that such techniques are only workable on a large scale. Thus, just as cost factors prove that greater volume is required in incineration to create an economical disposal system, the cost of recycling is diminished sufficiently to permit economic justification only when the scale of output is high.

Based on the data and observations of solid waste disposal made in this chapter, several trends, though not proven, are indicated by a bulk of the evidence. The use of sanitary landfill operations to dispose of solid waste from densely populated urban centers will decrease in the years ahead. Decreased availability of land for fill sites, higher prices for land near the cities, increased transportation costs to distant sites, and lower incineration net costs will result in a decline in the use of landfill techniques. Landfill advantages in the form of valuable reclaimed land will not outweigh the increased success of incineration facilities in the large urban centers.

Incineration facilities, especially with the many antipollution and recovery systems employed, will see a widespread increase in the large urban areas. These new incineration systems offer close proximity to collection apparatus (creating lower transportation costs), pollution-free and noise-free operation reducing the public distaste for them that has existed in the past, and lower costs (if operated on a large scale, and with the employment of recovery systems as cost offsets).

In the suburban and rural areas of the United States, however, the forecast is much different. Cities with under 400,000 population, generating less than 1,000 tons of refuse per day (based on an average of 4.5 pounds of refuse per person per day) will not find the high construction and operating costs of modern incineration and recovery systems economical. These systems do not function on a low net cost basis, unless a volume of about 1,000 tons per day of refuse processed is maintained. The only way suburban areas could operate such facilities would be to transport refuse from other suburbs to their facility. This would incur the high costs of transportation, however, and defeat the economies of scale inherent in these systems.

Therefore, it is the conclusion of this study that in most cases, cities with populations over 400,000 will find the new low-polluting, high-volume, recovery-geared incineration systems to be most economical. This assumes that the ever-increasing land shortages, and steadily rising transportation costs for sanitary landfilling will persist in urban areas. Suburban and rural areas, especially under 100,000 population, will find that the new incineration facilities are uneconomical because of the high volume of refuse required, and because cheap land is more readily available for sanitary landfilling operations.

Other factors may, however, alter the above estimates. The cost factors outlined above may be altered drastically by governmental regulations and/or aid programs. The increasing effort by various interests to get federal, state, and local pollution control laws enacted could result in increased costs to both incineration and landfilling facilities. Impending governmental regulations seem not to favor one method over the other.

The potentially more important, though presently less predictable, area of governmental action involves subsidization and government aid programs. Should the ultimate question of optimum utilization of available resources become vital, such as caused by the potential for large shortages in many raw materials, the areas of resource recovery through incinerator recovery systems, and the use of compost, may warrant governmental action. One governmental measure that may be crucial for the success of advanced incinerator systems that permit the recovery of salable energy and by-products may be devices to effect the cooperation of local administrative units. This was suggested by the experience of Connecticut now launching a statewide solid waste program with recovery features to process the solid waste of 190 municipalities, few of which would have generated enough individually to justify such measures.

Although composting has not been successfully employed by municipal disposal facilities, governmental subsidization could result in the recovery of an otherwise discarded resource. While current attempts at recovering metals, paper, glass, heat, and energy from incineration facilities have had only limited success, governmental funding might result in development of economical resource recovery systems. Such systems could prove invaluable in case of material shortages caused by depletion of resources, or economic boycott by producing nations toward consumer nations, most important of which is the United States.

Finally, it must be emphasized that the municipal solid waste disposal system to be employed in an area depends on the population (and resultant volume of refuse generated), availability of cheap land, and governmental restrictions and funding available. These factors will basically determine the net cost for the solid waste disposal systems available; and hence, the most economical system for the municipality.

Bibliography

1. Federal Solid Waste Management Research Staff under the direction of Andrew W. Breidenbach. *Composting of Municipal Solid Wastes in the United States.* Washington, D.C.: Environmental Protection Agency, 1971, pp. x and 103.
2. National Industrial Pollution Control Council. *Deep Ocean Dumping of Baled Refuse.* Washington, D.C.: Commerce Department, 1971, p. 15.
3. Hearings Before the Subcommittee on Fiscal Policy of the Joint Economic Committee of the Ninety-Second Congress of the United States, 1st Session. *The Economics of Recycling Waste Materials.* Washington, D.C.: United States Congress, 1971, p. 198.
4. Council on Environmental Quality. *Environmental Quality.* Washington, D.C.: President's Council on Environmental Quality, 1972, pp. xxvi and 450.
5. Van Tassel, Alfred J. (ed.). *Environmental Side Effects of Rising Industrial Output.* Lexington, Massachusetts: Lexington Books, D.C. Heath and Company, 1970, pp. xx and 548.
6. National Industrial Pollution Control Council. *Junk Car Disposal.* Washington, D.C.: Commerce Department, 1970, p. 54.
7. Institute for Solid Wastes of the American Public Works Association. *Municipal Refuse Disposal.* Chicago, Illinois: Public Administration Service, 1970, pp. xvii and 538.
8. Jensen, Michael E. *Observations of Continental European Solid Waste Management Practices.* Washington, D.C.: Public Health Service, 1969, pp. v and 46.
9. Smith, David D. and Brown, Robert P. *Ocean Disposal of Barge-Delivered Liquid and Solid Wastes from U.S. Coastal Cities.* Washington, D.C.: Environmental Protection Agency, 1971, pp. ix and 119.

10. National Industrial Pollution Control Council. *Plastics in Solid Waste.* Washington, D.C.: Commerce Department, 1971, p. 20.
11. American Public Works Association. *Proceedings of the Second Annual Meeting of the Institute for Solid Wastes of the American Public Works Association.* Chicago, Illinois: American Public Works Association, 1967, pp. viii and 67.
12. Committee on Solid Wastes of American Public Works Association. *Refuse Collection Practice, 3rd ed.* Chicago, Illinois: Public Administration Service, 1966, pp. xviii and 525.
13. Working Party on Refuse Disposal. *Refuse Disposal.* London, England: Her Majesty's Stationery Office, 1971, pp. vi and 198.
14. American Public Works Association Research Foundation in cooperation with American Public Works Association Institute for Solid Wastes. *Resource Recovery from Incinerator Residue, Volume 1–Findings and Conclusions.* Chicago, Illinois: American Public Works Association, 1969, pp. viii and 35.
15. Brunner, Dirk R. and Keller, Daniel J. *Sanitary Landfill Design & Operation.* Washington, D.C.: Environmental Protection Agency, 1972, pp. viii and 59.
16. Delaney, James E. *Satellite Vehicle Waste Collection Systems.* Washington, D.C.: Environmental Protection Agency, 1972, pp. iii and 14.
17. Kiefer, Irene. *A Study of Solid Waste Collection Systems Comparing One-Man With Multi-Man Crews–A Condensation.* Washington, D.C.: Environmental Protection Agency, 1972, p. 32.
18. Golueke, C.G. and staff of the College of Engineering of the University of California. *Solid Waste Management: Abstracts and Excerpts from the Literature, Volumes 1 and 2.* Washington, D.C.: Public Health Service, 1970, pp. xi and 308; vii and 147.
19. Environmental Protection Agency. *Solid Waste Management Glossary.* Washington, D.C.: Environmental Protection Agency, 1972, pp. iii and 20.
20. Journal of Commerce Special. "In St. Louis Power Plant Solid Waste Is Used as Fuel." *Journal of Commerce* (April 5, 1973), p. 10.
21. "Cash in Trash? Maybe," *Forbes* 105 (January 1970): 18-22 and 24.
22. "High temperature garbage disposal turns up iron, glass," *Industry Week* 166 (March 2, 1970): 16-17.
23. Whitney, Charles Allen. "Profit in Solid Waste–for Whom?" *The Exchange* 33, 11 (November 1972): 6-11.
24. Darnay, Arsen, and Franklin, William E. *Salvage Markets for materials in solid wastes.* Washington, D.C.: Environmental Protection Agency, 1972.
25. Van Tassel, Alfred J. "Methodological Problems in the Economic Study of the Environment." Paper presented at a meeting of the New York State Economic Association on April 28, 1973, pp. 1-20.
26. Van Tassel, Alfred J. (ed.). *Our Environment: The Outlook For 1980.* Lexington, Massachusetts: Lexington Books, D.C. Heath and Company, 1973, pp. vii and 589.
27. Cowart, Clarence R., Jr. "Dumps Are Signs of Poor Administration." *American City* 86 (December 1971): 54 and 56.

28. Sexton, John. "How to Make a Landfill Attractive." *American City* 81 (November 1966): 92-93.
29. Davis, Keith. "Planned Landfills Cut Costs and Complaints." *American City* 83 (December 1968): 102-104.
30. Golueke, C.G. and McGauhey, P.H. *Comprehensive Studies Of Solid Waste Management, First and Second Annual Reports.* Washington, D.C.: Public Health Service, 1970, pp. xx and 202; pp. xvii and 245.
31. "Chicago and London Take the Financial Sting Out of Garbage." *Engineering-News Record* 183 (December 4, 1969): 38-39.
32. "Making Garbage Glitter." *Fortune* (January 1973), p. 8.
33. Ashkinaze, Carole. "Green Light for Novel Waste Plant." *Newsday* (July 18, 1972).
34. Cynar, Ken. "Black Clawson: Resource Recovery System." *Environs*, published by the Nassau County Environmental Management Council, September 1973.
35. Vaughan, Richard D. "Solid Waste: Problems and Possible Solutions." *Urban and Social Change Review* 3, 2 (Spring 1970): 22-27.
36. Knight, Michael. "Plants to Make Fuel of Garbage." *New York Times* (May 17, 1974), p. 11.
37. Hertzberg, Dan. "2 Firms Submit Offers on Recycling Plant." *Newsday* (March 8, 1974), p. 26.
38. Thompson, Art. "Hempstead Town Accepts Resource Recovery Bid." *Town of Hempstead News* (May 21, 1974).
39. Bird, David. "Jersey Law Could Raise Costs of Garbage Disposal for the City." *New York Times* (May 8, 1973), p. 47.
40. Personal letters to the editor updating information in reference 38. From Robert C. Williams, Counsel, Town of Hempstead, December 18, 1974, and from H.R. Creelman, Commercial Manager, Black Clawson Fibreclaim, Inc., 200 Park Avenue, New York, N.Y.

6 Consumer Appliances: The Real Cost

Massachusetts Institute of Technology Center for Policy Alternatives with the Charles Stark Draper Laboratory, Inc.

Very few investigators have looked at the way in which American consumers use energy for daily life and entertainment. Here's a report that looks at these questions.

Introduction

During the past decade, American society has witnessed a consumer advocacy movement that is unprecedented since the rise of public concern over business practices in the 1930s.

Between 1966 and 1969 alone more than 25 consumer-oriented legislative acts were passed by Congress, and the number of federal government-sponsored consumer-protection programs jumped from 250 to more than 400; the number of state and local provisions are immeasurable. An entirely new sector of public advocacy has been institutionalized, taking the form of consumer-action groups organized by private citizens to redress alleged business abuses. And both the business literature and the general press carry daily charges of misleading advertising, defective products, and unethical buyer treatment.

Today, perhaps the largest single area of consumer concern—from the individual housewife to state and federal legislators—is with the cost and quality of product service. Manufacturers are blamed for poor product design, faulty manufacturing, and deceptive warranty practices. Repair service is singled out as incomplete and excessively expensive. And increasingly the American consumer is concerned not only with what he gets for what he pays but also with why he pays what he does.

Among the most frequently asked questions are, how efficient is industry's system for designing, manufacturing, distributing, and servicing consumer products? How do the various elements of this system affect what the U.S. consumer pays for a product? Where are the problem areas and how can they be improved?

Precisely to answer these and related questions, the Massachusetts Institute of Technology Center for Policy Alternatives and the Charles Stark Draper Laboratory, Inc. were commissioned by the National Science Foundation to

This chapter is the text of a 32 page summary that appears in a 325 page technical report on the subject of *Consumer Appliances: The Real Cost.*

undertake a major, two-year study of the consumer appliance industry. The study's major purpose was to examine and evaluate the servicing industry, and to find alternatives for increasing the productivity of service in the context of what the consumer pays for a product during its useable life.

Appliance Industry Chosen

The appliance industry was selected because it represents a substantial investment by American consumers, who in 1972 alone paid $5.4 billion for home electronics products and another $7.5 billion for other major appliances. (That same year, an estimated 330 million major appliances and 115 million television sets were in use in the United States.)

Less apparent to consumers, but of major importance in the context of the study, were the costs of servicing and operating these products. Consumers in 1972 paid an estimated $1.5 billion to $3.2 billion for radio and television repair and approximately $900 million for repair of other appliances. Moreover, the cost of electrical energy consumed by these products is estimated at $5 billion, a particularly significant figure at present given rapidly escalating energy costs (projected to increase approximately 75% from 1970 to 1980) and the national priority on conserving energy resources.

While these statistics underscore the economic magnitude of the appliance industry, they also illustrate the major implication of the MIT study: that the public generally is not aware of the total cost of appliances, i.e., what is termed the "life-cycle cost" of the product, which includes purchase price, cost of energy used in operating the product, the cost of maintenance and repair, and even the disposal cost. Least understood is servicing cost, which significantly increases the total price of a product and may even amount to more than the purchase price. Too, the sharply increasing cost of energy is of great significance currently.

Color TV and Refrigerator

These frequently unrecognized factors were the focus of the MIT study, initiated by the Research Applied to National Needs (RANN) program of the National Science Foundation. While the scope of the study ranged widely from general household appliances to entertainment and communications products, it concentrated specifically on two—the color television receiver and the refrigerator—as representative of the entire appliance industry.

In 1972, consumers paid $1.7 billion for the purchase of refrigerators and $4 billion for purchase of color television sets. In terms of aggregate annual expenditures for purchase, service, and energy, these products rank first and third, respectively, in order of economic importance to the U.S. consumer. (In ranking of dollars expended: refrigerators, black-and-white television, color television, washing machines, electric ranges, electric dryers, air conditioners, freezers, vacuum cleaners, and dish washers.)

Moreover, they are representative of two very different product categories:

1. *White Goods*—such major household appliances as ranges, clothes washers and dryers, freezers and refrigerators. The refrigerator, of which almost 7,000,000 per year are produced domestically, is perhaps the most essential product of this type in the American home.
2. *Brown Goods*—home electronics products characterized by their high technology which are subject to more rapid technological change than white goods. The color television set is typical of this group. It is approaching the 10,000,000-per-year production level but has grown much more rapidly than refrigerators.

Findings

The refrigerator and color television industries provided excellent opportunities to evaluate how the life-cycle costs of two different household products are affected by the various elements of the total servicing system.

The MIT study's findings included:

1. Servicing costs account for 35% of the total dollars spent on a color television set during its useful life, while the purchase price and electrical power costs account for 53% and 12%, respectively. This means that the owner of a $400 color TV can expect to spend another $400 during its usable life.[a]

2. Servicing costs account for only 6% of the refrigerator life-cycle cost, with electrical power cost accounting for the largest portion, 58%, and purchase cost totaling 36%. The owner of a $300 refrigerator will spend another $530[b] over the life of the product.

3. Product reliability in both has increased substantially over the past decade, as evidenced by the dramatic decline in the measured need for service in the first year (a 50% decline in the past 14 years for refrigerators, a 50% decline in the past 8 years for color television). However, the cost of such service—labor and parts—has increased so sharply as to offset what would have been a reduced life-cycle cost of each during those periods.

4. Inclusive product warranties, covering labor and parts for the first year of

[a]In terms of actual dollars spent in the future, which are subject to inflationary pressures, this figure could actually be much higher. The life-cycle costs for refrigerators and TV sets are calculated in terms of present dollars and do not reflect the future influence of inflation. Discounting and deflating future expenditures is explained on p. 151.

[b]See footnote a.

the products' lives, were initiated by manufacturers as marketing tools but subsequently have had a decidedly positive influence on design, reliability and serviceability. Too, warranties have provided new information about the causes for service calls that has increased manufacturers' efforts to educate consumers on the use of their products.

5. Manufacturers, consumers, and repair services were found to be jointly responsible for increasing the cost of warranty service. Fully 30% of all warranty service calls are unnecessary, requiring only minor set adjustment the owner could have performed. Manufacturers' rules regarding service calls paid under warranty can be misleading and sometimes wrongly place the repair-cost burden on service establishments and consumers. And repair services can abuse the warranty system by fraudulent billing.

6. Reduced service costs will be achieved most readily not from a more efficient service industry, as has been proposed, but from greater product reliability. The number of service calls by a repairman has remained at five to seven per day for the past several years and is unlikely to increase significantly. However, the need for service calls is projected to decline by 10% to 15% for color television and 11% for refrigerators in the next seven years, based on the current rate of reliability improvement.

7. Manufacturers have not performed as well as they could regarding product reliability, consumer information, and servicing during and beyond the warranty period. For example, products could be designed for greater long-term reliability, product life-cycle cost information could be made available at the point of purchase, and service warranties could be extended beyond the one-year norm.

Lastly, the MIT study of consumer appliances found that the products have more features, cost less and function better than their predecessors as a result of the industry system for design, manufacturing, distribution, sales and servicing. These and other findings are discussed at length on the following pages. While they apply specifically to color television sets and refrigerators, they are believed to apply generally to most consumer appliances.

Industry Background

The industry system examined by the MIT study team encompasses design activity, manufacturing, distribution and sales, servicing, and disposal. While both the color television and refrigerator industries operate along these lines, the differences in both their history and life-cycle cost components are significant. (Figure 6-1 shows graphically the overall system studied.)

Color Television

First introduced in 1954, color television has been subject to rapid growth and technological change. Compared to present day receivers, early color sets

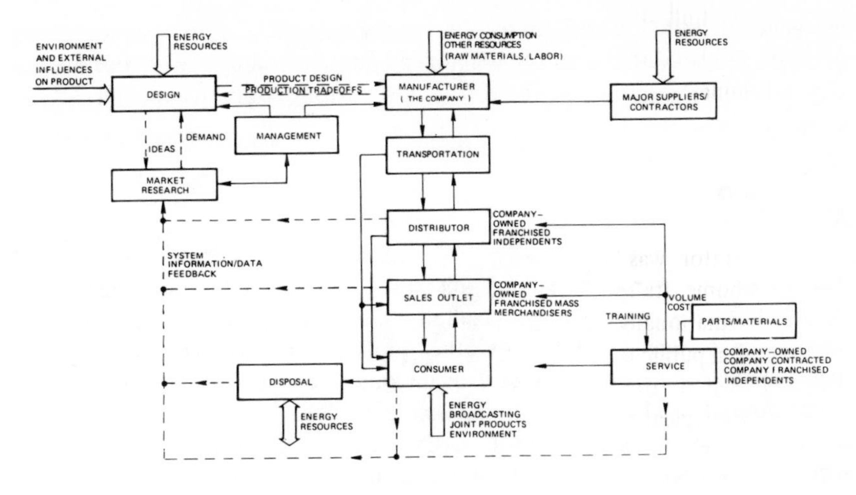

Figure 6-1. Consumer Appliance Product System

provided a poor picture, were difficult to adjust and required frequent degaussing, i.e., neutralization of built-up magnetism.

Significantly, the many innovations and advances in the color TV industry have been brought to the American consumer at steadily decreasing prices. Since 1962, the average retail price has dropped from over $600 to approximately $500, despite inflation and increasing labor and materials costs, while this same period has seen significant improvement in picture tubes, conversion to solid state, extensive use of special-purpose integrated circuits, the introduction of modularization, and increasing emphasis on reliability.

In 1972, 95% of all American households had at least one TV set compared to 87% in 1960 and 9% in 1950. Today, a typical color TV set fails approximately once per year as opposed to six times per year in 1954 and about three times per year in 1960. Moreover, the power consumption of color TV has been reduced significantly, from 300-400 watts in the earlier all-tube sets to as a few as 140 watts in some of today's solid-state sets.

The manufacturers of color TV in 1972 produced an aggregate retail dollar volume of over $4 billion, with the four largest companies controlling approximately 64% of the market and the 20 largest companies controlling 99%.

Receiving informational inputs from distributors and retailers and feedback from the service industry to supplement their own market analysis, manufacturers decide on product and market strategies which determine the price and features of the products they sell. They, in turn, provide marketing help and training aids to distributors and retailers, and in many cases empower the distributors to handle warranty claims.

Sold at the retail level by mass merchandisers and franchised and independent retailers, TV sets are serviced by company-owned, company-franchised, or independent agents as well as by the retailer himself. The independent agent performs the bulk of this service both in and out of warranty; owner-operated or one-man shops account for 76% of all TV service establishments and 31% of service volume.

Refrigerators

The refrigerator was selected for this study because of its prevalence in the American home, its technological stability relative to the TV, and its economic importance nationally relative to other products. The refrigerator represents higher annual public expenditures than any other major consumer appliance.

Introduced in the 1920s, the electromechanical refrigerator underwent its rapid technological change much earlier than the color television, although innovation has continued in terms of increased functions, greater efficiencies and such improved features as frost-free freezer space. Since 1962, the industry has been relatively stable (i.e., subject to few product engineering and design changes) and, like color TV, has offered the consumer increasingly lower retail prices, the average refrigerator cost dropping from nearly $350 in 1960 to approximately $300 in 1972.

Manufacturers of refrigerators in 1972 produced an aggregate retail dollar volume of $1.7 billion, with the four top manufacturers accounting for roughly 65% of production, as they have for nearly 20 years. Most refrigerators flow from manufacturer to distributor to mass merchandiser, franchised dealer, or independent retailer.

Refrigerator manufacturers rely more heavily than TV manufacturers on the large, company-owned service organization for warranty servicing. (These organizations typically account for about 60% of such service.) After warranties expire, however, independent service dealers perform most of the work. Additionally, servicing is provided by mass-merchandiser-owned service centers.

While both the refrigerator and the color TV are subject to the same influences on their life-cycle costs, these influences affect each industry in notably different ways, in many cases because of the varying product and technology characteristics discussed previously. The following sections, which contain the bulk of the study's implications for the American consumer, assess these varying cost influences during each product's usable life in terms of the following: Purchase Cost, Energy Cost, and Servicing Cost.

Product "Life-Cycle Costs"

The fact that consumer appliances carry unrecognized costs that can exceed their purchase price has significant implications for consumers, manufacturers, and service organizations. Additionally, it raises questions of product design, service productivity, warranty practices, content and comprehensiveness of product labeling and consumer information. Particularly significant is the study's identification of electrical energy as a major influence on life-cycle cost during a nationally declared "energy crisis" that shows little promise of abating for many years to come, if ever.

What, then, does the consumer actually pay for these products? Why does he pay it? And what are the prospects for greater efficiency and cost reductions?

Before examining total life-cycle costs—i.e., the sum of all dollars paid during a product's useful life—it is necessary to understand how these costs were arrived at. The next paragraph explains two techniques used by the MIT study team to *discount* and *deflate* life-cycle costs to constant dollars[c] for more accurate comparisons in different years of manufacture. While these techniques are necessary for projecting trends in life-cycle costs, had they not been used the costs quoted in this section would be much higher (for example, one of the techniques factors out the influence of inflation).

Calculation of Life-Cycle Costs. To account for the fact that power costs and servicing costs are spread over the life of a product, a technique called "discounting of future expenditures" was used to obtain the present value of these future expenses (i.e., the amount of money one would have to put in the bank today at a given interest rate in order to have enough money to pay service and power expenses in the future).

Additionally, for comparative purposes, it was necessary to create a unit of cost that is not affected by inflation. Thus the life-cycle costs were deflated to *constant* dollars or *real* costs, a standard technique used by economists to exclude inflation from cost comparisons over time. However, while this technique allows more accurate comparisons, the costs quoted in this section would be much higher if future inflation were taken into account.

Life-Cycle Cost. Based on industry information, the MIT study team established that the color TV has a life of about 10 years, the refrigerator a life of about 14. In constant dollar terms, the average life-cycle cost of a color television is today about $800. Purchase cost accounts for about one-half, service for one-third, and power for one-sixth.

Figure 6-2 shows the trend of these costs over time. Note that the figure shows an overall cost reduction of 30% in constant dollars since 1964, reflecting primarily the decreasing cost of purchase and service.

Figure 6-2 also shows the real life-cycle cost of refrigerators, discounted and deflated, from 1954 to 1972. While the total cost has declined more slowly than for color TV, this slight decline—from about $630 to $600 in constant dollar terms—occurred at the same time that convenience, performance, and unit size

[c]1967 was the base year for all constant dollar calculations.

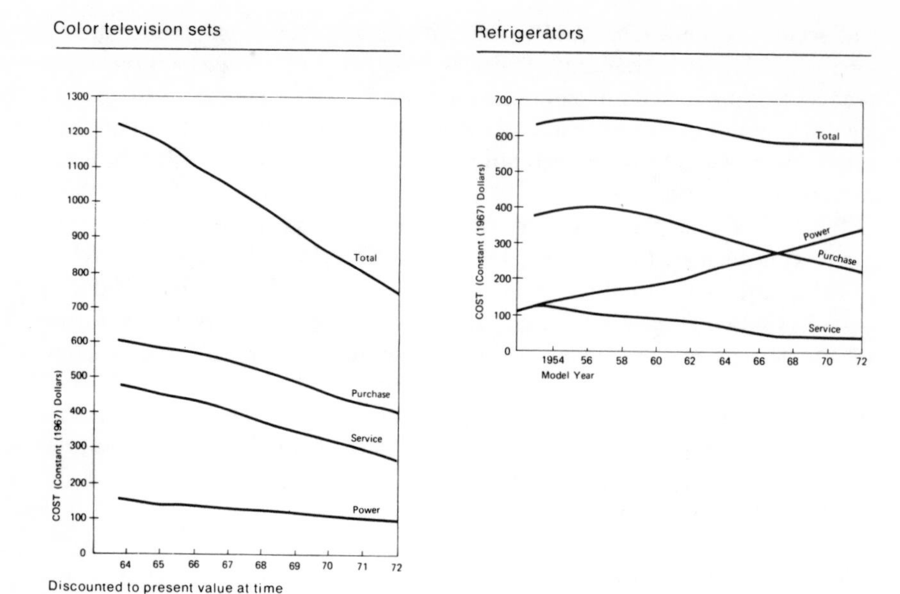

Figure 6-2. Trends in Life-Cycle Cost

increased dramatically. Too, note the increasing influence of power, nearly tripling since 1954 to 58% of total cost, while purchase and service costs declined relatively slightly to 36% and 6% respectively.

These percentage relationships as well as those for color television sets are shown in Figure 6-3. The sections that follow examine each of the major cost components in depth.

Purchase Cost—The Price of Acquisition

To the average consumer, the most immediate and obvious portion of life-cycle cost is the purchase price. For the average color TV, he pays today approximately \$500 or 53% of the total life-cycle cost, and the refrigerator today costs him about \$300 or 36% of life-cycle cost. Moreover, these prices have declined steadily over time, bringing the consumer more versatile and more efficient products at a lower purchase price (in 1960, the average color TV set sold for about \$600, while the average refrigerator sold for close to \$350).

The Influence of Design

Of all the influences on the purchase price of a product, a manufacturer's design philosophy is one of the most important, especially for an appliance as

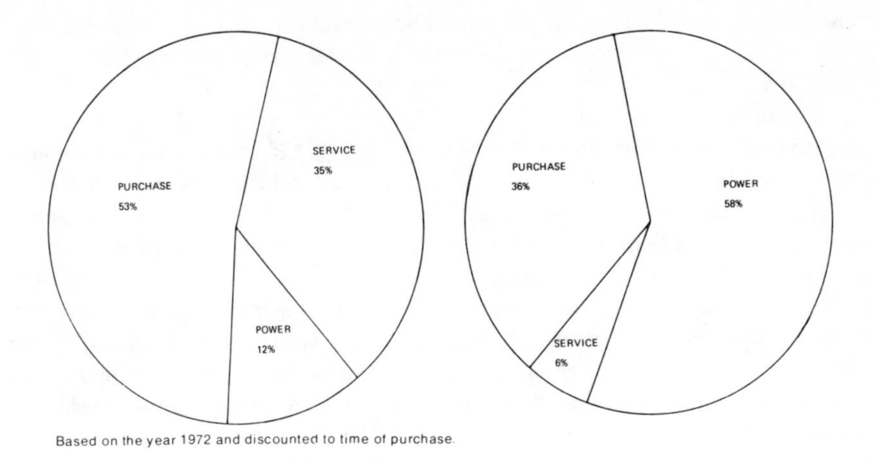

Figure 6-3. Components of Life-Cycle Costs

technologically complex as the color television. Because of warranty costs and the intensity of competition among TV manufacturers, a TV set is designed with the objective of getting the greatest reliability and most performance features for a given manufacturing cost.

Likewise, the refrigerator, considerably more complex than its counterpart of 10 years ago, with frost free freezer space, icemaker options, and other features, owes much of its purchase cost to the influence of design. Yet, unlike the TV industry where design must still cope with technological change, refrigerator design deals with an established basic product and concentrates more on taking advantage of new materials and better components than on total product design. Thus the manufacturing cost of the refrigerator, and ultimately the sales price, is largely due to the materials, parts, and components which comprise it, involving constant design tradeoffs between function, capacity, features, and reliability versus price in the marketplace.

Design tradeoffs for the color TV, however, are considerably more complex, involving engineering and design for the total product rather than just for materials and parts, and focusing heavily on balancing the cost of manufacturing against the cost of subsequent service during the warranty period.

The Warranty and Color TV

Since the advent of the consumer advocacy movement in the early 1960s, the television industry has seen the one-year labor and parts warranty practically become an industry standard, primarily due to consumer and competitive pressures that caused manufacturers to assume greater responsibility for early failures of their products. This is a vast change from prior warranty practices which had manufacturers typically assuming responsibility only for parts failures during the first 90 days of ownership, putting a much heavier service-cost burden on the consumer.

As might be expected, manufacturer's warranty service costs have risen accordingly. Under the 90-day, parts-only warranty, the cost per set was about $3 or roughly 1% of the retail price, representing a total cost to the TV industry of approximately $18 million annually. Since 1969, however, the one-year labor and parts warranty has increased this figure by a factor of 7 or 8 to as much as $25 per set or 5% to 10% of retail price.

In order to lower these costs, manufacturers are increasingly seeking to increase the reliability of their sets during the warranty period. Moreover, a cost/performance battle among manufacturers, spurred primarily by increased competition from Japanese firms, is causing added pressures for improved product reliability.

In short, the tradeoff between designing a TV set for lowest possible purchase cost and designing it for minimum servicing during the warranty period has become a major industry concern.

Manufacturers, seeking to gain a competitive edge by increasing reliability and hence lowering warranty service cost, are increasingly applying sophisticated systems to obtain warranty failure information. Through computer analysis of early failures in new models, they forecast service requirements that ultimately will be used to improve manufacturing quality, reliability and design of future sets as well as to reduce future warranty expenses.

All of these developments bode well for the consumer, at least while his set is covered by a warranty, and perhaps longer. For instance, the design engineer is likely to have more flexibility in his design alternatives with regard to reliability. If it can be shown that by using a more complex circuit or more reliable part warranty costs will be reduced, the manufacturer will likely go ahead with the design improvement. In turn, designer improvements that reduce a manufacturer's warranty cost should ultimately be passed on to the consumer in terms of a lower purchase price. The result: an increasingly more reliable product for the consumer at a better price.

It should be noted, however, that the manufacturer's motivation to design for reliability is purely an economic one. He wants to lower his warranty service cost, and thus makes design improvements primarily for the warranty period. There does appear, however, to be a trend toward longer warranties, and this, too, should benefit the consumer.

The inclusive warranty is also of substantial benefit to the consumer because it ensures good service. And the service is easy to obtain because the manufacturer typically pays warranty rates that are competitive with customer-paid servicing. While the labor warranty costs of $15 to $25 per set must be passed to the consumer through the purchase price, in the long run the inclusive warranty should lead to lower life-cycle costs through manufacturers' efforts to remain competitive by increasing reliability.

Warranty Problems

Although the warranty system has resulted in many positive developments that benefit not only the consumer but also the manufacturer and serviceman, it has brought with it a number of problems and abuses as well. For example, the Presidential Task Force on Appliance Warranties and Service in 1969 concluded that "it is fair to state that in some instances the exclusions, disclaimers, and exceptions so diminished the obligations of the manufacturer that it was deceptive to designate the document as a warranty." Two other problems deserve particular attention: fraudulent claims and consumer "education-calls."

The Claims Problem. When a serviceman performs warranty service, he sends a claim to the local distributor detailing the name and address of the customer, the model number of the set, the type of repair, and the charge. The distributor sends the claim to the manufacturer who reviews and approves it, sending the distributor a check. The distributor in turn reimburses the serviceman.

The transaction process, which can take six weeks to complete, often strains the serviceman's cash reserves. Additionally, it creates ready opportunities for fraudulent claims. While the warranty system shifts the costs of fraudulent claims to the manufacturer, these costs will ultimately be passed on to the consumer in the form of higher purchase prices.

Fraudulent claims can occur for a number of reasons. For example, a service center determines that to make a profit its labor charge for each repair must be at least $16, while the manufacturer sets a standard rate of $12 per repair for warranty work. The dealer appeals but the manufacturer argues that the figure is representative nationally and, additionally, the warranty work gives the service center an edge for later non-warranty business. Rather than take a loss on the warranty business, however, the serviceman might simply submit one false claim for every three real claims to make up the difference. The indirect nature of the claim process separates the serviceman from the manufacturer, making it unlikely that he will be caught.

Another practice is simply for the serviceman to claim major repair work when only minor repair work has been done, thus fraudulently inflating the cost of repair. Too, he may generate an unnecessary warranty service call by offering to perform a last-minute tune-up just before a customer's warranty expires.

Consumer Education Calls. These are service calls caused by consumer ignorance rather than failure of the TV set. They often result simply from failure to adjust the set properly or even to ensure that it is plugged in the wall socket. These problems are estimated to account for fully 30% of all warranty service. Typically the manufacturer will refuse to pay for such service calls. The consumer, who has been assured that his warranty covers all service costs, will also refuse to pay. The serviceman is usually left with the charge, but not for long. He simply replaces a part or claims that he replaced a part, and seeks warranty costs from the manufacturer.

Energy Cost—The Price of Power

While the magnitude of the U.S. energy crisis is widely debated, the cost of electrical energy promises to become an increasingly important concern both nationally and for the individual consumer. In fact, electricity is already a key contributor to the life-cycle costs of the refrigerator and other home appliances, in some cases accounting for over 50% of a product's total cost to the consumer.

But before the magnitude of these costs can be understood, they must first be placed in the perspective of overall residential electrical consumption. In 1971, the residential sector consumed 32.7% of all electrical energy sold to ultimate consumers in the United States, representing a growth rate of 8.9% per year from 1960 to 1971 and making the residential sector the fastest growing user of electrical energy nationally. Moreover, while future projections are extremely uncertain in today's energy climate, the cost of this energy is projected to double by 1980.

Within the residential sector, the large users of electrical energy are refrigerators, water heaters, and lights, followed by air conditioners, ranges, and clothes dryers. Figure 6-4 presents the composite breakdown of residential energy consumption in these categories. The section of the figure marked "Other" includes small appliances, motors for heating plants, and electricity-users that are unaccounted for.

Looking at Figure 6-4, note that TV accounts for only 4.5% of total residential energy consumption. This figure represents a declining power require-

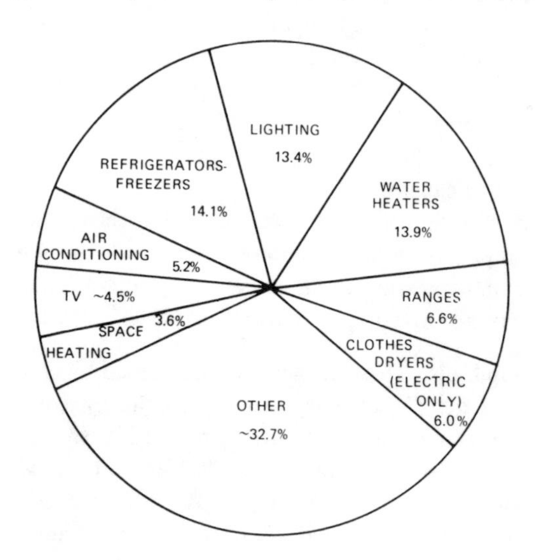

Estimated breakdown for U.S.A. as of 1967.

Figure 6-4. Residential Electrical Energy Use

ment based on increased energy efficiency in color TV design. However, the rising cost of power has tended to counteract the reduced requirement for it, thus keeping the life-cycle power cost nearly constant for the past 10 years (see Figure 6-2).

On the other hand, refrigerator power requirements have increased dramatically by comparison, nearly quadrupling in the past 20 years to over 4 kilowatt hours per day because of increased size, frost-free freezer space (which means that the freezer must be heated, then cooled), and other electricity-consuming innovations. In terms of 1972 dollars, this means that the consumer may be paying as much as $700 for power over the life of the average refrigerator. Moreover, the refrigerator/freezer is the largest single electrical energy consuming product in the residential sector, accounting for fully 14.1% of the total (see Figure 6-4), a figure that is increasing slightly faster than the 8.9% consumption rate for all residential electricity.

Prospects for Savings

For a number of reasons, including finite energy resources, the high growth rate of residential energy consumption appears impossible to sustain over the long term. Unfortunately, the American consumer is still more concerned with such features as size, appearance, function, and price than with the energy consumption of refrigerators and other home appliances. Already, however, many states are preparing legislation that would require some form of energy-use labeling on leading appliances, and manufacturers are exploring the possibilities of making energy-consumption information available to the consumer at the point of sale. In short, as the magnitude of home electricity consumption becomes more apparent, consumers hopefully will begin reversing the trend that has seen the growth of increased convenience at the cost of increased energy consumption.

There are certain appliances for which manufacturers themselves could achieve reduced energy consumption, in many cases without sacrificing functions and with a decrease in total life-cycle cost.

The Refrigerator. While redesign of the modern, no-frost refrigerator involves a number of complex tradeoffs, it is possible to reduce this product's energy consumption now through several rather simple changes:

1. Change the insulation material from fiberglass to foamed polyurethane to reduce heat leakage. This involves a simple tradeoff between the increased purchase price due to more expensive insulation and reduced energy cost associated with less running time for the compressor unit. Such a change is estimated to reduce total life-cycle costs by $85, slightly more than 10%. (Some refrigerators now incorporate this feature.)

2. Improve the operating efficiency of the electric motor that powers the

compressor unit. By using more copper or aluminum to increase the cross-section of the motor's conductors, it should be possible to increase its efficiency from about 65% to at least 75%. While this change would also increase the purchase price, it could result in a discounted net saving (through reduced electrical consumption) of $35, or about 5%, over the life of the refrigerator.

Other Appliances. The four appliances using electricity primarily for heating purposes–space heating, clothes drying, cooking, and water heating–together account for 30% of residential consumption (see Figure 6-4) and represent an obvious target for energy reduction. Unfortunately, the potential for such reduction is not great–water heaters could be improved slightly through better insulation, and ranges would take somewhat less energy if self-cleaning ovens were not used; minor savings could be achieved with the clothes dryer by improving such "process parameters" as temperature and time; and space heaters offer potential reduction through using a "heat pump" in conjunction with solar energy, but this remains only a possibility subject to current research.

Homeowners, however, could obtain substantial improvement in the electricity they use for lighting simply by trading their incandescent lamps for the highly efficient mercury vapor and fluorescent lamps that are currently available.

With regard to television, which today may account for 5% to 6% of residential electrical consumption, there does not appear to be much room for improvement. Solid state circuits have already reduced color TV power consumption significantly, and further reduction would probably require major modifications.

Any energy conserving design changes implemented by manufacturers are likely to increase the purchase price of major home appliances. Understandably, industry will be reluctant to make such changes unless the consumer can be persuaded, through such vehicles as sales literature and labeling, to look at a product from the point of view of its total life-cycle cost. It is, after all, in the consumer's best interest to select the more expensive machine if it will cost him less in the long run. Moreover, if the appliance industry is unable to encourage the adoption of energy conserving appliances, federal, state, and local government will likely do so through regulation.

Service Cost–The Price of Repair

The subject of consumerist concern in a wide range of industries, servicing is one of the most important components of life-cycle costs, and perhaps the least understood. Especially for technologically complex products, it accounts for a hefty percentage of the buying public's life-cycle dollar (e.g., it consumes fully 35% of a color TV set's 10-year cost to the consumer) as well as for a large proportion of all consumer complaints.

Among all consumer products, the TV set stands out as one of the most frequently mentioned sources of consumer servicing problems, competing with automobiles and refrigerators as the major source of complaint in a number of consumer surveys. Additionally, a 1972 Harris Poll found that 34% of TV owners expected trouble getting their sets repaired, while nearly 60% felt that, when they did get repairs, the bill would be too high. In fact, part of the motivation for selecting color TV for special emphasis in the MIT study was the expectation that its product service requirements would be high.

Refrigerators are also a major source of consumer servicing problems, though they require considerably less service than color TV. Too, they tend to cause more consumer distress due to the potential for food loss during a breakdown. In another Harris Poll, while 86% of refrigerator owners were pleased with their product, 10% experienced trouble getting repairs over a three-year period, more than for any other major appliance except TV sets and food waste disposers.

Before examining the service issue in greater depth, a definition of terms is needed:

Service call—a visit by the service technician to the location of the television set or refrigerator, usually the consumer's home.

Consumer education call—a service call prompted by consumer ignorance, e.g., the appliance may be improperly adjusted or the plug may be pulled.

Carry-in-repair—a completed repair prompted by the consumer bringing the product to the shop.

Product Service Requirements—the total of all requests for service on a product.

Service incident—any request for service on a product.

Service incidence rate—the rate at which requests for service occur, usually measured on an annual basis and expressed as an *average*, i.e., some products will fail several times and others not at all.

Life-cycle service cost—the total amount paid by an appliance owner to servicemen, including consumer-education costs but excluding warranty service, which is covered by the purchase price.

The Need for Service

How often can the average color TV and refrigerator owner expect to call the serviceman? Do these products require less service today than in the past? And how much does all of this cost?

According to industry sources, color television service requirements currently stand at about one service incident per year for the first year in the life of an average set, while the rate for the average refrigerator is 0.2 to 0.35 for the same period.

More important, the trend of service requirements for these two products has run steadily downward over time. Figure 6-5 shows this trend graphically for both products during their first year of ownership by the consumer. Note that in 1965 the average color TV required two service calls per year and by 1972 required only one; moreover, a similar reduction of 50% occurred for refrigerators between 1958 and 1972.

Apparently due to the need for consumer education, installation adjustments, production defects and related factors, the service incidence rate for both products tends to be higher in the first few months of service than for the remainder of the year.

The period after the first year, when the consumer assumes the responsibility for service cost at warranty expiration, is considerably more difficult to measure. However, allowing for scanty information, preliminary estimates are that service requirements for color TV sets are relatively constant during at least the first two years of ownership. For refrigerators, the service-incidence rate is higher during the first year than during the second through fifth years of ownership, and it increases in later years due to age and wear of parts.

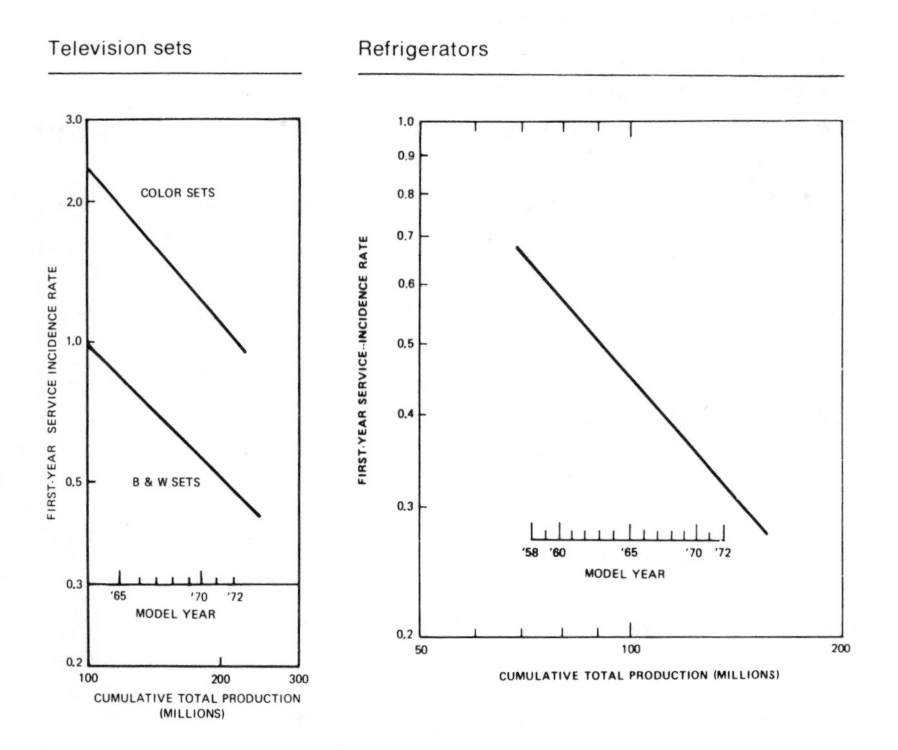

Figure 6-5. First-Year Service-Incidence Rates

Life-Cycle Service Cost. As mentioned earlier, this cost includes all money paid the serviceman by the consumer, excluding warranty service. Thus, it is directly tied to the service incidence rate.

First consider color TV. From 1965 to 1972 the cost for each service call rose from $20 to $30, an increase of 50%. Additionally, this figure is expected to rise another 30% to nearly $50 per repair by 1980. But now consider these costs in the context of the declining trend of service incidents. The need for service is *declining* faster than the cost of service is rising, producing a slight but clear reduction in the color TV's life-cycle service cost, a reassuring fact given its 35% hold on total life-cycle costs in the color TV industry.

The situation is much the same for refrigerators. While the cost of each repair is slightly less than for color TV, it is projected to increase nearly as dramatically by 1980. This cost, too, tends to be cancelled by the reduced need for repair, producing a slightly declining trend in the life-cycle service cost of refrigerators.

Productivity of Service

Given the sizable influence of the service industry on the total dollars the consumer pays during the life of a major appliance, it is natural to seek ways of decreasing this cost through greater productivity in the service sector. (Productivity is generally defined as the efficiency with which some resource is being used to produce some final product. For example, using a wheelbarrow to transport bricks from one point to another instead of carrying them by hand decreases the time to do the job and increases the efficiency of the operation.)

Both in individual industries and in the total economy, productivity is of growing national concern. For one thing, productivity growth is related to inflation—when productivity is high, inflationary pressures are usually reduced. For another, the U.S. economy is fast shifting from the production of goods to the production of services, a sector where productivity growth is widely believed to lag behind the rest of the economy.

The MIT study focused on repair service for consumer appliances, concentrating on the productivity of repair-service labor, since labor is a major component of the consumer's product repair costs. (According to industry data, labor charges account for 60% to 70% of gross service-center revenues.)

Measuring Productivity

Since appliance and television repair shops "produce" repairs, a good measure of repair-service productivity is the number of service calls that a service technician completes in a day, including in-home repairs, consumer education calls, and carry-in repairs.

According to industry sources, a service technician is able to complete between five and seven service calls in a typical eight-hour day, a figure that does not vary much between television and major appliance servicing, and one that does not appear to have changed during the past several years. Moreover, an analysis of the service incidence rate and the growth in numbers of service technicians during the past seven years indicates that repair-service productivity is actually declining slightly. The service labor force is increasing slightly faster than the total number of service incidents, implying a small decrease in the number of service incidents per technician day.

Service-Industry Characteristics

In order to understand the productivity of repair service and the prospects for improving it, one must first understand the nature of the repair business.

The majority of service performed on major appliances is handled by highly dispersed small-business establishments. In the television and radio industry, for example, repair shops of less than 10 employes predominate, and 76% of these are owner-operated. Furthermore, increased size does not appear to result in a corresponding increase in gross receipts per employe, an indication that there is little economy of scale.

While the refrigerator and television repair businesses vary widely, they tend to fall into three general categories:

1. *The one-man shop*—The owner-operator is primarily a serviceman, not a businessman, though those who succeed do so more through business acumen than through service ability. Productivity for this type of business is hard to estimate because of the flexible work day and the owner's tendency to charge on an hourly basis.

2. *The manufacturer or merchandiser-owned company*—These repair centers are identified with a manufacturer's brand names. They typically charge more for their services and claim to provide a better level of service to the consumer, tending to bill a set amount for in-home-repair labor and a set amount for shop-repair labor.

3. *The large independent service company*—This type is neither owned nor exclusively franchised by a manufacturer or merchandiser, but may act as an authorized service center for one or more brands. It differs from the company-owned shop in that it services many brands instead of just one and does not receive management support from the manufacturer.

In all three types of shops, the service process is a complex one. Its primary variable factor is the serviceman's time; the rest is overhead, parts, and profit.

Figure 6-6 shows how the serviceman spends his time from the point that a customer calls for service to the point where service is completed satisfactorily, a process that on average takes about one hour and 15 minutes (spread over several days) to complete.

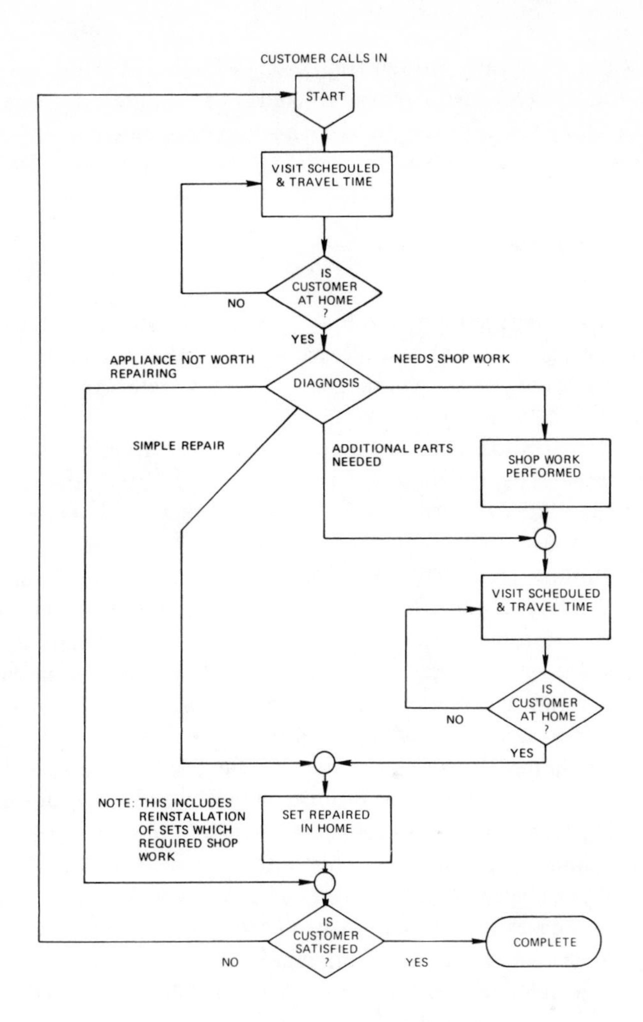

Figure 6-6. The Service-Call Process

Furthermore, the high labor intensity of this process creates several problems. For one thing, the consumer pays for the serviceman's time throughout, including travel time. For another, if the billing method is for actual time used, there is a temptation for some serviceman to take extra repair time, especially if they are having a slack day. This problem is avoided somewhat by flat-rate billing, which is becoming increasingly prevalent, but the serviceman could still take in sets unnecessarily for shop repair.

Too, consumers sometimes are the cause of unnecessary charges. It was noted that some customers would balk at a bill for $25 in labor and $5 in parts, yet would accept one for $5 in labor and $25 in parts, more readily believing this was evidence of something substantially wrong with the set. Thus, a serviceman, aware of this psychological pressure, could be tempted to replace parts unnecessarily; they typically carry a 100% markup from cost.

Can Service Be Improved?

The MIT study team used Figure 6-6 to construct a "Service Call Model" that assessed the average time and cost of each step of the serving process. The team found that the most time-consuming service tasks were (a) travel time, and (b) diagnosis and in-home repair time.

The area of greatest potential improvement was travel time. Television technicians spend approximately 42% and refrigerator technicians about 50% of total repair time simply scheduling a call and traveling to a customer's home. The cost, of course, is borne by the consumer. Of all the alternatives for reducing travel time, geographical specialization appears to be the most promising. Interviews with service companies indicate that service centers could save about 10% of the total time per call by limiting the geographical area each technician must cover in a day. Simply, this would involve a planning process whereby the service technician would make as many calls as possible in the smallest possible area.

Regarding in-home diagnosis and repair, the television service technician typically spends 44% of his time meeting with the customer, diagnosing the problem, and making the repair (as opposed to 14% for in-shop repair), while the refrigerator technician spends 50% of his time on these activities. Given the fact that most TV (and virtually all refrigerator) repair is completed in the home, these high percentages also appear to be likely targets for reduction. Unfortunately, the labor-intensive nature of the repair process itself limits the potential savings for both in-home and travel time—even a reduction of 50% in average actual repair time represents only an $18 saving in the life-cycle service cost of a set.

There doesn't appear to be much room for lowering costs by raising the level of service technician skills, either, according to the study. In terms of customer satisfaction and good public relations, the technician represents the manufacturer's only tangible contact with the consumer. Both manufacturers and industry associations are active in raising the occupational status of servicing, making it a career more attractive to a higher caliber of worker and to young people. Job-training is provided in vocational and high schools, and manufacturers and large service companies often provide on-the-job training. Too, all manufacturers in the study had good systems for maintaining and updating service technician skills through published materials and retraining sessions.

In other words, service, like preventive medicine or routine machine maintenance, is an activity where an ounce of prevention is worth a pound of cure.

While there are ways to improve the repair business, they appear difficult to achieve, given the prevalence of small shops and the time consuming elements of the repair process.

So why not reduce the need for service? As the section on service incidence pointed out, the main counterbalance to rising service costs is the declining need for service in the appliance industry. Further evidence is shown in Table 6-1. It is based on the assumption that the number of products in use per technician reflects both the efficiency of the serviceman and the reliability (service requirements) of the product. Since technician productivity hasn't changed much, the substantial 10-year increase in products per technician shown in the table is almost entirely due to a reduction in service requirements.

In short, when it comes to reducing life-cycle service costs, much more is to be gained by increasing reliability than by increasing serviceability. (The consumer, of course, could also help considerably by reducing the need for consumer-education calls.)

Service in the Future

If reducing the need for service holds the most promise for lowering its cost, what are the future prospects for such reduction? Using mathematical models and computer analysis, the MIT study team attempted to predict the future service requirements for television and refrigerators. While the projections were made only for these two products, they should give some indication of what lies ahead for other major home appliances and electronics products.

Television. Table 6-2 shows the number of service incidents per year for 1960 and 1970 and the projection for 1980, as well as the number of service calls per day for each technician in these years. The table also shows the estimates for the number of technicians needed in 1980. The conclusions:

1. In 1970, television servicing employed approximately 137,000 technicians, a figure expected to drop to 126,000 by 1980, reflecting a slight decline in the need for service and a slight increase in the number of calls a technician can make per day.

2. While color TV accounted for 40% of all TV set failures in 1970, it is expected to account for 67% by 1980.

3. Television carry-in service is expected to rise from 57% of all service in 1970 to 74% in 1980.

4. In sum, the productivity of TV service is not expected to change much in the coming years and the television service business in 1980 will be somewhat smaller, leaning heavily toward color TV and carry-in service (the latter due to smaller sets and modularization). While the need for technicians is expected to decline, other home electronics products such as cable TV and video tape systems may take up the slack.

Table 6-1
Products in Use per Technician

Year	TV Sets in Use (millions)	TV Technicians (thousands)	TV Sets per Technician	Appliances in Use (millions)	Appliance Technicians (thousands)	Appliances per Technician
1960	57.8	105.4	548	158.5	89.7	1767
1970	99.0	137.4	721	286.6	122.8	2334

Table 6-2
Estimated Number of Technicians Required for Television Service in 1980

Year	Total Number of Service Calls per Year (millions)	Total Number of Technicians (thousands)	Number of Service Calls per Day per Technician
1960	137	105	6.7
1970	144	137	6.4
1980	136	125	6.5

Refrigerators. The forecast for the rest of this decade is a decreasing need for refrigerator service, dropping from 18 million service calls in 1973 to 16 million by 1980, an 11% decline. This reduction, however, will not be as rapid as over the past 15 years.

Conclusion

The sole purpose of the MIT study has been to foster a better understanding of the system for maintaining and servicing major home appliances and electronics products. Approaching the problem from a systems point of view, the study attempted to shed new light on the total costs paid by consumers during a product's useable life.

While a number of problems were uncovered–notably the abuses in service billing practices and warranty claims–the industry system for designing, manufacturing, distribution and sales, and servicing has provided the American buying public with increasingly reliable products that do a better job. Nevertheless, the consumer, who looks first to purchase cost, is not aware of the substantial magnitude of servicing and energy costs. As a result, manufacturers have not paid as much attention as they should to reducing these costs.

There are a number of steps that consumers, manufacturers, and government agencies can take to improve problem areas and correct the most glaring abuses.

Consumers

1. Use the concept of life-cycle cost to appraise purchase options, actively seeking information on not just the purchase price but also the energy and service costs of competing products.
2. Reduce servicing costs by ensuring that a product is truly malfunctioning, e.g., read the instruction manual carefully before calling the serviceman.
3. Help correct warranty and servicing abuses and appliance-performance prob-

lems by using available channels of complaint—e.g., customer service departments of major manufacturers or the Major Appliance Consumer Action Panel (MACAP).

4. Given the substantial and growing influence of energy cost, carefully evaluate added features and conveniences of products against the energy they consume.

Manufacturers

1. Develop ways to correct warranty abuses, which add to cost and distort product performance information. These might include more efficient warranty billing systems and alternate methods for fulfilling warranty contracts (e.g., a minimum charge for each service call might reduce consumer education costs).
2. Improve the warranty system itself by (a) extending it to cover a greater portion of the product's life and (b) developing accurate feedback mechanisms to determine appliance performance during the warranty period.
3. Respond to the life-cycle cost concept by designing and manufacturing products that represent an optimal balance between acquisition, operation, and service costs during their lifetime.
4. Make life-cycle cost data available to consumers at the point of purchase.

Government

1. Provide incentives that would encourage reduced life-cycle costs and resource consumption.
2. Monitor future service technician needs and aid in development of training and education functions to meet shifts in product service requirements.
3. Study in greater depth how warranties and service contracts affect the entire system of manufacturer, retailer, serviceman, and consumer, and develop more effective, less costly means of dealing with appliance failure.

Part II: The Environmental Price of New Sources of Energy

What will be the impact on the environment if America turns to new sources to meet its energy needs?

7 Social Institutions and Nuclear Energy

Alvin M. Weinberg

To many, nuclear power is the obvious answer to America's expanding energy needs. What problems are involved?

Fifty-two years have passed since Ernest Rutherford observed the nuclear disintegration of nitrogen when it was bombarded with alpha particles. This was the beginning of modern nuclear physics. In its wake came speculation as to the possibility of releasing nuclear energy on a large scale: By 1921 Rutherford was saying "The race may date its development from the day of the discovery of a method of utilizing atomic energy" [1].

Despite the advances in nuclear physics beginning with the discovery of the neutron by Chadwick in 1932 and Cockcroft and Walton's method for electrically accelerating charged particles, Rutherford later became a pessimist about nuclear energy. Addressing the British Association for the Advancement of Science in 1933, he said: "We cannot control atomic energy to an extent which would be of any value commercially, and I believe we are not likely ever to be able to do so" [2]. Yet Rutherford did recognize the great significance of the neutron in this connection. In 1936, after Fermi's remarkable experiments with slow neutrons, Rutherford wrote ". . . the recent discovery of the neutron and the proof of its extraordinary effectiveness in producing transmutations at very low velocities opens up new possibilities, if only a method could be found of producing slow neutrons in quantity with little expenditure of energy" [3].

Today the United States is committed to over 100×10^6 kilowatts of nuclear power, and the rest of the world to an equal amount. Rather plausible estimates suggest that by 2000 the United States may be generating electricity at a rate of 1000×10^6 kilowatts with nuclear reactors. Much more speculative estimates visualize an ultimate world of 15 billion people, living at something like the current U.S. standard: nuclear fission might then generate power at the rate of some 300×10^9 kilowatts of heat, which represents 1/400 of the flux of solar energy absorbed and reradiated by the earth [4].

This large commitment to nuclear energy has forced many of us in the nuclear community to ask with the utmost seriousness questions which, when

This chapter is the text of the Rutherford Centennial lecture presented at the annual meeting of the American Association for the Advancement of Science, Philadelphia, December 27, 1971. It first appeared in *Science* 7 (July 1972), volume no. 177, pp. 27-34.

first raised, had a tone of unreality. When nuclear energy was small and experimental and unimportant, the intricate moral and institutional demands of a full commitment to it could be ignored or not taken seriously. Now that nuclear energy is on the verge of becoming our dominant form of energy, such questions as the adequacy of human institutions to deal with this marvelous new kind of fire must be asked, and answered, soberly and responsibly. In these remarks I review in broadest outline where the nuclear energy enterprise stands and what I think are its most troublesome problems; and I shall then speculate on some of the new and peculiar demands mankind's commitment to nuclear energy may impose on our human institutions.

Nuclear Burners–Catalytic and Noncatalytic

Even before Fermi's experiment at Stagg Field on December 2, 1942, reactor designing had captured the imagination of many physicists, chemists, and engineers at the Chicago Metallurgical Laboratory. Almost without exception, each of the two dozen main reactor types developed during the following 30 years had been discussed and argued over during those frenzied war years. Of these various reactor types, about five, moderated by light water, heavy water, or graphite, have survived. In addition, breeders, most notably the sodium-cooled plutonium breeder, are now under active development.

Today the dominant reactor type uses enriched uranium oxide fuel, and is moderated and cooled by water at pressures of 100 to 200 atmospheres. The water may generate steam directly in the reactor [so-called boiling water reactor (BWR)] or may transfer its heat to an external steam generator [pressurized water reactor (PWR)]. These light water reactors (LWR) require enriched uranium and therefore at first could be built only in countries such as the United States and the U.S.S.R., which had large plants for separating uranium isotopes.

In countries where enriched uranium was unavailable, or was much more expensive than in the United States, reactor development went along directions that utilized natural uranium: for example, reactors developed in the United Kingdom and France were based mostly on the use of graphite as moderator; those developed in Canada used D_2O as moderator. Both D_2O and graphite absorb fewer neutrons than does H_2O, and therefore such reactors can be fueled with natural uranium. However, as enriched uranium has become more generally available (of the uranium above ground, probably more by now has had its normal isotopic ratio altered than not), the importance of the natural ^{235}U isotopic abundance of 0.71 percent has faded. All reactor systems now tend to use at least slightly enriched uranium since its use gives the designer more leeway with respect to materials of construction and configuration of the reactor.

The PWR was developed originally for submarine propulsion where compactness and simplicity were the overriding considerations. As one who was closely involved in the very early thinking about the use of pressurized water for submarine propulsion (I still remember the spirited discussions we used to have in 1946 with Captain Rickover at Oak Ridge over the advantages of the pressurized water system), I am still a bit surprised at the enormous vogue of this reactor type for civilian power. Compact, and in a sense simple, these reactors were; but in the early days we hardly imagined that separated ^{235}U would ever be cheap enough to make such reactors really economical as sources of central station power.

Four developments proved us to be wrong. First, separated ^{235}U which at the time of *Nautilus* cost around $100 per gram fell to $12 per gram. Second, the price of coal rose from around $5 per ton to $8 per ton. Third, oxide fuel elements, which use slightly enriched fuel rather than the highly enriched fuel of the original LWR, were developed. This meant that the cost of fuel in an LWR could be, say, 1.9 mills per kilowatt hour (compared with around 3 mills per electric kilowatt hour for a coal-burning plant with coal at $8 per ton). Fourth, pressure vessels of a size that would have boggled our minds in 1946 were common by 1970: the pressure vessel for a large PWR may be as much as 8½ inches thick and 44 feet tall. Development of these large pressure vessels made possible reactors of 1000 megawatts electric (Mwe) or more, compared with 60 Mwe at the original Shippingport reactor. Since per unit of output a large power plant is cheaper than a small one, this increase in reactor size was largely responsible for the economic breakthrough of nuclear power.

Although the unit cost of water reactors has not fallen as much as optimists such as I had estimated, present costs are still low enough to make nuclear power competitive. I compare the relative position of a 1000-Mwe LWR and of a coal-fired plant of the same size (Table 7-1).

Water-moderated reactors burn ^{235}U, which is the only naturally occurring fissile isotope. But the full promise of nuclear fission will be achieved only with successful breeders. These are reactors that, essentially, burn the very abundant isotopes ^{238}U or ^{232}Th; in the process, fissile ^{239}Pu or ^{233}U acts as regenerating catalyst—that is, these isotopes are burned and regenerated. I therefore like to call reactors of this type *catalytic nuclear burners.* Since ^{238}U and ^{232}Th are immensely abundant (though in dilute form) in the granitic rocks, the basic fuel for such catalytic nuclear burners is, for all practical purposes, inexhaustible. Mankind will have a permanent source of energy once such catalytic nuclear burners are developed.

Most of the world's development of a breeder is centered around the sodium-cooled, ^{238}U burner in which ^{239}Pu is the catalyst and in which the energy of the neutrons is above 100×10^3 electron volts. No fewer than 12 reactors of this liquid metal fast breeder reactor (LMFBR) type are being worked on actively, and the United Kingdom plans to start a commercial 1000-Mwe fast breeder by 1975. Some work continues on alternatives. In the

Table 7-1
Estimated Total Cost of Power from 1000-Mwe Power Plants (mils per electric kilowatt hour)

	PWR Plants		Coal Plants			
			No SO_2 System		With SO_2 System	
	Run-of-river	With cooling towers	Run-of-river	Cooling towers	Run-of-river	Cooling towers
Capital cost ($/kwe)	365	382	297	311	344	358
Fixed charges	7.8	8.2	6.4	6.6	7.4	7.7
Fuel cost	1.9	1.9	3.9	3.9	3.9	3.9
Operation and maintenance cost	0.6	0.6	0.5	0.5	0.8	0.8
Total power cost (mils/kwhe)	10.3	10.7	10.8	11.0	12.1	12.4

Note: The costs include escalation to 1978. Nuclear fuel costs were taken from (8). The coal plant fuel costs are based on average delivered coal price of about $8 per ton in 1971, with escalation to 1978 at 5 percent per year. This leads to about $10.5 to $10.7 per ton in 1978. Estimates for costs of operating SO_2-removal equipment range from zero to about 2×10^6 dollars per year.

^{233}U-^{232}Th cycle, on the light water breeder and the molten salt reactor; in the ^{239}Pu-^{238}U cycle, on the gas-cooled fast breeder. But these systems are, at least at the present, viewed as backups for the main line which is the LMFBR.

Nuclear Power and Environment

The great surge to nuclear power is easy to understand. In the short run, nuclear power is cheaper than coal power in most parts of the United States; in the long run, nuclear breeders assure us of an all but inexhaustible source of energy. Moreover, a *properly* operating nuclear power plant and its subsystems (including transport, waste disposal, chemical plants, and even mining) are, except for the heat load, far less damaging to the environment than a coal-fired plant would be.

The most important emissions from a routinely operating reactor are heat and a trace of radioactivity. Heat emissions can be summarized quickly. The thermal efficiency of a PWR is 32 percent; that of a modern coal-fired power plant is around 40 percent. For the same electrical output the nuclear plant emits about 40 percent more waste heat than the coal plant does; in this one respect, present-day nuclear plants are more polluting than coal-fired plants. However, the higher temperature nuclear plants, such as the gas-cooled, the molten salt breeder, and the liquid metal fast breeder, operate at about the same efficiency as does a modern coal-fired plant. Thus, nuclear reactors of the future ought to emit no more heat than do other sources of thermal energy.

As for routine emission of radioactivity, even when the allowable maximum exposure to an individual at the plant boundary was set at 500 millirems (mrem) per year, the hazard, if any, was extremely small. But for practical purposes, technological advances have all but eliminated routine radioactive emission. These improvements are taken into account in the newly proposed regulations of the Atomic Energy Commission (AEC) requiring, in effect, that the dose imposed on any individual living near the plant boundary either by liquid or by gaseous effluents from LWR's should not exceed 5 mrem per year. This is to be compared with the natural background which is around 100 to 200 mrem per year, depending on location, or the medical dose which now averages around 60 mrem per year.

As for emissions from chemical reprocessing plants, data are relatively scant since but one commercial plant, the Nuclear Services Plant at West Valley, New York, has been operating, and this only since 1966. During this time, liquid discharges have imposed an average dose of 75 mrem per year at the boundary. Essentially no ^{131}I has been emitted. As for the other main gaseous effluents, all the ^{85}Kr and ^{3}H contained in the fuel has been released. This has amounted to an average dose from gaseous discharge of about 50 mrem per year.

Technology is now available for reducing liquid discharges, and processes for retaining ^{85}Kr and ^{3}H are being developed at AEC laboratories. There is every reason to expect these processes to be successful. Properly operating radiochemical plants in the future should emit no more radioactivity than do properly operating reactors—that is, less than 10 percent of the natural background at the plant boundary.

There are some who maintain that even 5 mrem per year represents an unreasonable hazard. Obviously there is no way to decide whether there is any hazard at this level. For example, if one assumes a linear dose-response for genetic effects, then to find, with 95 percent confidence, the predicted 0.5 percent increase in genetic effect in mice at a dose of, say, 150 mrem would require 8 billion animals. At this stage the argument passes from science into the realm of what I call trans-science, and one can only leave it at that.

My main point is that nuclear plants are indeed relatively innocuous, large-scale power generators if they and their subsystems work properly. The entire controversy that now surrounds the whole nuclear power enterprise therefore hangs on the answer to the question of whether nuclear systems can be made to work properly; or, if faults develop, whether the various safety systems can be relied upon to guarantee that no harm will befall the public.

The question has only one answer: there is no way to guarantee that a nuclear fire and all of its subsystems will never cause harm. But I shall try to show why I believe the measures that have been taken, and are being taken, have reduced to an acceptably low level the probability of damage.

I have already discussed low-level radiation and the thermal emissions from nuclear systems. Of the remaining possible causes of concern, I shall dwell on the three that I regard as most important: reactor safety, transport of radioactive materials, and permanent disposal of radioactive wastes.

Avoiding Large Reactor Accidents

One cannot say categorically that a catastrophic failure of a large PWR or a BWR and its containment is impossible. The most elaborate measures are taken to make the probability of such occurrence extremely small. One of the prime jobs of the nuclear community is to consider all events that could lead to accident, and by proper design to keep reducing their probability however small it may be. On the other hand, there is some danger that in mentioning the matter one's remarks may be misinterpreted as implying that the event is likely to occur.

Assessment of the safety reactors depends upon two rather separate considerations: prevention of the initiating incident that would require emergency safety measures; and assurance that the emergency measures, such as the emergency core cooling, if ever called upon, would work as planned. In much of the discussion and controversy that has been generated over the safety of nuclear reactors, emphasis has been placed on what would happen if the emergency measures were called upon and failed to work. But to most of us in the reactor community, this is secondary to the question: How certain can we be that a drastic accident that calls into play the emergency systems will never happen? What one primarily is counting upon for the safety of a reactor is the integrity of the primary cooling system: that is, on the integrity of the pressure vessel and the pressure piping. Excruciating pains are taken to assure the integrity of these vessels and pipes. The watchword throughout the nuclear reactor industry is *quality assurance*: every piece of hardware in the primary system is examined, and reexamined, to guarantee insofar as possible that there are no flaws.

Nevertheless, we must deal with the remote contingency that might call the emergency systems into action. How certain can one be that these will work as planned? To better understand the analysis of the emergency system, Figures 7-1 and 7-2 show, schematically, a large BWR and a PWR.

Three barriers prevent radioactivity from being released: fuel element cladding, primary pressure system, and containment shell. In addition to the regular safety system consisting primarily of the control and safety rods, there are elaborate provisions for preventing the residual radioactive heat from melting the fuel in the event of a loss of coolant. In the BWR there are sprays that spring into action within 30 seconds of an accident. In both the PWR and BWR, water is injected under pressure from gas-pressurized accumulators. In both reactors there are additional systems for circulating water after the system has come to low pressure, as well as means for reducing the pressure of steam in the containment vessel. This latter system also washes down or otherwise helps remove any fission products that may become airborne.

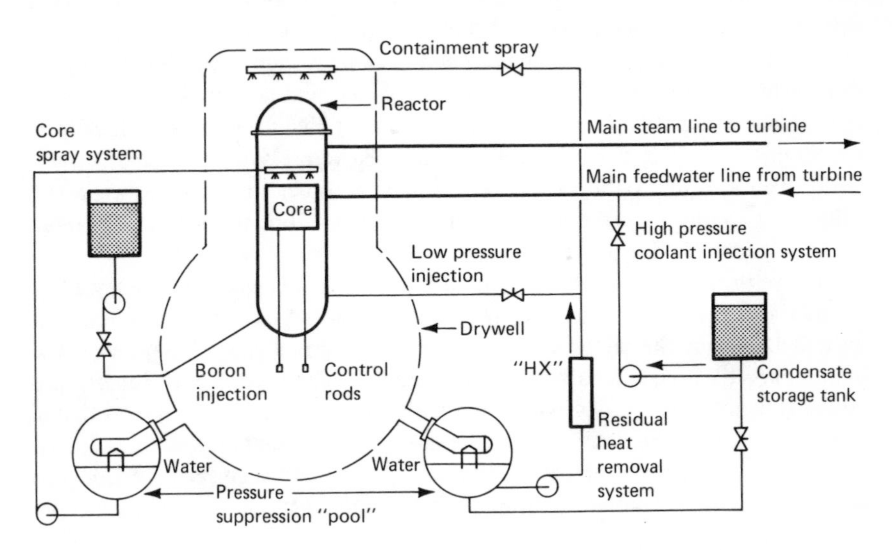

Figure 7-1. Boiling Water Reactor Emergency Cooling Systems

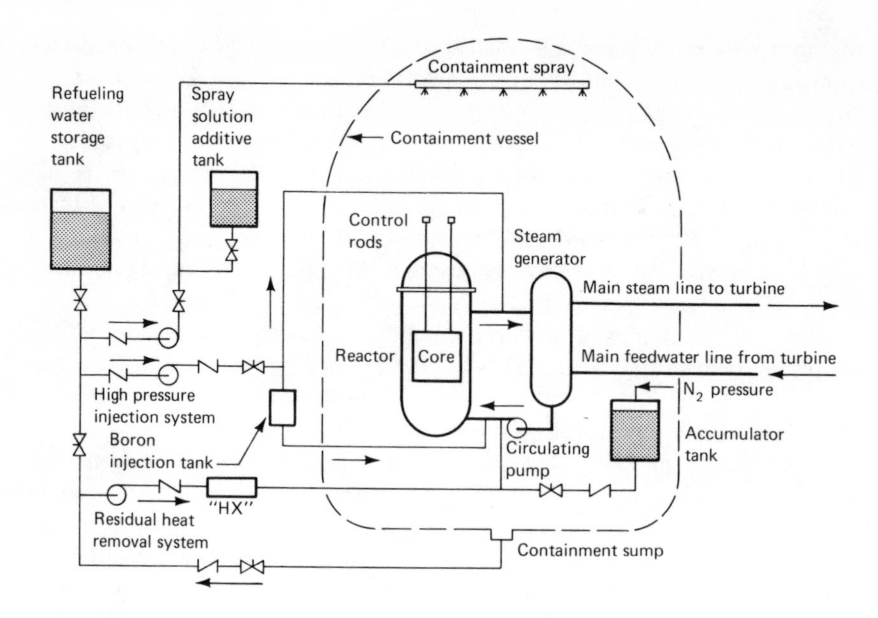

Figure 7-2. Pressurized Water Reactor Emergency Cooling Systems

In analyzing the ultimate safety of a LWR, one tries to construct scenarios—improbable as they may be—of how a catastrophe might occur; and then one tries to provide reliable countermeasures for each step in the chain of failures that could lead to catastrophe. The chain conceivably could go like this. First, a pipe might break, or the safety system might fail to respond when called upon in an emergency. Second, the emergency core cooling system might fail. Third, the fuel might melt, might react also with the water, and conceivably might melt through the containment. Fourth, the containment might fail catastrophically, if not from the melt itself, then from missiles or overpressurization, and activity might then spread to the public. There may be other modes of catastrophic failure—for example, earthquakes or acts of violence—but the above is the more commonly identified sequence.

To give the flavor of how the analysis of an accident is made, let me say a few words about the first and second steps of this chain. As a first step, one might imagine failure of the safety system to respond in an emergency, say, when the bubbles in a BWR collapse after a fairly routine turbine trip. Here the question is not that some safety rods will work and some will not, but rather that a common mode failure might render the entire safety system inoperable. Thus if all the electrical cables actuating the safety rods were damaged by fire, this would be a common mode failure. Such a common mode failure is generally regarded as impossible, since the actuating cables are carefully segregated, as are groups of safety rods, so as to avoid such an accident. But one cannot *prove* that a common mode failure is impossible. It is noteworthy that on September 30, 1970, the entire safety system of the Hanford-N reactor (a one-of-a-kind water-cooled, graphite-moderated reactor) did fail when called upon; however, the backup samarium balls dropped precisely as planned and shut off the reactor. One goes a long way toward making such a failure incredible if each big reactor, as in the case of the Hanford-N reactor, has two entirely independent safety systems that work on totally different principles. In the case of BWR, shutoff of the recirculation pumps in the all but incredible event the rods fail to drop constitutes an independent shutoff mechanism, and automatic pump shutoff is being incorporated in the design of modern BWR's.

The other step in the chain that I shall discuss is the failure of the emergency core cooling system. At the moment, there is some controversy whether the initial surge of emergency core cooling water would bypass the reactor or would in fact cool it. The issue was raised recently by experiments on a very small scale (9-inch-diameter pot) which indeed suggested that the water in that case would bypass the core during the blowdown phase of the accident. However, there is a fair body of experts within the reactor community who hold that these experiments were not sufficiently accurate simulations of an actual PWR to bear on the reliability or lack of reliability of the emergency core cooling in a large reactor.

Obviously the events following a catastrophic loss of coolant and injections of emergency coolant are complex. For example, one must ask whether the fuel rods will balloon and block coolant channels, whether significant chemical reactions will take place, or whether the fuel cladding will crumble and allow radioactive fuel pellets to fall out.

Such complex sequences are hardly susceptible to a complete analysis. We shall never be able to estimate everything that will happen in a loss-of-coolant accident with the same kind of certainty with which we can compute the Balmer series or even the course of the ammonia synthesis reaction in a fertilizer plant. The best that we can do as knowledgeable and concerned technologists is to present the evidence we have, and to expect policy to be based upon informed—not uninformed—opinion.

Faced with questions of this weight, which in a most basic sense are not fully susceptible to a yes or no scientific answer, the AEC has invoked the adjudicatory process. The issue of the reliability of the emergency core cooling system is being taken up in hearings before a special board drawn from the Atomic Safety and Licensing Board Panel. The record of the hearings is expected to contain all that is known about emergency core cooling systems and to provide the basis for setting the criteria for design of such systems.

Transport of Radioactive Materials

If, by the year 2000, we have 10^6 megawatts of nuclear power, of which two-thirds are liquid metal fast breeders, then there will be 7,000 to 12,000 annual shipments of spent fuel from reactors to chemical plants, with an average of 60 to 100 loaded casks in transit at all times. Projected shipments might contain 1.5 tons of core fuel which has decayed for as little as 30 days, in which case each shipment would generate 300 kilowatts of thermal power and 75 megacuries of radioactivity. By comparison, present casks from LWR's might produce 30 kilowatts and contain 7 megacuries.

Design of a completely reliable shipping cask for such a radioactive load is a formidable job. At Oak Ridge our engineers have designed a cask that looks very promising. As now conceived, the heat would be transferred to air by liquid metal or molten salt; and the cask would be provided with rugged shields which would resist deformation that might be caused by a train wreck. To be acceptable the shipping casks must be shown to withstand a 30-minute fire and a drop from 30 feet onto an unyielding surface (Figure 7-3).

Can we estimate the hazard associated with transport of these materials? The derailment rate in rail transport (in the United States) is 10^{-6} per car mile. Thus, if there were 12,000 shipments per year, each of a distance of 1000 miles, we would expect 12 derailments annually. However, the number of serious accidents would be perhaps 10^{-4}- to 10^{-6}-fold less frequent; and shipping casks are designed to withstand all but the most serious accident (the train wreck near an oil refinery that goes into flames as a result of the crash). Thus the statistics—between 1.2×10^{-3} and 1.2×10^{-5} serious accidents per year—at least until the

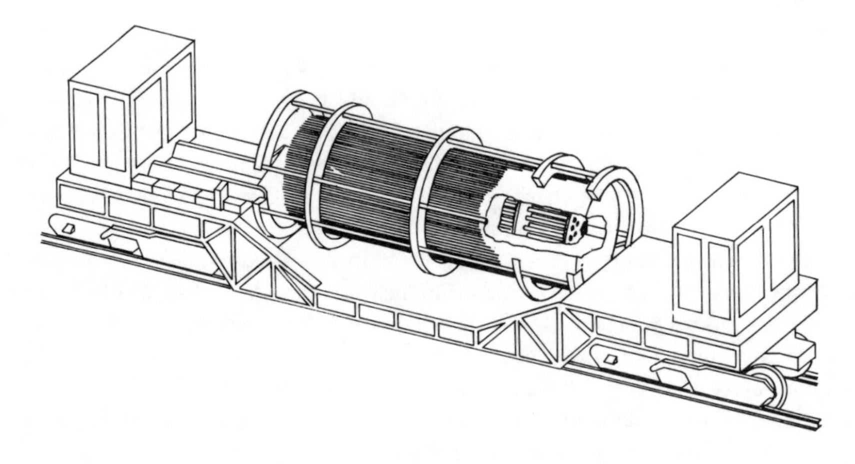

Figure 7-3. Liquid Metal Fast Breeder Reactor Spent Fuel Shipping Cask (18 Assemblies)

year 2000, look quite good. Nevertheless the shipping problem is a difficult one and may force a change in basic strategy. For example, we may decide to cool fuel from LMFBR's in place for 360 days before shipping: this reduces the heat load sixfold, and increases the cost of power by only around 0.2 mill per electric kilowatt hour. Or a solution that I personally prefer is to cluster fast breeders in nuclear power parks which have their own on-site reprocessing facilities [5]. Clustering reactors in this way would make both cooling and transmission of power difficult; also such parks would be more vulnerable to common mode failure, such as acts of war or earthquakes. These difficulties must be balanced against the advantage of not shipping spent fuel off-site, and of simplifying control of fissile material against diversion. To my mind, the advantages of clustering outweigh its disadvantages; but this again is a trans-scientific question which can only be adjudicated by a legal or political process, rather than by scientific exchange among peers.

Waste Disposal

By the year 2000, according to present projections, we shall have to sequester about 27,000 megacuries of radioactive wastes in the United States; these wastes will be generating 100,000 kilowatts of heat at that time. The composition of these wastes is summarized in Table 7-2.

The wastes will include about 400 megacuries of transuranic alpha emitters. Of these, the ^{239}Pu with a half-life of 24,400 years will be dangerous for perhaps 200,000 years.

Can we see a way of dealing with these unprecedentedly treacherous materials? I believe we can, but not without complication.

There are two basically different approaches to handling the wastes. The first, urged by W. Bennett Lewis of Chalk River [6], argues that once man has opted for nuclear power he has committed himself to essentially perpetual surveillance of the apparatus of nuclear power, such as the reactors, the chemical plants, and others. Therefore, so the argument goes, there will be spots on the earth where radioactive operations will be continued in perpetuity. The wastes then would be stored at these spots, say in concrete vaults. Lewis further refines his ideas by suggesting that the wastes be recycled so as to limit their volume. As fission products decay, they are removed and thrown away as innocuous nonradioactive species; the transuranics are sent back to the reactors to be burned. The essence of the scheme is to keep the wastes under perpetual, active surveillance and even processing. This is deemed possible because the original commitment to nuclear energy is considered to be a commitment in perpetuity.

There is merit in these ideas; and indeed permanent storage in vaults is a valid proposal. However, if one wishes to perpetually rework the wastes as Lewis suggests, chemical separations would be required that are much sharper than

Table 7-2
Projected Waste Inventories at the Permanent Repository

	Calendar Year		
	1980	1990	2000
Number of Annual Shipments			
High-level waste*	23	240	590
Alpha waste†	420	1,200	0
Accumulated High-Level Waste			
Volume of waste (cubic feet)	3,170	74,200	319,000
Salt area used (acres)	9	200	900
Total thermal power (megawatts)	1.17	24.4	94.9
Total activity (megacuries)	329	7,030	27,700
^{90}Sr (megacuries)	59.0	1,310	5,290
^{137}Cs (megacuries)	83.1	1,850	7,500
^{238}Pu (megacuries)	0.102	2.34	9.88
^{239}Pu (megacuries)	0.00157	0.0368	0.158
^{240}Pu (megacuries)	0.00400	0.101	0.470
^{241}Am (megacuries)	0.151	3.54	15.3
^{244}Cm (megacuries)	1.58	34.1	133.3
Accumulated Alpha Waste‡ §			
Volume of waste (10^6 cubic feet)	2.1	10.3	19.3
Salt area used (acres)	20	96	180
Total thermal power (megawatts)	0.0142	0.170	0.476
Total activity (megacuries)	14.2	151	300
Total mass of actinides (metric tons)	1.40	15.8	38.3
^{238}Pu (megacuries)	0.232	2.57	6.02
^{239}Pu (megacuries)	0.0515	0.580	1.41
^{240}Pu (megacuries)	0.0741	0.834	2.02
^{241}Pu (megacuries)	13.8	146	286
^{241}Am (megacuries)	0.0617	1.03	4.74

*Each shipment consists of 57.6 cubic feet of waste in 36 cylinders (6 inches in diameter). Each cubic foot of waste represents 10,000 megawatt days (thermal) of reactor operation. Half of the waste is aged 5 years, and half is aged 10 years at the time of its shipment. Last shipments are assumed to be made in the year 2000.

†Shipments are made in ATMX railcars; each shipment contains 832 cubic feet of waste. Last shipments are assumed to be made in the year 1999.

‡At end of year.

§The isotopic composition of Pu at the time of its receipt is 1 percent ^{238}Pu, 60 percent ^{239}Pu, 24 percent ^{240}Pu, 11 percent ^{241}Pu, and 4 percent ^{242}Pu.

those we now know how to do; otherwise at every stage in the recycling we would be creating additional low-level wastes. We probably can eventually develop such sharp separation methods; but these, at least with currently visualized techniques, would be very expensive. It is on this account that I like better the other approach which is to find some spot in the universe where the wastes can be placed forever out of contact with the biosphere. Now the only place where we know absolutely the wastes will never interact with man is in far outer space. But the roughly estimated cost of sending wastes into permanent orbit with foreseeable rocket technology is in the range of 0.2 to 2 mills per electric kilowatt hour, not to speak of the hazard of an abortive launch. For both these reasons I do not count on rocketing the wastes into space.

This pretty much leaves us with disposal in geologic strata. Of the many possibilities—deep rock caverns, deep wells, bedded salt—the latter has been chosen, at least on an experimental basis, by the United States and West Germany. The main advantages of bedded salt are primarily that, because salt dissolves in water, the existence of a stratum of bedded salt is evidence that the salt has not been in contact with circulating water during geologic time. Moreover, salt flows plastically; if radioactive wastes are placed in the salt, eventually the salt ought to envelop the wastes and sequester them completely.

These arguments were adduced by the National Academy of Sciences Committee on Radioactive Waste Management [7] in recommending that the United States investigate bedded salt (which underlies 500,000 square miles in our country) for permanent disposal of radioactive wastes. And, after 15 years of discussion and research, the AEC about a year ago decided to try large-scale waste disposal in an abandoned salt mine in Lyons, Kansas (Figure 7-4). If all goes as planned, the Kansas mine is to be used until A.D. 2000. What one does after A.D. 2000 would of course depend on our experience during the next 30 years (1970 to 2000). In any event, the mine is to be designed so as to allow the wastes to be retrieved during this time.

The salt mine is 1000 feet deep, and the salt beds are around 300 feet thick. The beds were laid down in Permian times and had been undisturbed, until man himself intruded, for 200 million years. Experiments in which radioactive fuel elements were placed in the salt have clarified details of the temperature distribution around the wastes, the effect of radiation on salt, the migration of water of crystallization within the salt, and so on.

The general plan is first to calcine the liquid wastes to a dry solid. The solid is then placed in metal cans, and the cans are buried in the floor of a gallery excavated in the salt mine. After the floor of the gallery is filled with wastes, the gallery is backfilled with loose salt. Eventually this loose salt will consolidate under the pressure of the overburden, and the entire mine will be resealed. The wastes will have been sequestered, it is hoped, forever.

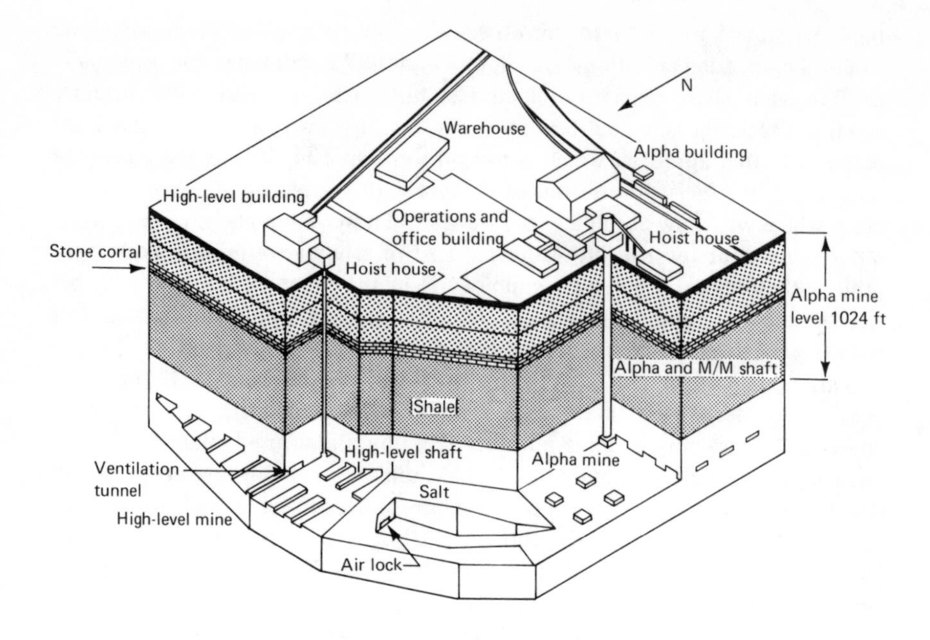

Figure 7-4. Federal Repository

Much discussion has centered around the question of just how certain we are that the events will happen exactly as we predict. For example, is it possible that the mine will cave in and that this will crack the very thick layers of shale lying between the mine and an aquifer at 200 feet below the surface? There is evidence to suggest that this will not happen, and I believe most, though not all, geologists who have studied the matter agree that the 500-foot-thick layer of shale above the salt is too strong to crack so completely that water could enter the mine from above.

But man's interventions are not so easily disposed of. In Kansas there are some 100,000 oil wells and dry holes that have been drilled through these salt formations. These holes penetrate aquifers; and in principle they can let water into the mine. For the salt mine to be acceptable, one must plug all such holes. At the originally proposed site there were 30 such holes; in addition, solution mining was practiced nearby. For this reason, the AEC recently authorized the Kansas State Geological Survey to study other sites that were not peppered with man-made holes. The AEC also announced recently its intention to store solidified wastes in concrete vaults, pending resolution of these questions concerning permanent disposal in geologic formations.

Man's intervention complicates the use of salt for waste disposal; yet by no means does this imply that we must give up the idea of using salt. In the first place, such holes can be plugged, though this is costly and requires development. In the second place, let us assume the all but incredible event that the mine is flooded–let us say 10,000 years hence. By that time, since no new waste will be placed in the mine after A.D. 2000, all the highly radioactive beta decaying species, notably ^{90}Sr and ^{137}Cs, would have decayed. The main radioactivity would then come from the alpha emitters. The mine would contain 38 tons of ^{239}Pu mixed with about a million tons of nonradioactive material. The plutonium in the cans is thus diluted to 38 parts per million; since plutonium is, per gram, 10,000 times more hazardous than natural uranium in equilibrium with its daughters, these diluted waste materials would present a hazard of the same order as an equal amount of pitchblende. Actually, the 38 tons of ^{239}Pu is spread over 200 acres. If all the salt associated with the ^{239}Pu were dissolved in water, as conceivably could result from total flooding of the mine, the concentration of plutonium in the resulting salt solution would be well below maximum permissible concentrations. In other words, by virtue of having spread the plutonium over an area of 200 acres, we have to a degree ameliorated the residual risk in the most unlikely event that the mines are flooded.

Despite such assurances, the mines must not be allowed to flood, especially before the ^{137}Cs and ^{90}Sr decay. We must prevent man from intruding–and this can be assured only by man himself. Thus we again come back to the great desirability, if not absolute necessity in this case, of keeping the wastes under some kind of surveillance in perpetuity. The great advantage of the salt method over, say, the perpetual reworking method, or even the aboveground concrete vaults without reworking, is that our commitment to surveillance in the case of salt is minimal. All we have to do is prevent man from intruding, rather than keeping a priesthood that forever reworks the wastes or guards the vaults. And if the civilization should falter, which would mean, among other things, that we abandon nuclear power altogether, we can be almost (but not totally) assured that no harm would befall our recidivist descendants of the distant future.

Social Institutions–Nuclear Energy

We nuclear people have made a Faustian bargain with society. On the one hand, we offer–in the catalytic nuclear burner–an inexhaustible source of energy. Even in the short range, when we use ordinary reactors, we offer energy that is cheaper than energy from fossil fuel. Moreover, this source of energy, when properly handled, is almost nonpolluting. Whereas fossil fuel burners must emit oxides of carbon and nitrogen, and probably will always emit some sulfur dioxide, there is no intrinsic reason why nuclear systems must emit any pollutant–except heat and traces of radioactivity.

But the price that we demand of society for this magical energy source is both a vigilance and a longevity of our social institutions that we are quite unaccustomed to. In a way, all of this was anticipated during the old debates over nuclear weapons. As matters have turned out, nuclear weapons have stabilized at least the relations between the superpowers. The prospects of an all-out third world war seem to recede. In exchange for this atomic peace we have had to manage and control nuclear weapons. In a sense, we have established a military priesthood which guards against inadvertent use of nuclear weapons, which maintains what a priori seems to be a precarious balance between readiness to go to war and vigilance against human errors that would precipitate war. Moreover, this is not something that will go away, at least not soon. The discovery of the bomb has imposed an additional demand on our social institutions. It has called forth this military priesthood upon which in a way we all depend for our survival.

It seems to me (and in this I repeat some views expressed very well by Atomic Energy Commissioner Wilfrid Johnson) that peaceful nuclear energy probably will make demands of the same sort on our society, and possibly of even longer duration. To be sure, we shall steadily improve the technology of nuclear energy; but, short of developing a truly successful thermonuclear reactor, we shall never be totally free of concern over reactor safety, transport of radioactive materials, and waste disposal. And even if thermonuclear energy proves to be successful, we shall still have to handle a good deal of radioactivity.

We make two demands. The first, which I think is the easier to manage, is that we exercise in nuclear technology the very best techniques and that we use people of high expertise and purpose. Quality assurance is the phrase that permeates much of the nuclear community these days. It connotes using the highest standards of engineering design and execution; of maintaining proper discipline in the operation of nuclear plants in the face of the natural tendency to relax as a plant becomes older and more familiar; and perhaps of managing and operating our nuclear power plants with people of higher qualification than were necessary for managing and operating nonnuclear power plants: in short, of creating a continuing tradition of meticulous attention to detail.

The second demand is less clear, and I hope it may prove to be unnecessary. This is the demand for longevity in human institutions. We have relatively little problem dealing with wastes if we can assume always that there will be intelligent people around to cope with eventualities we have not thought of. If the nuclear parks that I mention are permanent features of our civilization, then we presumably have the social apparatus, and possibly the sites, for dealing with our wastes indefinitely. But even our salt mine may require some small measure of surveillance if only to prevent men in the future from drilling holes into the burial grounds.

Eugene Wigner has drawn an analogy between this commitment to a permanent social order that may be implied in nuclear energy and our commitment to a stable, year-in and year-out social order when man moved from hunting and gathering to agriculture. Before agriculture, social institutions hardly required the long-lived stability that we now take so much for granted. And the commitment imposed by agriculture in a sense was forever: the land had to be tilled and irrigated every year in perpetuity; the expertise required to accomplish this task could not be allowed to perish or man would perish; his numbers could not be sustained by hunting and gathering. In the same sense, though on a much more highly sophisticated plane, the knowledge and care that goes into the proper building and operation of nuclear power plants and their subsystems is something that we are committed to forever, so long as we find no other practical energy source of infinite extent.[a]

Let me close on a somewhat different note. The issues I have discussed here—reactor safety, waste disposal, transport of radioactive materials—are complex matters about which little can be said with absolute certainty. When we say that the probability of a serious reactor incident is perhaps 10^{-8} or even 10^{-4} per reactor per year, or that the failure of all safety rods simultaneously is incredible, we are speaking of matters that simply do not admit of the same order of scientific certainty as when we say it is incredible for heat to flow against a temperature gradient or for a perpetuum mobile to be built. As I have said earlier, these matters have trans-scientific elements. We claim to be responsible technologists, and as responsible technologists we give as our judgment that these probabilities are extremely—almost vanishingly—small; but we can never represent these things as certainties. The society must then make the choice, and this is a choice that we nuclear people cannot dictate. We can only participate in making it. Is mankind prepared to exert the eternal vigilance needed to ensure proper and safe operation of its nuclear energy system? This admittedly is a signifcant commitment that we ask of society. What we offer in return, an all but infinite source of relatively cheap and clean energy, seems to me to be well worth the price.

References and Notes

1. "50 and 100 Years Ago," *Sci. Amer.* **225**, 10 (Nov. 1971).
2. J. Bartlett, *Familiar Quotations* (Boston: Little, Brown, ed. **14**, 1968).
3. E.N. da C. Andrade, *Rutherford and the Nature of the Atom* (Garden City, N.Y.: Doubleday, 1964), p. 210.
4. A.M. Weinberg and R.P. Hammond, in *Proceedings of the Fourth International Conference on the Peaceful Uses of Atomic Energy* (New York: United Nations, in press); *Bull. Atom. Sci.* **28**, 5, 43 (March 1972).
5. A.M. Weinberg, "Demographic Policy and Power Plant Siting," Senate Interior and Insular Affairs Committee, Symposium on Energy Policy and National Goals, Washington, D.C., October 20, 1971.

[a]Professor Friedrich Schmidt-Bleek of the University of Tennessee pointed out to me that the dikes of Holland require a similar institutional commitment in perpetuity.

6. W.B. Lewis, *Radioactive Waste Management in the Long Term* (DM-123, Atomic Energy of Canada Limited, Chalk River, July 13, 1971).
7. *Disposal of Solid Radioactive Wastes in Bedded Salt Deposits* (National Academy of Sciences–National Research Council, Washington, D.C., 1970); *Disposal of Radioactive Wastes on Land* (Publication 519, National Academy of Sciences–National Research Council, Washington, D.C., 1957); *Report to the U.S. Atomic Energy Commission* (Washington, D.C.: National Academy of Sciences–National Research Council, Committee on Geologic Aspects of Radioactive Waste Disposal, May 1966).
8. L.G. Hauser and R.F. Potter, "The effect of escalation on future electric utility costs" (report issued by Nuclear Fuel Division, Westinghouse Electric Corporation, Pittsburgh, Pa., 1971).

8 Fission Energy and Other Sources of Energy

Hannes Alfven

Should we opt for nuclear energy as our principal source of energy in the years to come? What are the alternatives?

The problem of how to satisfy the avalanching demand for energy in the world is attracting rapidly increasing interest. The energy crisis in the United States—whether real or manipulated—has stimulated the already lively discussion. The problem has a scientific-technical aspect. What energy sources are available now and in the future, and what are the ecological consequences of their use? It has an economic aspect. What price do we have to pay for energy produced in different ways? Finally, it is associated with at least two important world policy problems, one concerning the international competition for energy sources and the other concerning the relations between atomic energy and atomic warfare. Of special importance is that the large-scale deployment of fission reactors is creating an abundance of nuclear material in the world which deserves much attention from Pugwash. During the 1972 Pugwash conference at Oxford, the energy problem in general was discussed, and the Continuing Committee recommended the setting up of an international institute for the study of the scientific, technological, political and economic aspects of energy problems in the whole world.

When the atom bombs exploded over Hiroshima and Nagasaki, many of the scientists who had taken part in the Manhattan Project became frightened by the result of their work. They tried to satisfy their conscience in two ways. Some claimed that the horror of the bomb should put an end to all wars—the same thought Alfred Nobel expressed when he invented dynamite. The Vietnam war has taught us that this was not true. Others claimed that the development of fission had given mankind the ideal source of energy—the fission reactor—which should be of such benefit that it would overshadow the curse of the bomb.

In the United States the development of the atomic energy reactor began under extremely good conditions. A competent team of scientists, trained by their work on the bomb, worked in excellent laboratories which were well financed by the U.S. government. Big industries established for bomb manu-

This chapter is based on a paper presented to the 23rd Pugwash Conference on Science and World Affairs at Aulank, Finland, fall 1973. It first appeared in *Science and Public Affairs*, January 1974.

facturing switched, at least in part, to the development and manufacture of reactors, a profitable area because the U.S. government paid and took all the risks.

Development of the atomic energy reactor spread rapidly to other countries. This fast reaction was largely due however to military considerations. In some countries, namely, the Soviet Union, the United Kingdom, France and later China, atomic bombs were actually manufactured, whereas in other countries only atomic reactors for peaceful use were built. However, in several countries the motivation for building reactors was, at least initially, a desire to keep open an option for making atomic bombs sooner or later. For different reasons—technical, economical or political—no other country has yet made atomic bombs but some may "go nuclear" in the not too distant future. The strong international opinion against atomic bombs (to some extent because of Pugwash discussions and the Non-Proliferation Treaty) and the activity of the International Atomic Energy Agency (IAEA) have been and still are a rather efficient brake on the spread of nuclear weapons.

Energy Policy Decisions

Atomic energy thus received a flying start and development proceeded. A strong motivation for further investment in nuclear reactors was that the knowledge and technology already built-up must be utilized. The result was that in many countries energy policy decisions were distorted because the primary goal was not how to cover the energy need of the country but how to find an application for atomic energy.

From what has been said, it is clear that the nuclear industry partially received its internationally powerful position due to its association with the atomic bomb. In fact, the nuclear industry has been and probably still is supported by military subventions. The enriched uranium on which the reactors in most countries depend is a by-product of atomic bomb production, and a great deal of the development work for its manufacturing has been charged to military accounts. It would be interesting to obtain an objective clarification of whether atomic energy can be considered a cheap and competitive energy source without the more or less hidden subventions.

During the period of the development of nuclear reactors some opposition to this technology appeared; but from a technical, scientific point of view these objections seemed to be unwarranted. Certainly there were some unsolved problems, but these did not appear very serious. The prospects for atomic energy looked very promising. (As a personal declaration, up to a few years ago I was convinced that fission energy was the solution to the energy problem until fusion energy was ready.)

Objections to Fission Technology

The optimistic period for fission technology ended around 1970. There were several reasons for this.

1. It became increasingly obvious that plutonium and several of the waste products, especially radioactive strontium, are perhaps the most poisonous elements we know. For example, if introduced into the human body some of them tend to be deposited in the skeleton which they irradiate for a long time, increasing the risks of cancer even if the total quantity is only a small fraction of a milligram.

2. There are in nature a number of complicated biologic processes which enrich some of the radioactive waste products by a factor of 1,000 or 100,000. It is possible that even more efficient concentrating processes exist. Therefore, it is dangerous to deposit radioactive waste anywhere in the biosphere, even if highly diluted.

3. Development of the breeder reactor was proceeding and the uranium reactors already in use began to be considered as a transition to the breeder. The breeder technology, which at least at present is mainly based on the uranium-plutonium cycle, means an enormous increase in the production of plutonium.

4. Hitherto the discussions of the waste problem referred to one or a few reactors. Now plans were made to use atomic energy to satisfy a substantial part of the world's energy needs, and that meant necessarily the mass production of radioactive waste and plutonium.

5. The waste products from one or a few reactors can be taken care of; but when a huge amount of such products is accumulated, a very serious problem appears because these products cannot be destroyed by any applicable technology. This has been demonstrated, for example, by the project to make a "nuclear repository" in a salt mine in the United States in Kansas, which had to be stopped because of leakage risks. Supported by competent geologists, the state of Kansas refused to accept the project.

At present there does not seem to be any existing, realistic project on how to deposit radioactive waste; but there are a multitude of optimistic speculations on how to do so. The problem is how to keep radioactive waste in storage until it decays after hundreds or thousands of years. The deposit must be absolutely reliable as the quantities of poison are tremendous. It is very difficult to satisfy these requirements for the simple reason that we have had no practical experience with such a long-term project. Moreover, permanently guarded storage requires a society with unprecedented stability.

6. The ecological awakening has changed the basic view on technology. Hitherto a new technology has been allowed to take its short-term advantages, leaving the long time disadvantages to posterity. The essence of the ecological debate is that such a procedure cannot be allowed any longer. Applied to our case, the fission reactor produces both energy and radioactive waste: we want to

use the energy now and leave the radioactive waste for our children and grandchildren to take care of. This is against the ecological imperative: Thou shalt not leave a polluted and poisoned world to future generations.

Fission Reactors and Ecology

Because burning coal and oil with present methods produces much air pollution, it has been claimed that fission energy is much cleaner, and hence from an ecological point of view is preferable. Some ecologists, however, call fission energy "the most dirty of all energy sources."

A single reactor or a few reactors which are carefully controlled are not likely to constitute a very serious ecological threat. Research reactors, either for scientific purposes or as technical "prototypes," are rather innocent. However, if nuclear technology spreads in such a way that a considerable fraction of the energy consumption of a country or of the whole world comes from nuclear reactors, the picture changes completely. The reason is that the production of nuclear energy is necessarily associated with the production of radioactive elements; and a very large production of nuclear energy necessarily means the mass production of radioactive poisons in quantities which are terrifying.

This is the basic reason for the opposition against the use of atomic energy, which has become a worldwide controversy.

On one hand, everybody must have the deepest admiration for all the ingenious precautions the reactor constructors have made in order to contain the radioactive products and to prevent them from reaching the biosphere. On the other hand, one must also respect the objectors, who are not driven by some "nuclear hysteria" but by a very well motivated fear of a new threat to the lives and health of their generation and future generations.

With reference to some arguments which have been stated or published with regard to this controversy, it seems legitimate to state:

1. It is not correct to claim that reactors are absolutely safe—because no technological product can ever be safe and no operator is absolutely reliable.

2. It is not fair to claim that reactor accidents should be accepted in the same way as train and airplane accidents—because of the much more serious consequences which a reactor accident may lead to.

3. It is not correct to claim that long time deposit of radioactive waste is not a serious problem—because this problem has not been solved as yet and, further, no one knows how to solve it on the required large scale if nuclear technology spreads and provides a considerable fraction of the energy consumed by a country or the world. On the other hand, one cannot exclude the possibility that future research may lead to acceptable solutions of these difficulties. There may be some chance that future discoveries may make fission energy acceptable; but we have not reached this state yet and no guarantee can be given that we will ever reach it.

Other Energy Sources

From what has been said, it seems obvious that with the present state of development fission energy should be accepted as a large scale energy source only if the need of energy is desperate and no other sources of energy exist.

How much energy *really* is needed in an acceptable society will not be discussed here. It is obvious that technological civilization is on its way to a limitation. We have reached a new stage in the development where our actions can no longer be dictated by a desire to increase the population or the consumption of a country. We shall confine ourselves here to the technical-scientific problem of how to solve the energy problem.

When comparing different sources of energy we must observe that a large amount of competent work and much money have been invested in fission energy, whereas very modest efforts have been made to develop other forms of energy. As stated above, the reason for this is because in many countries the problem has not been how to solve the energy problem, but how to give fission energy not only military but also civilian use. Accordingly, we must start thinking in a radically new way: *We must imagine how other sources of energy would have appeared today if research and development had been concentrated on them.* And what we could have expected of them in the future.

Apart from fission energy, the following energy sources are discussed as serious alternatives:[a]

Fossil Fuels. It is often claimed that oil and natural gas sources will suffice only for the next 20 years; but in view of the fact that large new sources have been discovered, they may last much longer. In any case there is coal sufficient for centuries. At present the environmental objections to fossil fuels are serious. However, one may hope that if research of the same quality and quantity that has been devoted to fission energy is directed to a non-polluting handling of fossil fuels, these sources may supply the world with energy for a long time and in a way that is tolerable for the environment. There are several new ideas of how to utilize the huge coal reserves in a clean way.

Fusion Energy. A decisive difference between fission and fusion energy is that the fusion processes of interest result in nonradioactive end products. However, the intense neutron flux from a fusion reactor necessarily produces some radioactivity in the structure of the reactor. Also, a fusion reactor contains tritium, as an intermediate product which is radioactive, and this causes some leakage risk. There is no doubt that from an ecological point of view the fusion reactor is much less objectionable than the fission reactor. It is often claimed that technical fusion reactors will not be developed before year 2000 and,

[a]Hydropower and wind and tidal energy are not considered here; hydropower is geographically limited, and wind and tidal energy do not seem to be available in large quantities as to be of importance except in special cases.

therefore, fusion energy should not be mentioned in the present debate. The causality chain may be the reverse: as the breeder reactor lobby does not like the competition with the fusion alternative, this is eliminated by the claim that it belongs to a very distant future.

Solar Energy. As each square kilometer of the Earth's surface receives as much energy from the Sun as a big fission reactor delivers (about one GW), we have here an inexhaustible and completely clean source of energy. It is, however, at present very expensive. New research results give reasons for optimism about the future economy of solar energy.

Geothermal Energy. Geothermal energy means energy from the hot interior of the Earth. Out-streaming vapor and hot water in volcanic areas have long been used, for example, in Iceland, Italy and the Soviet Union. A new method, called the "hot rocks" method, has been suggested recently. Two holes close to each other are drilled until a hot region is reached, perhaps about 5 kilometers downward. The rock between the holes is cracked by some method. There, water is poured down through one of the holes, and when it comes in contact with the hot rock it is vaporized and comes up through the other hole as steam. This may constitute an almost inexhaustible source of energy for all countries, especially for those countries in which the rocks have a high thermal gradient.

The Human Factor

Fusion, solar and geothermal energy sources are not yet sufficiently developed to ensure the solution of our energy problem; nor can we be sure that fossil fuels can be handled in a clean way as to satisfy the environmentalists. But, as both the Manhattan project and the Apollo program have shown, our science and technology are so powerful that if an intense effort is made, we can do almost anything we want in, say, 10 years—provided we are not in conflict with the laws of Nature!

It is obvious that the present international debate is creating new ideas; and it is very likely that energy sources will be found which may make fission energy unnecessary. It is, therefore, a mistake to concentrate energy policy on a line which initially seems attractive, but which in the near future may be considered obsolete and dangerous. As we have learned from history, it is a normal process that a technology which at a certain time seemed attractive is substituted with a better one at a later time.

As stated above, we cannot rule out that new discoveries will make fission technology acceptable by solving the safety and the waste disposal problems. However, as so much highly qualified work has already been devoted to these fields, this is not very likely. We now recognize that these problems are not the usual scientific-technological problems but are ones closely connected with the "human factor." To what extent can we trust that operators will really do what they are instructed to do? Do the social systems in different countries and in the whole world possess the unprecedented stability which the fission technology requires? This means that the basic questions fall outside the field of competence of the fission technologists.

Inseparable Twins?

Are the military and the peaceful atomic energy programs inseparable twins? Another strong objection to fission energy is derived from its close relation to atomic bombs. This relation is clarified in a number of reports of the IAEA and in the Stockholm International Peace Research Institute's *Yearbooks* as well as at a June 1973 SIPRI symposium on "Review of Nuclear Proliferation Problems." There is no doubt that the IAEA has had remarkable success in constructing a system of international inspection of nuclear material, according to the plans of the Non-Proliferation Treaty. However, not all countries have signed the NPT, and even for those countries which have signed it, IAEA authority is confined to inspection. There is no guarantee that efficient international sanctions would be applied to a country which breaks the Non-Proliferation Treaty.

According to existing plans, within 10 years the production of fissile material for peaceful purposes will be sufficient for the production of about 10,000 atomic bombs a year.

It has been claimed that the large plutonium quantities produced by the reactors cannot easily be used for manufacturing bombs because of the difference between "reactor plutonium," which is normally produced in the reactors, and "weapon-grade plutonium," which is used for atomic bombs. The former is obtained when a reactor is run in the economically most favorable way, with change of fuel elements about once every 18 months; the latter is obtained if the burn-out is limited to a few months. However, there is no serious technical difficulty for an establishment, possessing complete fission energy equipment, to change to the production of weapon-grade plutonium. Moreover, even ordinary reactor plutonium can be used for making bombs, certainly somewhat clumsy and with less than maximum yield and less accuracy of performance, but anyhow terrible enough. Hence by stealing some 20 kilograms of ordinary reactor plutonium, a guerrilla or a criminal could obtain possession of the material for making atomic bombs.

It has also been claimed that the technology of manufacturing bombs is a very difficult process and requires knowledge of "atomic secrets." Today this is not true. Certainly, fabrication is not so easy that anyone could make an atomic bomb "in his garage" after having stolen a sufficient quantity of plutonium, as is said sometimes. But, according to what appear to be reliable reports, only a few engineers with a normal scientific and technical education and a good but not very fancy workshop are needed to make a bomb. This conclusion is quite reasonable from the point of view of the general experience of the technological development. What was a top scientific-technical achievement in 1945, when the first atomic bomb was made, may very well be an achievement that is rather easy to repeat in 1974. (However, the hydrogen bomb is still difficult to make!)

With all this in mind it is difficult to imagine how a proliferation of atomic bombs can be avoided in the future—when according to existing plans thousands or tens of thousands of atomic reactors will be working all over the world, and when the enormous production of plutonium from breeders has started. Indeed, it will be extremely difficult to prevent atomic bombs from falling into the hands of many groups of people who may like to use them for political or criminal purposes.

If we want to promote the spread of nuclear technology, the only way to avoid such a proliferation of nuclear bombs seems to be to impose very strict police control over the whole world. (Compare, Alvin Weinberg, "Social Institutions and Nuclear Energy," *Science*, 177 (1972):27.) This will be difficult to achieve and does not lead us to a very attractive future society.

What Could Pugwash Do?

It seems that the energy problem is a very important field which Pugwash should study carefully in order to make recommendations on how to direct the scientific and technological development. A preliminary suggestion is:

1. There should be increased research on how to make fission energy acceptable. Efforts should be made to increase reactor safety, and to find a method of satisfactory long time storage of the radioactive wastes.

2. Most important is to clarify whether there is any realistic chance of avoiding a proliferation of atomic bombs in a world in which a large part of the energy is generated by fission reactors.

3. The authority of IAEA should be strengthened and its safeguarding activity supported.

4. Because final acceptability of fission energy cannot be taken for granted, a warning should be issued that large-scale application of fission technology may not be a realistic solution to the world's energy problem. It is unwise to spend more work and money on the development and deployment of fission reactors, especially breeders, before the acceptability problem is clarified. A nuclear moratorium may be advisable in order to allow time for a detailed analysis of this question.

5. A major effort should be made to develop alternative energy sources which satisfy ecological demands.

6. An analysis should be made of how much energy is really needed for a high quality civilization. The *need* may be much smaller than the *demand.*

7. An international institute should be established, devoted to the international energy problem, as recommended during the 1972 Pugwash conference at Oxford.

9 Geothermal Energy

Ronald I. Axler

Geothermal power offers one possible solution to our energy needs. What is being done to exploit it and how important can it be?

The United States has reached a stage in its development when any further progress toward a more modern society must be measured by the possibly detrimental effects on the environment. How to cope with the massive increase in the future demand for energy output without producing even greater pollution problems, is one of the most vital questions society must face.

The United States is currently experiencing a serious energy crisis, evidenced by fuel shortages and accompanied by rising prices. The outlook for the future is not optimistic. It is estimated that by the year 2000 a 37 percent increase in our nation's population will be accompanied by more than a two and a half times increase in the use of all forms of energy [7:1]. With the predicted growth rate of 3.6 percent per annum, gross energy consumption will increase from the 1971 level of 69 quadrillion BTUs to 192 quadrillion BTUs by the end of the century.

As part of this increase, use of natural gas will grow from 22 trillion cubic feet to 33 trillion cubic feet, with 5½ trillion more being produced from three hundred million tons of coal and one hundred million barrels of oil. Petroleum use of five and a half billion barrels will grow to thirteen billion barrels, including the amount used to manufacture gas. Coal use will increase by more than two and a half times to over 1.3 billion tons. Nuclear power, which now contributes only 2 percent of our energy needs will provide over 50 percent by the year 2000 [30:II-3].

Accompanying the projected increase in energy demand is the grim realization that additional sources of energy are becoming more scarce. There are few feasible hydroelectric sites which have not already been developed to their fullest extent. The expansion of fossil-fueled plants has been limited by rising prices, dwindling supply, and governmentally enforced environmental restrictions. Although the reserve supplies of coal and oil are ample for our present needs, most reserves of both substances have high sulfur contents and when burned for fuel the sulfur oxides become a major pollutant in the air. Low sulfur fuels, presently in high demand for use in steam plants, are becoming increasingly more expensive to obtain. Natural gas resources are being depleted faster than those of other energy sources due to the increasing demand for what is considered the least polluting source of energy. Nuclear-fueled power, although it represents what some people believe to be the ultimate answer to the fuel crisis, meets heavy opposition due to its potentially hazardous radiation and thermal pollution effects.

Despite these problems the energy outlook need not be one of utter despair. In recent years an increasing amount of research and development has been directed toward new energy sources. One of the most promising of these relatively newly found sources is geothermal power.

Types of Geothermal Formations

Geothermal energy emanates from the natural heat of the earth, which is derived from radioactive decay, friction caused by tidal and crustal plate motion, and possible primal heat. The heat is concentrated locally in the earth's crust by volcanism and tectonism. A layer of molten rock, called magma, which is located at a distance of approximately twenty miles beneath the earth's crust, transmits the heat upwards. Where fissures have appeared in the upper layers of rock, this magma is forced close to the surface, evidencing its presence in the form of volcanic activity. The heat is often transferred by convection to cells of circulating hot water above the magma chambers. Geysers, fumaroles, and hot springs may result.

The use of the pressurized steam and heated water from these underground systems to produce energy is the principle upon which geothermal power is based.

There are four basic classes of geothermal energy sources: dry steam systems, hot water systems, hot dry rock systems, and geopressured reservoir systems. Dry steam systems consist of steam and some hot water under extreme pressure and high temperatures. When a well is drilled it provides an outlet for the pressurized fluid. When the pressure decreases the heat from the surrounding rock can dry and superheat the steam.

Hot water systems are mostly filled with water that is kept at temperatures well above its boiling point, but at extremely high pressure that maintains it in liquid form despite its high temperature. This "super-heated" water has typical temperatures ranging from about 350 to 700 degrees Fahrenheit. Fifteen to 25 percent of the discharge of the wet field flashes into steam at the wellhead. The rest is hot water. The steam is separated at the surface and the water is either discharged on the surface or reinjected back into the ground.

The next class of geothermal system is a hot dry rock system. Here there is an impermeable layer of rocks over the heat source, i.e., magma chamber, so that there is no natural supply of water circulating near the surface. Water must be introduced to provide a circulating heat source if productive use is to be made of this occurrence.

The last type of geothermal system is a geopressured reservoir system. This is one of highly porous sands saturated with brines of high temperature and very high pressure. This type of field is found along the Louisiana Coast and offshore Southern Texas. A geopressured reservoir system is thought to result from a heat flow trapped by under-compacted clays which act as an insulating layer. Water from the compacting and dehydrating of the clays accumulate in the sand and add to the fluid reserve. More research is needed to develop processes which could make use of this source on a more economical level.

Many countries have joined the search for this underutilized source of energy. There are now geothermal fields being utilized in Italy, the United States, New Zealand, Japan, Iceland, Mexico, the U.S.S.R., Hungary, and Chile.

Geothermal fields are employed in the production of electric power in ten locations throughout the world. Larderello, Italy, is still the largest power-generating location with 365 megawatts (mw). Others are Monte Amiata, Italy, with 26 mw; Wairakei, New Zealand, with 192 mw; Matsukawa, Japan, with 20 mw; Otake, Japan, with 12 mw; Cerro Prieto, Mexico, with 75 mw; Pathe, Mexico, with 1/2 mw; Namafjall, Iceland, with 3 mw; and Pauzhetsk, U.S.S.R., with 5 mw.

The Geysers in California is the only United States location currently generating electric power from geothermal resources. But there are Known Geothermal Resource Areas (KGRAs) scattered throughout all states in the continental United States west of the Continental Divide, plus Alaska and South Dakota. The terrains of these locations is diverse, including the Great Basin, the Rockies, the mountains along the Pacific Coast, the Cascade Range, and the Aleutians. The lands include forests, grasslands, high and low deserts, tundra, and agricultural lands.

The harnessing of geothermal power has already reached an aggregate capacity of over 1,000 mw in plants around the world. At the present rate of development, this capacity would quadruple by the end of the decade.

Many people expect a good deal of development of government-owned land in the United States for geothermal exploitation. A National Science Foundation report has estimated that by 1985 our geothermal resources will have the potential of developing 132,000 mw of electricity; by the year 2000 they could provide 395,000 mw. Currently the total production capacity in the United States is approximately 350,000 mw for all methods of electrical power production combined [40:1].

According to Russian estimates, the earth's crust to a depth of 7 to 10 kilometers has a total amount of stored up heat five thousand times greater than the heat-producing capacity of all types of conventional fuels on the Earth. Within Soviet boundaries one-third of the land is estimated to contain subterranean hot waters, most of which is located in the sparsely populated region of Siberia. So far, only the Pauzhetka power station uses geothermal energy from the Soviet sources. There are plans, though, to study the potentialities of the

Avacha volcano. They estimate that to a depth of only 3 to 5 kilometers it contains stored heat capable of powering a one million kilowatt power station for five hundred years [22:3].

The Federal Power Commission estimates a dramatic increase in the requirements for thermal power for the future, as detailed in Table 9-1. The country and much of the world is currently experiencing a fuel shortage. The projections in Table 9-1 indicate the extent that new resources must be found to provide the fuel needed in the future. This is why new sources of energy need to be investigated. Oil companies extol the advantages of offshore drilling and the Alaskan pipeline. Many scientists advocate the massive use of nuclear power plants as a long-term answer. Geothermal power can possibly serve as a partial alternative to some of these methods which pose serious environmental problems.

Tables 9-2 and 9-3 demonstrate the replacement capabilities estimated for geothermal energy through the year 1985. Table 9-2, for example, shows that by 1980 geothermal power is expected to reach a generating capacity between 1,000 and 2,000 mw. If geothermal power is not developed, thirteen to twenty-six million barrels of oil would be needed during the year to provide the needed electricity. The use of geothermal power allows those millions of barrels of oil to be used elsewhere where there may be a need, or to be saved for some future time, thus extending the time in which oil resources will be available. Table 9-3 provides a similar comparison between geothermal energy and natural gas usage. By 1985 as much as 524 billion cubic feet of gas can be saved each year. When the replacement figures are compared with actual fuel use, striking facts appear. In 1970 the annual total domestic consumption of natural gas was

Table 9-1
Projected Annual Fuel Requirements For Thermal Power Generation

Fuel	Year 1970	1980	1990
Coal (millions of tons)	322	500	700
Natural Gas (billions of cubic feet)	3,600	3,800	4,200
Residual Fuel Oil (millions of barrels)	331	640	800
Uranium Ore (tons)			
With plutonium recycle	7,500	41,000	127,000
Without plutonium recycle	7,500	38,000	108,000

Source: Federal Power Commission, *1970 National Power Survey* (Washington, D.C.: Government Printing Office, December 1971), Table 1.5, p. I-1-20.

Table 9-2

Oil Requirement Needed in Western United States to Replace Electricity Generated by Geothermal Means Should Geothermal Power Not Be Expanded as Projected[1]

Year	Geothermal Power[2]	Oil Replacement: Million Bbls. Per Year	Oil Replacement: Thousand Bbls. Per Day
1973	298 megawatts from The Geysers field, California	4	11
1980	1,000-2,000 megawatts	13-26	37-77
1985	7,000-20,000 megawatts from all Western U.S. sources (assuming new technology can be developed to use hot water)	91-260	250-710

[1] The average crude oil yields about 5.8 million BTU per barrel (42 gallons) and the average low-sulfur fuel oil, a refined product, yields about 6 million BTU per barrel. For this purpose, it is assumed that a low-sulfur crude oil yielding 5.8 million BTU could be used in power generating facilities even though lighter fractions would probably be refined off and used for other purposes.

[2] Predicted geothermal power from National Petroleum Council, U.S. Energy Outlook, v. 2, 1971.

Source: The United States Department of the Interior, *Final Environmental Statement for the Geothermal Leasing Program*, Ch. IV (Washington, D.C.: U.S. Government Printing Office, 1973), p. 71.

Table 9-3

Gas Consumption Alternatives at the Three Projected Levels of Geothermal Development Considered Herein

Year	Geothermal Capacity (mw)	Gas Equivalent Per Year (trillion BTU)	Gas Equivalent Per Year (billion cu. ft.)
1973	298	8	8
1980	1,000-2,000	27-54	26-52
1985	7,000-20,000	190-540	184-524

Source: The United States Department of the Interior, *Final Environmental Statement for the Geothermal Leasing Program*, Ch. IV (Washington D.C.: U.S. Government Printing Office, 1973), p. 136.

22,412 billion cubic feet [24:I-4-14]. The total domestic and utility consumption of fuel oil was 804.2 million barrels [24:I-4-17].

Geothermal Electricity

The world's first geothermal project was started in Larderello, Italy, in 1904. It is still the world's largest geothermal project, although it has remained at the same capacity of 365 mw for decades. The first geothermal electric plant in the United States is located at Geyserville, California, eighty miles north of San Francisco. For the past ten years men have been drilling more than a mile into the earth's crust at this location, where they reach a bed of natural steam. Then generators are attached and electricity is produced. By 1976 it is predicted that these wells will have a capacity of 750 mw, enough energy to supply all the power required by a city the size of San Francisco [38:2].

The Geysers power plant in California is the only commercial development of geothermal energy in the Western Hemisphere. It is an excellent example of the use of steam for power production. The power plant is actually a scattering of several separate generating units, each linked to individual fumaroles. It works basically the same way as a conventional steam plant, but without the man-made boiler. At the Geysers drillers dig as far as a mile and a half down to release the steam. The steam is filtered of impurities by being put through a centrifugal separator. There the steam is whirled so that minute bits of fine rocks will spin out so as not to damage the turbine blades. The steam is put through the turbine and then piped to the condenser, where it is turned into hot water. It is then pumped to a cooling tower. After being cooled, most of the water is sent back to the condenser to cool the incoming steam. This process provides a vacuum at the exhaust end of the turbine, which doubles the efficiency of the plant by increasing the speed of flow throughout the cycle. The cooled water contains impurities so it is reinjected into the ground. This keeps the impure water far enough away so that it does not affect surface water supplies. The final procedure also helps prevent environmental damage from a water loss, and permits the water to be exposed once again to the hot rocks and thereby converted into steam a second time.

Steam from dry fields, when harnessed and piped directly into a turbine, greatly simplifies plant requirements for the production of electric power. As is the case with most geothermal methods, pollution and waste of resources is minimal. The steam, after it has expended its useful energy for electric power generation, can readily be converted into fresh water for consumption.

According to discoveries made to date, hot water, or "wet" fields may be twenty times more abundant than "dry" fields. One of the prime locations of this type of system is the Imperial Valley in Southern California.

With a wet field the water does not become steam until the pressure is released by drilling into the field. The steam can be used for electric power production, while the water has a multitude of potential uses. The superheated water in this type of system usually contains a high concentration of dissolved minerals, which may permit by-product recovery, but also poses engineering problems of corrosion, scaling, and disposal.

Project Plowshare

All of the geothermal projects currently in existence use nature's own resources for the basis of their energy production. There are proposals which would eliminate the dependence on finding such rare naturally occurring conditions. The Atomic Energy Commission has been investigating the feasibility of utilizing nuclear explosions to help recover energy from dry geothermal anomalies for the production of electricity.

The concept behind what has been labeled "Project Plowshare" is to use a nuclear explosion deep underground to break up a large portion of the rock layer. This produces a chimney effect of broken rock above the actual blast site. Water can be piped into and then extracted heated from the hot rock. Recycling the water can maintain a constant heat flow to generators on the surface. The "Plowshare" technique eliminates the need for finding the natural condition of a geothermal gas or hot water deposit. The finely broken rock also allows for freer circulation of the water to absorb the heat. This serves to improve the quality and increase the quantity of the steam produced [39:119].

An alternative proposal to the one mentioned above was suggested by scientists at the research center in Los Alamos, New Mexico. Their idea is to hydraulically fracture the rock layer with water pressure, a process known as "hydrofracing." They would pump cold water into the earth at a pressure of approximately 7,000 pounds per square inch to break up the rock. Additional fracturing, thermal fracturing, would be caused by thermal stresses induced by the temperature changes brought about by the injection of cold water into the hot rock. This is a common technique used in oil drilling. The hot rock would then heat the water. The water would be drawn into a second pipe and rise to the surface by convection [37:18]. The entire procedure is illustrated in Figure 9-1.

The technology required to fulfill completely the geothermal energy potential is yet to be fully developed. But we already possess capabilities to make some use of a great deal of the geothermal energy. The use of even the most modern methods are relatively inefficient, though, considering the vast extent of the energy potentially recoverable from geothermal sources.

The fact that geothermal resources have been known to exist for quite some time makes it appear strange that more research has not been undertaken. History shows that as early as 1920 attempts to harness the hot water found in western states were abandoned due to the extreme salinity of the water. Interest was not revived again until other finds of heated water displayed a brine content about as great as that of sea water [40:1]. The subsequent reluctance to develop geothermal power has arisen from the tendency to stay away from an area where new technology may be needed to harvest energy of acceptable levels of efficiency. The Geothermal Steam Act of 1970 has removed one obstacle to geothermal energy production by opening up government land at reasonable leasing terms to allow use by private industry [34:1].

Alternative Uses of Geothermal Energy

The majority of the initial attempts at harnessing geothermal energy have had a single purpose, generation of electric power. The Wairakei hot water field in New

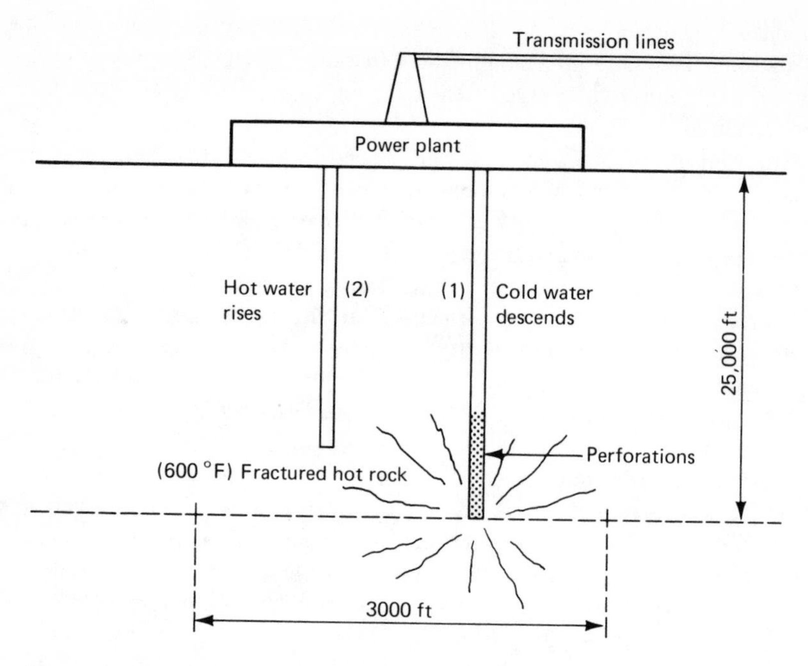

Figure 9-1. Creating New Sources of Geothermal Energy by Hydro-fracing. Source: Based upon description in John Noble Wilford, "New Plan Is Outlined for Tapping Geothermal Energy," *New York Times* (June 21, 1972), p. 18, Col. 3.

Zealand, for example, feeds the steams fraction of the discharge to a power generating plant and then just discards the hot water into a river. More recently there have been attempts to analyze the components and combinations of the wet fields to evolve diverse and even multipurpose projects. There are already installations which use the steam from the wet fields to produce power and then distill some of the hot water to make fresh water, without requiring the addition of more heat. At a field recently discovered in El Tatio, Chile, the government is using United Nations aid to investigate the development of a facility that will generate three different products. The steam will initially be used to generate electricity. Once the steam turns to hot water it will be put through a desalination plant to produce fresh water. The hot water which accompanies the gas in the discharge will be concentrated in a mineral-rich brine from which valuable minerals will be extracted in evaporation ponds. The United Nations has been undertaking a systematic search for mineral rich geothermal brines ever since there were accidental finds in Southern California and the Red Sea. Two potentially economically important discoveries have also been made in Chile and Ethiopia. Potentially valuable minerals contained in geothermal hot water are potassium, lithium, and calcium, among others.

Other uses for geothermal hot water are for manufacturing and processing operations, such as those being employed in Iceland, New Zealand, and the U.S.S.R. Russia also has experimented with geothermal energy for refrigeration, and with the recovery of by-product chemicals [30:II-15]. At Kawerau in New Zealand a paper and pulp company applies geothermal water to heating for industrial processes. Applications in Japan include use in experimental fish farming projects, cleaning, cooking, soil heating, and bathing. A system which uses the water to provide heat in winter and refrigeration in summer is being used in a hotel in New Zealand. They report that the energy cost is only a tenth of the cost of a system using electrically-run compressors [12:35].

Geothermal hot water is most efficiently utilized in direct space heating. Reykjavik, the capital city of Iceland, has 75 percent of its district heating and hot water supplied by geothermal fields. There are sixteen bore holes which are used to draw forth the super-heated water. The water temperature at the smaller fields is about 190 degrees Fahrenheit, while a larger field supplies water at 218 degrees and the fourth and largest field emits water at 260 degrees. Fourteen-inch steel pipes carry the water from as far away as ten miles to the city's central storage tanks, which are maintained at a level of approximately 20 percent of the highest twenty-four hour load. The drop in the water temperature of the effluent from the fields most distant from the city to the point of use is only nine degrees Fahrenheit. From the storage tanks and the main boosters the water is pumped to nine district pumping stations. The supply mains then pump the water to the consumers through either single or double pipe distribution systems. Double pipe systems collect the water which has cooled, and mix it with high temperature supply water to regain its heat. The distribution system itself has a heat loss of approximately 11 percent, which permits the maintenance of a set temperature of 176 degrees for water provided to the houses.

Gas and oil heating systems are much less efficient than the one used by the District Heating Service in Reykjavik. The geothermal system has only a 16 percent heat loss from the farthest wells to the point it enters the homes [16]. The average overall efficiency of heating from the power provided through electric power plants is only 30 percent, with much of the loss occurring at the plant itself. Gas or oil burning home heating systems have about a 60 percent end-use efficiency. Therefore, the fuel and energy savings from the use of geothermal hot water is considerable [15:1301].

Experimental projects using geothermal hot water for space heating are also being tested in New Zealand, the U.S.S.R., and Hungary, as well as Klamath Falls, Oregon, and Boise, Idaho, where over four hundred buildings and two hundred homes, respectively, are being supplied.

Cost Comparisons

It would appear that geothermal power appears to be of some importance to help provide the future needs of energy. But merely because it is an available source of power, and its use would not substantially adversely affect the environment, does not insure that its use will be sought. One of the prime prerequisites for any new commercial undertaking in our economic system is that it should provide profits. Thus, costs are often the determinative factor as to whether prospective projects are undertaken. It appears as though the developmental costs of geothermal energy, while considerable, are not prohibitive and, in the long run, can be competitive with the least expensive power producing sources.

The best source of financial information on production of electricity from geothermal steam is The Geysers, as it is the fastest growing field of its kind and will soon be the largest in the world. Tables 9-4 through 9-7 compare costs of different types of electrical power production.

Table 9-4 shows that at an assumed capacity of 1,500 mw electricity may be produced from geothermal steam less expensively than from any other energy source. Greater capacity would reduce the fixed unit cost and make it less important as compared with the variable costs. Lower capacity has the opposite effect of increasing the importance of the fixed unit costs relative to the variable costs. Based on this reasoning the fuels with lower variable costs will usually have lower total unit costs at high capacities. At lower capacity hydroelectric power might have a lower total unit cost because it has a lower fixed factor than geothermal energy. But, all of the other sources have greater fixed and variable cost factors and so would have higher total costs at any capacity level.

Tables 9-5 and 9-6 show estimated costs of electricity from geothermal energy and various other plants. Generally the costs of hot water plants are comparable to hydroelectric, nuclear and oil fired plants. Dry steam plants have

Table 9-4
Cost Comparison of Electrical Energy Production (mils per kilowatt hour)

Fuel	Fixed Load Factors[1]	Variable Load Factors[1]
Coal	5.22	6.14
Geothermal Power	2.96	2.96
Hydroelectric Power		
By private industry	3.45	4.79
By government	2.60	3.60
Natural Gas	4.82	6.19
Nuclear Power	5.42	5.49
Oil	4.87	6.27

[1] All calculations assume a plant capacity of 1,500 mw.

Source: Alvin Kaufman, "An Economic Appraisal of Geothermal Energy," *Public Utilities Fortnightly* 88, 7 (September 30, 1971): 22.

Table 9-5
Estimated Cost of Electricity from Variously Fueled Plants

Item	Gas-fired* 1	Coal-fired* 2	Hydro-electric	Nuclear**	Geothermal* (Hot water)	Oil-fired* 1
Unit investment cost of plant $/kw[3]	105	225	390	300	290[4]	115
Annual fixed charge, percentage of investment[5]	17	17	17	17	[4]	17
Kilowatt-hours generated per year per kw capacity[6]	7,000	7,000	7,000	7,000	7,000	7,000
Cost of fuels, cents/million BTU	34.8	20.3	–	17.5	37.6[7]	61.5
Cost of electricity, mills/kwh						
Plant investment	2.6	5.5	9.5	7.3	2.3	2.8
Operation & maintenance	0.6	0.8	0.1	0.4	1.5	0.7
Fuel	3.5	1.9	–	1.9	5.9	6.5
Total	6.7	8.2	9.6	9.6	9.7[8]	10.0

*Derived from 17th steam station cost survey, *Electrical World* 176 (Nov. 1, 1971).

**Derived from Hottel and Howard, *New Energy Technology*; article by Benedict, "Electric Power for Nuclear Fusion," *Proceedings of National Academy of Science* 68.

[1] Outdoor type plant.

[2] Indoor type plant. All figures valid only for western states.

[3] Includes land, structures, boilers, turbine generators, electrical equipment, miscellaneous plant equipment. Excludes switchyard.

[4] See Table 9-6 for explanation of investment costs and annual fixed charges.

[5] Includes cost of money, depreciation, interim replacements, insurance, and taxes.

[6] The 80 percent operating factor used here is applicable only to base load plants. Hydro is seldom a base load plant.

[7] Cost of fuel based on capital and operating costs of steam-winning system.

[8] Comparative cost for dry steam approximately 5.3 mils.

Source: The United States Department of the Interior, *Final Environmental Statement for the Geothermal Leasing Program*, Ch. III (Washington, D.C.: U.S. Government Printing Office, 1973), p. 91.

Table 9-6
Geothermal Plant Investment and Annual Fixed Charges[1]

Item	Unit Investment Costs	
	10-Year Life	30-Year Life
Production well system	$ 48	$ 25
Injection well system	75	32
Make-up water system	7	8
Subtotal, steam winning system	$130	$ 65
Generating plant[2]	0	95
Total	$130	$160
Annual fixed charges, percent of investment[3]	23	17

[1] Steam-winning costs based on Geothermal Resources Investigation, January 1972, by Bureau of Reclamation. Costs escalated to reflect inflationary trends in construction industry.

[2] Includes structures, turbine generators, electrical equipment, miscellaneous. Excludes land, steam-winning system, switchyard.

[3] Includes cost of money, depreciation, interim replacement, insurance and taxes.

Source: The United States Department of the Interior, *Final Environmental Statement for the Geothermal Leasing Program*, Ch. III (Washington, D.C.: U.S. Government Printing Office, 1973), p. 92.

a lower cost of electricity than any other plant (5.3 mills—as shown in Table 9-5, Note #8). But geothermal steam fields are very rare. Gas fired plants also cost considerably less than the others, but gas is also becoming a scarce source. Coal fired plants currently have a mild cost advantage, but it is unlikely at the present that they can keep the advantage and still adhere to increasingly stringent air quality standards.

Table 9-7 gives estimates for costs of producing new electrical power in the western states. It suggests the cost advantages mentioned above, and shows the extent of such differences for that portion of the nation.

Costs at The Geysers may not be representative even for steam fields throughout the country. The steam produced at The Geysers is relatively clean and can be tapped at shallow depths. Purification of the steam and water from sites producing discharges with greater amounts of dissolved substances will be more costly, though in most cases not prohibitively so. Drilling deeper wells will also increase costs. Transportation costs are an important factor limiting use of geothermal energy. Costs involved are substantial to transport energy any distance from geothermal sites. Other fuel sources, such as gas and oil, lend themselves to transportation by trucks or pipelines. Coal is most often shipped by rail, which is comparatively inexpensive.

Table 9-7
Broad Estimates of the Cost of Producing New Electrical Power from Various Energy Sources in the Western States

Power Plant Type	Power Costs (Mils/kwh)
Geothermal (dry steam)	5.25
Gas fired–steam	6.7
Coal fired–steam	8.2
Hydroelectric	9.6
Nuclear–steam	9.6
Geothermal (hot water)	9.7
Oil fired–steam	10.0

Source: The United States Department of the Interior, *Final Environmental Statement for the Geothermal Leasing Program,* Ch. II (Washington, D.C.: U.S. Government Printing Office, 1973), p. 19.

Geothermal plants use two and a half times the amount of steam as do fossil-fueled steam plants to produce the same amount of electricity. This is due to the relatively low temperature and pressure of the geothermal steam. A modern steam electric plant unit will have an average capacity of 800 mw, and a total plant size of several thousand megawatts. The Geysers uses generating units with a maximum size of 55 megawatts. Two such units are combined into one generating station. Thus, the conventional systems have advantages of economy of scale. Geothermal plants somewhat balance this by their incremental nature. Most power plants need to be planned for future requirements. This is because they take so long to build and start into operation that the demand for the stations output has increased by the time of completion. Geothermal fields are usually comprised of a number of small wells which combine to contribute to overall capacity. When the requirements of the geothermal plant are increased, a few new wells can be added to the existing system with only minor additional facilities needed.

In the United States, the principal and almost sole use of geothermal power has been as a source of steam for electric power generation; not so in Iceland. The average costs to the consumer in Reykjavik for the direct heating services are shown in Table 9-8. The 79,000 people, or 96 percent of the city's population, who use the system are metered and billed by volume. Table 9-8 demonstrates that the cost of the district heating system is about 60 percent that of fuel oil heating and 78 percent that of the least expensive method of electric power heating, which uses hydroelectric power [16:1].

Table 9-8
Reykjavik, Iceland Prices for Heat Production Based on Fuel Costs, May 1972

	Per MBTU
District Heating (Geothermal)	$1.04
Fuel Oil	1.73
Night Electricity	1.33

Source: "Hitaveita Reykjavikur," The City of Reykjavik, received from the Icelandic Consulate, New York, March 1973.

Environmental Considerations

The most important aspect of the use of geothermal power sources is that, ". . . it appears to have the potential of being less environmentally damaging than other power generating systems using coal, oil, or nuclear energy sources" [30:III-85]. The effect that the production of any form of energy might have on the environment is dependent upon the biological, geographic, geologic, physical, climatological, and demographic characteristics of the area involved. One must also consider mineral resources, esthetic, scenic, recreational, agricultural, industrial, and other potential land uses, the physical and chemical characteristics of the stream or associated fluids, relationships between the geothermal reservoir and fresh water reservoirs, and the extent and energy content of the geothermal resource.

Within the Known Geothermal Resource Areas in the United States, the Department of the Interior's 1973 edition of Threatened Wildlife indicates that approximately thirty-one fish species, ten reptile and amphibious species, twenty-one bird species, and seventeen mammal species are listed as threatened and could directly or indirectly be impacted by geothermal operations. In addition an estimated sixty-five peripheral species and sixty-nine species with presently undetermined status could be impinged upon [30:II-40].

In general, vapor-dominated systems yield relatively pure steam with minor amounts of other gases and minerals (e.g., boron, carbon dioxide, hydrogen, methane, nitrogen, hydrogen sulfide, mercury, radon, and ammonia). The condensates of the vapor leave water with predominantly dissolved ammonium and bicarbonate ions. Hot water systems may yield mineralized or saline waters with a wide variety of metallic salts and silica. The Salton Sea KGRA fluid is a highly concentrated brine which has proven too corrosive for use with presently available equipment. Geothermal fields in Oregon and Idaho, on the other hand, may yield sufficiently pure water for direct use for irrigation or other fresh water uses.

Probably the most troublesome of the pollution problems caused by geothermal development is that of gaseous emissions. Table 9-9 shows the break-

Table 9-9
Gases Comprising Geothermal Steam at The Geysers, California

	Volume Percentage
Water as steam	98,045
Carbon dioxide	1.242
Hydrogen	0.287
Methane	0.299
Nitrogen	0.069
Hydrogen sulfide	0.033
Ammonia	0.025
Phosphoric acid	0.002

Source: The United States Department of the Interior *Final Environment Statement for the Geothermal Leasing Program* Ch. V (Washington, D.C.: U.S. Government Printing Office, 1973), p. 57.

down of gases found in geothermal steam at The Geysers, while Table 9-10 shows the range of concentrations of the accompanying gases over a three-year period. The main hazard among the gases is from hydrogen sulfide. The natural production processes of periodically bleeding and venting the steam wells release gas into the atmosphere. Condensers and cooling towers also release gas from gas

Table 9-10
Noncondensible Gases Identified in Steam from Wells at The Geysers Power Plant

Gas	Range of Concentrations Measured Percentage by Weight	
	Low	High
Carbon dioxide	0.0884	1.90
Hydrogen sulfide	0.0005	0.160*
Methane	0.0056	0.132
Ammonia	0.0056	0.106
Nitrogen	0.0016	0.0638
Hydrogen	0.0018	0.0190
Ethane	0.0003	0.0019
Total noncondensibles	0.120	2.19

*Overall average from steam in headers at Units 1-4 during the period 1968-70 was 0.0461%.

Source: F.T. Searls, John C. Morrissey, and Philip A. Crane, Jr., *Before the Public Utilities Commission of the State of California: Application No. 52325* (San Francisco, Calif.: Pacific Gas and Electric Company), Exh. K.

ejectors. This affects the air quality, and if noxious gases are at a sufficient percentile in the air they pose a health hazard for plant employees. Hydrogen sulfide can be found at sixteen times the toxic level at The Geysers, and twenty-five times at Larderello, Italy. Hydrogen sulfide and each of its oxidized forms are readily soluble in water and are washed out of the atmosphere by rain. The oxides form an acid solution in water and so create a slightly corrosive raindrop. But, because it is originally mixed with steam, the acidity is weak and usually falls far downwind from the plant site. Hydrogen sulfide itself is the greater problem due to its toxicity and nuisance odor of rotten eggs, which is detectable in quantities as small as .025 ppm (parts per million). During stagnant air and air inversion conditions it can accumulate locally from the geothermal operations to a high nuisance level and perhaps a mildly toxic level.

The next greatest hazard is ammonia (NH_3), which is found at over five times the toxic level at The Geysers and over six times at Larderello. Ammonia is lighter than air so it does not remain at the point of emission. It is also readily soluble in water where it forms ammonium hydroxide. Rain creates this new compound and brings it down to the ground where it combines with soil minerals and plant acids to aid in fertilization.

Methods are being studied and tested to combat some of these sources of air pollution. At The Geysers an inorganic compound is used in the circulating cooling waters to control emissions from the cooling tower by oxidizing dissolved hydrogen sulfide to elemental sulfur, which is retained in the water phase [30:III-34].

The development and production of geothermal resources involve six phases: exploration, test drilling, production testing, field development, power plant and powerline construction, and full-scale operations. Each phase has some slightly differing impacts on the environment. The last phase tends to have somewhat less of a polluting effect due to the stability of the situation.

During the first few phases, and to a lesser degree afterwards, particulate matter greater than wind blown dust or dust created by the present vehicular movement over untreated or unsurfaced roads will be added to by geothermal related activities and use of such roads, and from the earth-moving activities of construction of drilling pads and other construction projects. Dust degrades the air quality and its settling on surrounding plant life may influence their growth and survival.

Blowouts, where steam or hot water escapes uncontrolled, may present control problems because of the difficulty of handling the escaping hot fluid. This is similar to problems experienced at oil and gas wells, but with geothermal wells there is no fire hazard involved. At Wairakei, New Zealand, of one hundred early wells only two had to be abandoned due to blowouts, and a blowout occurred at a third well only after some years of service. There were never any deaths involved. Safety measures which have since been developed have greatly reduced the rate of incidents [30:III-8].

While these and other individual sources of air pollution may be satisfactorily controlled, the total effect upon the environment may be incompatible with the surrounding land uses, thereby precluding full resource development and use in such areas.

The environmental problem caused by geothermal resource development which has the greatest potentially adverse effects concerns the depletion of the geothermal water reservoirs. Eighty percent of the steam condensate is consumed by evaporation for condenser cooling during the energy production process. The remaining 20 percent must be discarded, unlike the practice with conventional thermal units, which return condensed steam to the boiler to be recycled. Condensate use at The Geysers averages about 45 acre-feet per year per megawatt of plant capacity. The current capacity of 298 mw represents 13,000 acre-feet per year of water consumption. Production of 600 mw would mean 27,000 acre-feet per year. Electric power generation in hot water fields presumably would require comparable amounts of water for condenser cooling [30:III-34]. Water withdrawal can decrease streamflow and adversely affect a drainage basin for many miles downstream. Subsidence from water drainage would change terrain features. There are theoretical arguments that subsidence in geothermal areas would increase seismic activity. Disposal of the cooled water would be likely to pollute nearby fresh water sources, as it would frequently contain a high concentration of minerals.

The most widely accepted solution is to reinject the used water back into deep wells. Reinjection could alleviate the nonreversible process of subsidence and prevent pollution of surface waters. This answer has problems of its own, though. An area such as the Salton Sea has surface and subsurface irrigation waste water as the present major source of water. Significant quantities of highly saline geothermal waters, if added, could contribute to the degradation of existing marginal water qualities, and have a potentially adverse impact upon fish and wildlife in the area.

Figure 9-2 shows three types of power plant systems. At The Geysers the condensate is all reinjected into the ground. The middle example shows how with hot water fields the brine is separated and either discarded or used for by-product production. The third system is a closed system where a recycled coolant, in this case iso butane, absorbs the heat transferred by use of a heat exchanger. The coolant then performs the role of the steam portion of the geothermal effluence. The geothermal water is reinjected into the ground with the only change in character being a loss of heat. This eliminates most environmental concerns, as the reinjected water is in relatively the same condition as when it was drawn from the ground.

Physical land disturbance and modification occurs with the activities of construction of roads, drainage ponds, drill sites, the drilling of wells, pipeline construction, and power plant and transmission line construction. These activities could disturb the wildlife habitat and result in a loss of wildlife for the area.

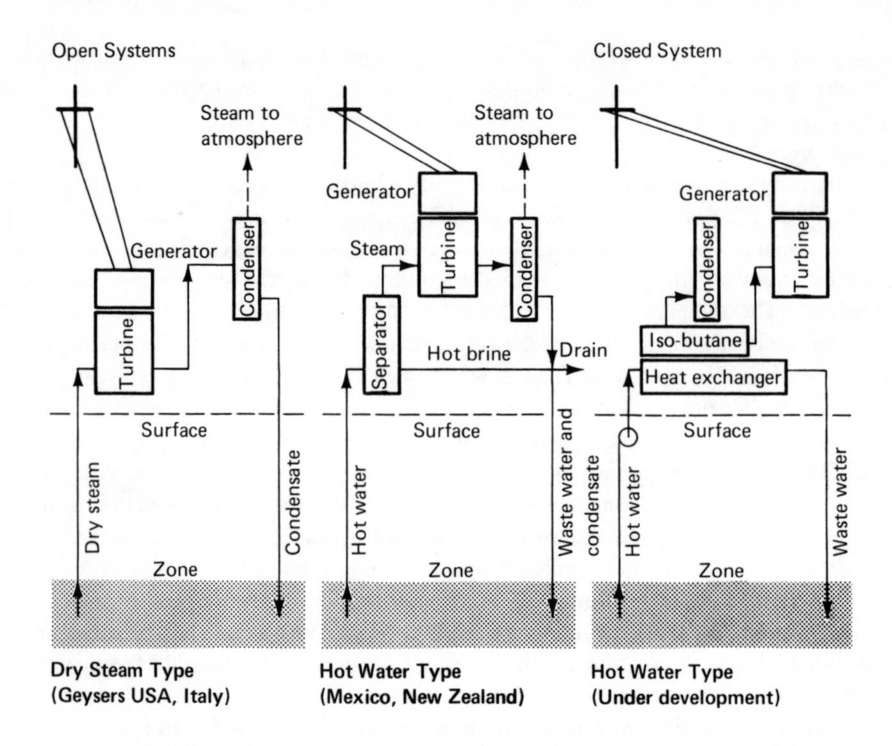

Figure 9-2. Types of Geothermal Power Plants. Source: The United States Department of the Interior, *Final Environmental Statement for the Geothermal Leasing Program*, Ch. V (Washington, D.C.: U.S. Government Printing Office, 1973), p. 156.

However, clearings, revegetated areas, roads, and trails could improve the wildlife habitat. There are also many adaptable animals. Adjacent areas should not be affected directly, but restriction of access roads could occur from the construction of power lines and other surface activities. This might work both ways as it would reduce hunting and angling also.

Painting the metal siding on the main plant structures a bayberry (olive-green) to blend with the surroundings, and painting the main stream line piping the same color aid in softening the impact of their existence. All yard equipment, piping, etc., are painted in compatible natural hues to match the surrounding terrain and vegetation.

Another form of pollution which must be considered is noise pollution. The movement of trucks and other vehicles, the drilling of wells, venting of steam and other ancillary sound sources raise the background noise levels. Excessive noise levels can pose a health hazard to employees, are objectionable to residents and visitors to the area and may disturb wildlife distribution and breeding patterns. Table 9-11 displays some comparisons of noise levels of various activities at The Geysers geothermal site with the levels at various other sources. It is plain that points near geothermal operations, the noise level can reach a dangerous stage.

The natural hot water used in Iceland has very unusual properties, which differ from the effects from conventional heating fuels. It is often expected that more dissolved solids will be found in hot spring water than in cold water. The water used by the District Heating Service in Reykjavik, though, contains very small amounts of dissolved solids, and none which are detrimental to health. The amount of dissolved solids varies between 200 and 300 miligrams per litre, depending upon which bore hole is analyzed. The comparable figure for cold tap water in Washington, D.C., is 140 miligrams per litre. It is noted that of the dissolved solids in Reykjavik, 30 to 35 percent is silicium oxide which is considered to be one of the least harmful solids to be found in drinking water. Another advantage of Reykjavik geothermal water is its low hardness or calcium content, which is several times less than that of Washington, D.C., and is accordingly more effective for washing.

The Reykjavik thermal water also contains exactly the recommended amount of flouride for the prevention of tooth decay. Dissolved gases, of which nitrogen is the most prevalent, are removed by gas separators as soon as the water is brought out of the bore holes. Oxygen is nonexistent in the water, which is an

Table 9-11
Comparison of Noise Levels Between The Geysers Area and Other Sources

Geysers Area Source	Level	Distance
Drilling Operation (air)	126 dB (A)	25 feet
Drilling Operation (air)	55 dB (A)	1,500 feet
Muffled testing well	100 dB (A)	25 feet
Muffled testing well	65 dB (A)	1,500 feet
Steam line vent	100 dB (A)	50 feet
Steam line vent	90 dB (A)	250 feet
Comparative Levels		
Jet aircraft takeoff	125 dB (A)	200 feet
Threshold of pain	120 dB (A)	
Unmuffled diesel truck	100 dB (A)	50 feet
Street corner in a large city (Average)	75 db (A)	
Residential area at night (Average)	40 dB (A)	

Source: The United States Department of the Interior, *Final Environmental Statement for the Geothermal Leasing Program*, Ch. V (Washington, D.C.: U.S. Government Printing Office, 1973), p. 56.

important factor in avoiding corrosion of the pipes. Most important is the fact that because of the extreme heat, geothermal water is completely free of bacteria [16:1].

A comparison of the effects of using geothermal resources in Reykjavik for space heating as opposed to the use of conventional fuels demonstrates favorable potentials for further development of the use of natural hot water sources. The total heating load for the entire District Heating Service in 1970 was the equivalent of burning 180,000 metric tons of fuel oil. This is the most readily noticeable difference created by the use of geothermal water. There is no dark cloud of pollution hanging over the city as there had been in the past. There is no smoke rising from chimneys of the homes and other buildings. Indeed, many of the newer buildings have no chimneys at all. It is estimated that the annual consumption of 180,000 tons of fuel oil, if burned in domestic burners, would have produced more than 2,000 tons of sulfur dioxide, 850 tons of nitrogen oxide and 210 tons of ash and soot to spread throughout the air of the city [16:1].

Depletion of the earth's natural heat and water in a local area is renewable in a short time relative to that of oil or gas. Although the resource will not renew itself within the lifespan of the individual project, it will replenish itself within a few hundred years as opposed to thousands or hundreds of thousands of years needed to renew an oil field.

There are about ten wells needed for each geothermal generating station. Besides the various individual disturbances created by the steps leading up to the final development of a site, the entire area becomes transformed in the end from its previous condition into an industrial complex. Cuts, fills, clearings, buildings, power lines, etc. represent permanent changes in the landscape and affect the esthetic qualities of the area. Many recreational activities will be curtailed, although the actual acreage involved may not be large. Overnight camping and such activities lose much of their quality if carried out within sight or sound of an industrial atmosphere. The effect is even more pronounced in open grassland and desert areas.

Even with controls a higher than natural noise level persists. Disturbances are caused by the emission of steam and other gases into the atmosphere, operational activities, additional vehicular traffic, and other such activities. Transmission lines will be a hazard to some wildlife and may result in minor levels of electrocution of eagles, hawks, and other birds which fly into the lines. Potential problems exist with land subsidence and increased seismic activity. The subsidence is non-reversible. There will be localized and regionally adverse impacts on habitat. Some loss of wildlife in the immediate vicinity may be expected. There are potential dangers to fish and other aquatic life from toxic fluids being added to the water. The long-range effects are fewer. Wildlife is likely to return to geothermal sites after operations are terminated.

Many of the environmental impacts of geothermal energy production would also apply to production of power from conventional fuels. In many cases the adverse impacts from the latter would be greater than they are with geothermal power.

Conclusions

In the past year the realization that our current sources of energy are not inexhaustible has burst upon the American scene with devastating impact, confounding the expectations of expert and ordinary citizen alike. A rationing program vigorous enough to bring about the necessary balancing of supply and demand could be instituted only with what most would regard as unacceptable damage to our institutions and customs.

The development of new and additional sources of energy would seem an obvious answer, but is beset with many difficulties. Many new sources such as offshore oil, shale oil, and strip mined coal involve major hazards to the environment.

Geothermal power development emerges as a potential partial solution helping meet energy needs without unacceptable damage to the environment. Even though the percentage of energy contribution from geothermal sources can only be small relative to the total energy generation from nuclear sources, oil, coal, gas, and hydroelectric power, the importance of such geothermal power development is increased manifold because of the alternative uses available. Besides generating electricity, geothermal energy can be utilized in direct heating, desalination, horticulture, and other direct immediate uses. These various extra employments for such resources can increase their potential tenfold [11:41]. This might very well make the geothermal contribution to our energy needs a significant one.

The only major criticism of geothermal power development is the limitation of potential locations for supply and use. It must be recognized, though, that geothermal power is for the most part a previously untapped natural resource. This means that geothermal resources represent a new, previously unaccounted for, addition to the total power available from all sources. This supplemental nature can be of considerable importance on the local or regional basis, and can easily affect the overall energy supply significantly.

Geothermal power may be severely limited to particular geographic locations. However, wherever geothermal resources are used some of the necessity of employing alternative energy sources is eliminated. The fuel which is saved in this manner can be transported to areas where shortages might exist. The local costs of geothermal energy have been shown to remain competitive with those of the conventional fuels currently in use. The fuel which is sent elsewhere will have only a nominal increase in cost added due to the transportation involved. Usually the original source location of the fuel that is thus replaced is a third location which may or may not be more distant from the new destination than it was from the old destination.

Instead of having the replaced amount of conventional fuel transported for current use, it can be stored and saved for the future. Or it need not be mined or drawn upon until it is needed. Thus geothermal energy may expand the useful life of any resource it replaces by reducing the annual demand for it. Whereas a specific oil field might have previously had an estimated useful life of thirty years before it would go dry, replacement of part of the load by geothermal energy could extend the life of the oil field to forty or fifty years.

It appears as well that the production of geothermal energy is likely to be less detrimental to the environment than any conventional power source except natural gas. The additional amount of energy gained from geothermal development will produce a disproportionately small amount of additional pollution. This is indeed a step in the right direction.

Geothermal energy may well be the most palatable of the hitherto overlooked natural resources which put relatively little strain on our environment. But acceptance may well be followed by extended research into other such sources as solar energy and wind energy. These may be the future answers to man's unceasing growth of power utilization.

Glossary

BTU (British Thermal Unit). Unit of heat required to raise the temperature of one pound of water one degree Fahrenheit at or near its maximum density (39.1 degrees Fahrenheit). .252 Calories.

Crustal Plate. The cool outer layer of the earth.

Fumarole. A hole in a volcanic region from which issue gases and vapors at high temperatures.

KGRA. Known Geothermal Resource Area.

Magma. Molten rock material originating within the earth.

Megawatts (mw). One million watts.

Tectonism. The structuring of rock layers beneath the earth due to the deformation of the earth's crust.

Volcanism. Volcanic action.

Bibliography

1. Anderson, Winslow. "The California Geysers." *Mineral Information Service* (San Francisco, Calif.: California Division of Mines and Geology) 22, 8 (Aug. 1969): 129-132.
2. "Chemistry of Broadlands' Geothermal Area, New Zealand." *American Journal of Science* 272, 1 (Jan. 1971): 48-69.

3. "Emerging Geo-thermal Power Source." *American Geophysics Transactions*, Apr. 12-16, 1971.
4. "Energy for the Future." *C.B.S. Morning News with John Hart*, Part 3 (Transcript), Feb. 7, 1973.
5. "Energy Resources of the Earth." *Scientific American* 22, 3 (Sept. 1971): 60-71.
6. "Energy Story, Part IV–Geothermal Power." San Francisco, Calif.: Pacific Gas and Electric Company, Aug. 1971, pp. 4-5.
7. "Energy Story." San Francisco, Calif.: Pacific Gas and Electric Company, Sept. 1972.
8. Finney, J.P.; Miller, F.J.; and Mills, D.B. *Geothermal Power Project of Pacific Gas and Electric Company at The Geysers, California.* Report to the Summer meeting of the IEEE Power Engineering Society. San Francisco, Calif.: Pacific Gas and Electric Company, July 1972.
9. Garrison, Lowell E. "Geothermal Steam in the Geysers–Clear Lake Region, California." *Geological Society of America Bulletin*, May 1972, pp. 1449-1468.
10. "Geo-thermal–Earth's Primordial Energy." *Technology Review* 74, 1 (Oct.-Nov. 1971): 42-49.
11. "Geo-thermal Energy." *Science Teacher* 39, 3 (March 1972): 40-44.
12. "Geo-thermal Resources Gather a Head of Steam." *Engineering News Record* 186, 18 (May 6, 1971): 30-34.
13. "Geo-thermal Stations Have Pollution Problems Too." *Power* 115, 5 (May 1971): 96-97.
14. "Heat Transfer, The Geysers Geothermal Plant." *Chemical Engineering Progress* 68, 7 (July 1972): 83-86.
15. Hirst, Eric and Moyers, John C. "Efficiency of Energy Use in the United States." *Science*, March 30, 1973, pp. 1299-1304.
16. *Hitaveita Reykjavikur*, The City of Reykjavik.
17. "Japan Moving Ahead in Geothermal Development." *Mineral Information Service*. San Francisco, Calif.: California Division of Mines and Geology, 22, 8 (Aug. 1969): 136-137.
18. Kaufman, Alvin. "An Economic Appraisal of Geothermal Energy." *Public Utilities Fortnightly* 88, 7 (Sept. 30, 1971): 19-23.
19. Koenig, James B. "The Geysers." *Mineral Information Service.* San Francisco, Calif.: California Division of Mines and Geology, 22, 8 (Aug. 1969): 123-128.
20. Large, Arlen J. "Free Electricity." *Wall Street Journal*, Dec. 10, 1970, p. 176, Col. 1.
21. Lessing, Lawrence. "Power From the Earth's Own Heat." *Fortune* 79 (June 1969): 138-141.
22. Mangushev, K. and Prikhodko, N. "Geothermal Energy in the U.S.S.R." *Current Digest of Soviet Press*, (Reprinted from *Pravda*, Mar. 30, 1972), Apr. 26, 1972, Vol. 24, p. 18.

23. McMillan, D.A., Jr. "Economics of The Geysers Geothermal Field, California." *United Nations Symposium on the Development and Utilization of Geothermal Resources*, 1970.
24. 1970 National Power Survey, Federal Power Commission, Washington, D.C.: Government Printing Office, 1971.
25. Searls, F.T.; Morissey, John C.; and Crane, Philip A., Jr. *Before the Public Utilities Commission of the State of California: Application No. 52325.* San Francisco, Calif.: Pacific Gas and Electric Company, Nov. 24, 1970.
26. "Subterranean Hot Water Oceans Seen As Power Source in West." *New York Times*, Aug. 6, 1970, p. 25, Col. 4.
27. "The Geysers." *PG&E Life*, Oct. 1972, pp. 8-9.
28. "The Potential for Geothermal Power." *Business Week*, March 17, 1973, pp. 74-75.
29. "The Roar from an Emerging Resource." *Reclamation Era* 57, 3 (Aug. 1971): 1-7.
30. The United States Department of the Interior. *Final Environmental Statement for the Geothermal Leasing Program*. Washington, D.C.: Government Printing Office, 1973.
31. "Too Good to Be True?" *Forbes*, Apr. 15, 1972, p. 74.
32. "Turning Turbines With Geo-thermal Steam." *Power Engineering* 76, 3 (March 1972): 36-41.
33. *Unit 12 The Geysers Power Plant Environmental Report January 1973.* San Francisco, Calif.: Pacific Gas and Electric Company, January, 1973.
34. *United States Congressional Hearings and Legislation on the Use of Geo-thermal Resources for Power.* United States Senate, 91st Cong., 2nd sess., Sept. 1970.
35. "Utilities Put Faith in Old Faithful." *Business Week*, Oct. 3, 1970, p. 50.
36. Wilford, John Noble. "Mexico–Underground Steam to Run Pollution Free Plant." *New York Times*, Aug. 22, 1970, p. 25, Col. 1.
37. Wilford, John Noble. "New Plan Is Outlined for Tapping Geo-thermal Energy." *New York Times*, June 21, 1972, p. 18, Col. 4.
38. "Geothermal Power Goes Commercial." *Seventy-Six*, Nov.-Dec., 1971.
39. Hearings before the Subcommittee on Minerals, Materials and Fuels of the Committee on Interior and Insular Affairs, "Statement of John S. Kelly, Director, Division of Peaceful Nuclear Explosives, Atomic Energy Commission," *A Bill to Authorize the Secretary of the Interior to Make Disposition of Geothermal Steam and Associated Geothermal Resources, and for Other Purposes*, United States Senate, 91st Cong., 2nd sess., July 17 and 28, 1970.
40. Gottschalk, Earl C., Jr. " 'Earth Power' Steam Below Ground Seen Giving Big Boost to U.S. Energy Supply." *The Wall Street Journal* 181, 55 (Mar. 20, 1973): p. 1, Col. 1.
41. Holles, Everett R. "Hot Ocean Defies Hunt for Energy." *New York Times* Feb. 17, 1974, p. 53B.

10 Strip Mining of Coal

Irwin Zurkowski

The vast resources of western coal are close enough to the surface to be mined by stripping. What has been the effect of strip mining in the past and what would be the impact in the west?

There has been a great deal of controversy about strip mining, particularly in recent years. Strip mining for coal and other minerals without reclamation may be one of the most grossly destructive practices of modern technology. Spurred by the shortage of coal for generating electricity, and an increase in the size and efficiency of machinery, strip mining is expanding at a fearsome rate.

This chapter will examine the impact that strip mining has had and will have in the United States. It will focus on the Appalachian and western regions. A vast amount of destruction has already occurred in Appalachia due to strip mining. Yet, up to this time reclamation has been minimal. The western region of the United States will become an increasingly important source of coal in the future. It has an abundance of low sulfur coal and the region is in store for an extensive amount of strip mining.

A major question must be answered. What are the costs of acceptable strip mining practice? This study has attempted to answer this question by assembling as much data as possible on the costs of reclamation.

However, before an examination of strip mining and reclamation is undertaken, it would be worthwhile to take an historical look at the United States energy picture. This historical examination will suggest the intense competition that the coal industry has faced from other energy sources.

Today, as in the past, the United States draws on several primary sources to satisfy its heavy energy demands. Table 10-1 shows that following World War II, coal accounted for about one-half the country's total energy consumption, oil for about one-third, and natural gas for slightly less than one-seventh. Going back even earlier to the period after World War I, coal's share of the total stood at close to three-quarters, and oil and natural gas together at not much more than 15 percent, with hydropower and fuel wood making up the remainder.

In 1970, however, petroleum and natural gas supplied 7 percent of the nation's energy needs, and coal about 19 percent. Energy demand in the United States has almost doubled in the last twenty years and is expected to double again in the next fifteen years. By and large, oil and gas rather than coal have satisfied the country's increased energy requirements in this era.

Table 10-1
United States Total Gross Consumption of Energy Resources by Major Sources (1947-1970—Selected Years)

	In Quadrillions of British Thermal Units						
Year	Anthracite	Bituminous & Lignite	Natural Gas Dry[1]	Petroleum[2]	Hydro-Power	Nuclear Power	Total Gross Energy
1947	1.2	14.6	4.5	11.4	1.3	–	33.0
1950	1.0	11.9	6.2	13.5	1.4	–	34.0
1955	.6	10.9	9.2	17.5	1.4	–	39.7
1960	.4	9.7	12.7	20.1	1.7	a	44.6
1965	.3	11.6	16.1	23.2	2.1	b	53.3
1970	.2	12.7	22.0	29.6	2.7	.2	67.4

[1] Exclude natural gas liquids.

[2] Petroleum products including still gas and liquefied natural gas and refinery gas.

[a] 6 trillion BTU

[b] 38 trillion BTU

Source: Table 10A-1.

For most of the components of Table 10-1, U.S. production of energy is the same as U.S. consumption of energy. Thus, for most of the period covered nearly all of the U.S. production of anthracite coal, dry natural gas, hydropower, and nuclear power was consumed at home. There were sizable, though varying, exports of U.S. produced bituminous coal. U.S. consumption of petroleum products had exceeded domestic production by a constantly increasing percentage since World War II. Thus, U.S. consumption of petroleum products exceeded domestic production by only 5.5 percent in 1947 and 18.4 percent in 1950. But by 1960 U.S. consumption exceeded production by over one-third and by 1970 by nearly one-half (Tables 10-1 and 10-2).

Liquid fuel has many advantages over solid fuel. It weighs about 30 percent less per unit volume than coal and occupies only 50 percent of the space required for an equivalent amount of coal. Liquid fuels are usually readily available, can be stored conveniently, produce high heats in less furnace space, begin to burn instantly and clearly, require little ash removal, and need relatively little attention.

The coal industry has been confronted with the most intense competition in our normally competitive society. After World War II it lost its major markets because of technological improvements that demanded different types of energy. The industry lost the railroad market to the diesel engine and the home-heating market to a large extent to natural gas and oil [16:115].

Diesel engine development began after World War I. The diesel engine displaced steam in many passenger vessels and freighters and also began to compete with the steam engine in powering railway locomotives. But, after World War II, United States' railroads entered into a program of dieselization, which largely relegated the steam locomotive to the past. At the same time, dieselization of motor trucks, tractors, and construction equipment occurred on a large scale.

The automatic oil burner for heating homes was introduced in the United States in 1917 and came into widespread use in the late 1920s. Its greatest gains, however, came after World War II. In addition there were millions of homes using space heaters which burned fuel oil distillate. Space heating was a big factor in the petroleum industry's phenomenal growth. Table 10-2 shows rate of growth in U.S. production of crude petroleum.

The importation of residual fuel oil along the eastern seaboard of the United States has also had a very serious impact upon the coal industry. Residual oil competes with coal in its primary marketing areas, which are electric generation and large industrial plants [16:87].

Residual oil is, as the name implies, what is left over after gasoline and other products have been refined from crude oil. Each year less and less residual oil is produced in this country, as shown in Table 10-3, as the United States domestic refineries become more efficient and yield a higher percentage of more valuable products.

Table 10-2
U.S. Production of Crude Petroleum (1947-1970–Selected Years)

Year	Billion Barrels
1947	1.9
1950	2.0
1955	2.5
1960	2.6
1965	2.8
1970	3.6

Source: Table 10A-2.

Table 10-3
U.S. Production of Residual Fuel Oil (1947-1970–Selected Years)

Year	Million Barrels
1947	448
1950	425
1955	420
1960	332
1965	269
1970	256

Source: Table 10A-3.

The distillate demand increase reflects principally the continuing efforts of consumers to get under the sulfur content standards imposed by various authorities, by the substitution of distillate for residual or its use as a blending agent to permit the higher sulfur residual oil to meet low sulfur specifications. The production of distillate fuel oil for the years 1945 to 1972 is shown in Table 10-4. This product is light in color, flows easily, and has good ignition and combustion characteristics, and generally has a sulfur content of less than three-tenths of one percent, which makes it suited ideally to residential and light commercial space heating applications with very minimal contributions to air pollution.

Heavy fuel oils are typically far less costly than distillates. They are dark in color and, because of their high viscosity, require heated tanks and piping to make them flow. Heavy fuel oils have a higher sulfur content, usually in the magnitude of 2 percent by weight. Nevertheless, heavy oils have provided an exceptional bargain to large commercial and industrial customers who have the more expensive equipment and storage facilities necessary to handle these fuels. The advantage of lower cost is enhanced by the fact that there are actually more heat units available per unit of volume from heavy oil than distillate.

Table 10-4
U.S. Production of Distillate Fuel Oil (1947-1970–Selected Years)

Year	Million Barrels
1947	312
1950	399
1955	603
1960	667
1965	765
1970	897

Source: Table 10A-4.

During the coal industry's lean period, every effort was made to keep the industry strong and competitive. Pinched between rapidly rising labor costs and low price competition from other fuels, billions of dollars were put into the development of new facilities, modernization, mechanization to improve the quality of the product as well as reduce its production costs. As a result, coal sold for less in 1965 than it had in 1960, 1955, and 1950 as shown in Table 10-5.

An important element in the revival of coal in the late 1960s is the high growth rate of output of electricity. However, in the late 1940s and the 1950s the growing use of coal by electric utilities could not offset the great tonnage losses experienced by coal in powering railroad locomotives and in space heating.

There are three main reasons why electric utilities have grown since World War II. First, there has been an increase in the population. Second, the increase in the consumption of manufactured products has caused new plants to open and old ones to expand. Third, there has been an increase in the use of electrical appliances, particularly for air conditioning [6:1].

Large electric generating plants powered by steam have been built to meet the demand of the electric utilities. These plants are designed to use the most

Table 10-5
U.S. Production of Bituminous Coal (1947-1970–Selected Years)

Year	Million Short Tons	Price (Dollars per ton)
1947	631	$4.16
1950	516	4.84
1955	465	4.50
1960	416	4.69
1965	512	4.44
1970	603	6.26

Source: Table 10A-5.

economical fuel available. The electric generating plants use coal, and are not turned away from it because of its bulk and chemical properties. The only requirement is that the cost per unit of heat is satisfactorily low. Many utilities are located in and near the coal field areas and provide ready markets for large amounts of strip mined coal. This market has been a major reason for the increase in strip mine coal production and the advance of strip mine technology [2:1].

According to the National Power Survey, coal is expected to continue as a major fuel for electric utilities throughout the foreseeable future. Each year sees increasing quantities of coal used for the thermal generation of power. Coal is the most abundant natural fossil fuel and currently provides the primary energy for about 56 percent of fossil fueled electric generation. This is slightly less than one-half of the total electric generation. In the ten years from 1961 to 1970, consumption of coal for electric generation increased steadily [34:I-4-2 and Figure 4-3].

The growing contribution of strip mined coal to the rapidly expanding United States energy needs is evidenced by the fact that strip mined production increased from 9.4 percent of total coal production in 1940, to 50 percent in 1972, as shown in Table 10-6. Today, as in 1940, strip mining is carried on in nearly every one of the major coal producing states, but it is virtually the only method of mining employed in some states. In fact, large reserves of western coals can only be extracted by strip mining.

The main reason strip mining has increased in importance is because of its economic and competitive advantages. Coal from strip mining is cheaper and easier to extract, and operates under fewer constraints than underground mining. Also, 90 percent of a coal seam can be recovered by strip mining compared to about 50 percent by underground mining.

Table 10-6
Strip Mining as a Percentage of Total Coal Production (1940-1970) (million-tons rounded)

Year	Total Production	Strip Mining Percentage of Total Production	Total Strip Mine Production
1940	460.80	9.4	43.1
1950	516.30	23.9	123.5
1955	464.60	24.8	115.1
1960	415.50	29.5	122.6
1965	512.08	32.3	165.2
1970	602.90	40.5	244.2

Source: Environmental Protection Agency. *Legal Problems of Coal Mine Reclamation* (Washington, D.C.: Government Printing Office, 1972), pp. 25, 26, Tables 1, 2.

Strip mined coal became increasingly important to the United States energy picture in the late 1960s and particularly so in 1970. Deep mined coal production declined by 1970, and natural gas appeared to be increasingly short in supply, while the domestic oil industry no longer readily met United States utility and industrial demands. Finally, construction of new nuclear plants fell behind schedule, in part because of limited domestic supplies of uranium and inadequate provision for disposal of radioactive wastes generated in operation. Opposition of environmentalists also slowed the approval and construction of nuclear plants. All of these circumstances left the door wide open for the coal industry. In 1970, strip coal mining proved it had not only the reserves but also the capacity to expand and meet, on short notice, a sharp increase in the demand for energy. The coal industry faces the future not only with reserves vastly greater than those of any other fuel, but with modern, high capacity production methods.

The coal industry has embarked on a twofold program of enlarging its present electric utility market, while seeking through research to open vast new markets in the conversion of coal into gas and gasoline. The coal industry's new marketing concept envisions the use of coal as a supplementary fuel for natural gas and petroleum as those fuels become less plentiful. It is estimated that if coal can capture only 10 percent of the gasoline market by 1980, a new coal market of 175 million tons a year will result. Coal gasification is an almost equally attractive prospect. If coal can pick up 10 percent of United States gas demand projected for 1980, coal markets will be increased by almost 140 million tons a year. From the year 1975 to the year 1990 there will be a continual rise in total coal production [3:234]. Obviously, with coal production on the rise, strip mining will also increase.

The Appalachian region contains some of the largest reserves of high quality coals in the United States. There are thirteen states in the Appalachian region. Nine of the states are coal producing states: Alabama, Kentucky, Maryland, Ohio, Pennsylvania, Tennessee, Virginia, West Virginia, and Georgia. The four non-producing states are New York, Mississippi, North Carolina, and South Carolina. The estimated total reserves of coal in the region is 271,000 million tons, approximately one-sixth of the total coal of all ranks in the United States. Ninety-five percent (258,000 million tons) of this reserve is bituminous coal and the remainder is Pennsylvania anthracite. The total bituminous reserves of the Appalachian region are distributed among states as follows: West Virginia, 40 percent (104,000 million tons); Pennsylvania, 23 percent (60,000 million tons); Ohio, 16 percent (41,000 million tons); Kentucky, 11 percent (28,000 million tons); and other states 10 percent (26,000 million tons). Table 10-7 shows the productivity of the different mining methods employed by the states in the Appalachian region.

Strip mining, in which the overburden is stripped away to expose and then remove the underlying coal deposit, can be divided into two general types. The two types are area and contour mining.

Table 10-7
Production of Coal in the Appalachian Region and Method of Mining (1969) (in millions of short tons rounded)

Appalachian Region	Underground	Strip	Auger	Total	Percentage Surface Mined
Alabama	9.3	8.1	.04	17.5	46.8
Kentucky	44.5	9.9	7.2	61.6	27.7
Maryland	.3	1.0	.1	1.4	76.4
Ohio	18.6	31.0	1.6	51.2	63.7
Pennsylvania	58.1	26.5	.6	84.1	31.9
Tennessee	4.5	3.4	.2	8.1	44.7
Virginia	30.4	3.6	1.6	35.6	14.6
West Virginia	121.6	14.5	5.0	141.0	13.7
Total	287.4	98.0	16.3	401.6	28.5

Source: U.S. Congress. Committee on Interior and Insular Affairs. *Regulation of Strip Mining.* Hearings, 92nd Congress, 1971, p. 245, Table 3.

Area strip mining is conducted on land that has a rather flat terrain. It is employed chiefly in the flatter sections of the Appalachian states of Alabama, Ohio, Pennsylvania, and West Virginia. An area stripping operation starts with the excavation of a long straight trench exposing a portion of a coal deposit. The coal deposit is then removed. A second trench is then made parallel to the first, and the overburden from the second trench is placed into the first. This is a continuous process and the final cut leaves an open trench as deep as the thickness of the overburden, plus the ore recovered. More often than not the final cut is a mile or more away from the first cut. If the land is not graded the land resembles a huge washboard. Without reclamation, area strip mining can preclude future productive land use, pollute water with siltation and acid mine drainage, and destroy the esthetic values in a large area. Other potential environmental impacts of area mining are changed surface water courses, ground water pollution, and temporary destruction of ground cover. With adequate reclamation the mined area need not be precluded from future productive land use [5:34].

Contour strip mining occurs most often in rolling or mountainous regions such as Appalachia. It begins with the removing of overburden by explosives and bulldozers to get at the coal seam. This type of mining creates a shelf on the hillside. The blasted earth is pushed over the shelf to create an unstabilized spoil bank. Unless such spoil material is stabilized it can cause severe erosion and landslides, and the mountain may be damaged before a ton of coal is removed. The next step is for shovels and bulldozers to come in and remove the coal. Acid mine drainage is another major problem caused by contour strip mining, since the material next to the coal deposit often contains pyrites and other acid forming substances. When such harmful materials are exposed to weathering, they are converted to soluble acids and minerals and are carried away to streams and ground water. About 12 percent of the acid mine drainage in Appalachian streams derives from abandoned coal surface mines and access roads, while the rest comes from the underground mines. Unstable highwalls are a hazard to life and property. Highwalls that crumble and erode from weathering ruin drainage patterns and significantly add to water pollution. Material falling off the highwall can retard surface water flow and thereby prolong the contact between water and acid producing materials [5:34].

Reclamation is the reconditioning or restoration of an area of land affected by strip mining operations, to a condition where its surface value is at least as great as it was prior to the beginning of the strip mining. Lands thus reclaimed may be used for the same purposes for which it was used prior to the beginning of the strip mining. In any reclamation process a prime consideration should be the maintenance to the maximum extent possible of land use values. As of 1972, 4 million acres of land had been disturbed by strip mining in the United States, yet, over half of the land was unreclaimed. Twenty thousand miles of highwalls remain, and the water quality of thousands of miles of streams and thousands of acres of lakes have been severely degraded. This disruption of wildlife habitat and impairment of esthetic and recreational value increases as mining continues to be inadequately controlled. Table 10-8 shows the acreage disturbed and the acreage reclaimed in the coal producing states of the Appalachian region.

Table 10-8
Acreage Disturbed and Reclaimed Due to Strip Mining in the Appalachian Region (1969) (in millions of short tons rounded)

State	Production	Acreage Disturbed	Acreage Reclaimed
Alabama	8.2	Not Reported	Not Reported
Kentucky	17.8	12.2	9.6
Maryland	1.0	.3	.5
Ohio	32.6	10.6	7.9
Pennsylvania	27.2	12.3	9.8
Tennessee	3.6	Not Reported	Not Reported
Virginia	5.2	2.3	2.3
West Virginia	19.4	15.7	17.1

Source: U.S. Congress. Committee On Interior and Insular Affairs, *Regulation of Strip Mining* Hearings, 92nd Congress, 1971, p. 245, Table 4.

There are several mining techniques which can provide for contour mining, concurrent reclamation with minimal disturbance and environmental impacts on adjacent land. The modified block cut is one such technique. It incorporated reclamation as an essential part of the mining operation, but it is not applicable to all sites. Lands are reclaimed during mining by backfilling the previously worked area with newly removed overburden. Except for the first cut, spoils are not deposited on the downslopes, and the land is almost instantly restored to its original contour. For this reason landslides, esthetic blight, water pollution, and other environmental effects are lessened. However, the destruction during the active mining operation cannot be fully avoided. The modified block cut is not widely used at the present time. It offers one promising approach to reduce environmental effects in many, although certainly not in all areas [7:21].

Another technique is contour mining with the shaping of the spoil bank. The potential for landslides and erosion can be lowered by spreading and stabilizing the spoil over a large area. This is done by removing all vegetation from the hillside below the cut, and the overburden is then spread over the downslope in compacted layers. A portion of the spoil material can be amassed along the edge of the bench. After the coal is removed, the spoil material is redistributed on the bench.

Slope reduction and parallel slope are two methods of shaping the spoil material on the downslope. In slope reduction, the spoil is graded to form a reduced slope angle on the spoil bank. In the parallel slope method, the spoil is spread over the downslope in layers parallel to the original slope of the hillside. However, creating a satisfactory vegetative cover on the resulting large spoil disposal area is usually difficult. Even though both methods lessen the potential of landslides, they may cause massive sheet and gully erosion and slumping of the slopes, particularly in a high rainfall area such as Appalachia.

Adequate vegetation should follow shortly after grading. A vegetative cover should be set up to hold the exposed surface soil in place. Planting or grading alone cannot satisfactorily solve the spoil bank problem, but a combination of the two can reduce much of the environmental damage [7:16].

Another technique is contour mining with the backfilling of the bench. The material in the spoils bank is moved back onto the bench and regraded to a designated shape, in the bench backfilling method. A majority of the spoil material is restored to the bench and regraded to nearly the original contour of the hillside. Since the volume of the spoil material is typically larger than the volume of the cut, a portion of the spoils is normally stabilized on the downslope [7:18].

A final technique is terrace backfilling. Terrace backfilling is used when the slope is too steep for other techniques or when the soil condition leads to excessive erosion. A portion of the spoil bank is used to cover the acid producing spoil in the face of the highwall and a portion is used to lessen the slope below the bench. The terrace backfilling method normally lessens the bench width by forming a series of terraces. However, terrace backfilling along with the other methods previously mentioned could lead to a substantial erosion and bank movement if it is applied in the wrong places. This occurs most often when materials are placed on the outslope. Nevertheless, it does get rid of the highwall and leaves the mined area in a shape roughly like the pre-mining condition [7: 19].

Costs of reclamation depend upon many factors such as the soil characteristics, local cost factors, coal seam, overburden thickness, rainfall, and the character of the desired reclamation. Table 10-9 contrasts total and incremental costs for the contour mining techniques. This analysis assumes a given slope and coal seam thickness. Under different conditions the total costs and incremental costs would not be the same.

Costs were developed by the Council on Environmental Quality for a mining operation on a slope of 20° with a 3-foot coal seam. The bench width was assumed to be 125 feet, with a 25-foot undisturbed barrier at the outer edge, resulting in a 9-foot low wall and a 55-foot highwall. These characteristics describe a hypothetical mining operation in southwestern Pennsylvania, but the relative cost difference between techniques are representative of other parts of the Appalachian region as well.

Using the assumptions set forth in the footnote to Table 10-9, the total costs for the contour mining techniques were broken down as follows: mining with no reclamation; mining with smoothing of the spoil bank; mining with terrace backfilling of the bench; and mining using a modified block cut method.

The first four contour mining techniques are the customary ones. All four

Table 10-9
Estimated Costs of Contour Strip Mining and Reclamation Approaches[1] (in dollars per ton)

Type and Degree of Reclamation	Production Costs	Incremental Reclamation Costs	Incremental Reclamation Costs Above the 1972 Minimum Required
No Reclamation	3.90		
Shaping of Spoil Bank[2]	4.29	.39	
Terrace Backfilling	4.59	.69	.30
Contour Backfilling	4.85	.95	.56
Modified Block Cut	4.46	.56	.17

[1] Costs estimated assuming a slope of 20° with a 3-foot-thick coal seam. Bench width assumed to be 125 feet with a 25-foot undisturbed barrier at the outer edge resulting in a 9-foot low wall and a 55-foot high wall.

[2] Shaping of spoil bank required in all major Appalachian mining states.

Source: Council On Environmental Quality, *Coal Surface Mining And Reclamation* (Washington, D.C.: Government Printing Office, 1973), p. 3, Table II.

techniques deposit overburden in a spoils pile on the edge of the bench or on the downslope, and then go to different levels of effort to recover, reshape, and revegetate it. Therefore, these techniques are similar as far as the unit operations and equipment are concerned.

A diesel powered dragline with a seven cubic yard bucket is used in developing costs. It is used to pile some of the spoils on the edge of the bench and drop the rest down the hillside. Prior to the use of the dragline, a bulldozer is used to clear the surface of vegetation and to remove topsoil. The bulldozer is also used to compact and shape the spoil bank in those cases where no spoils are returned to the bench. This lessens its steepness and thus reduces the probability of landslides. In the terrace backfilling and contour backfilling techniques, the dragline pulls part or all of the spoils back onto the bench, and the bulldozer smooths any remaining soil back on the downslope. It also shapes the spoils on the bench in terraces or in a shape close to the original contour.

These four techniques result in cost estimates which were arrived at by first estimating the cost for mining without reclamation, and then incorporating data on the proficiency of utilization of the various pieces of equipment and standard labor productivity rates. Costs for successively higher degrees of reclamation such as smoothing the spoil bank, terrace backfilling, and contour backfilling were then estimated based on the extra unit operations needed, lowered productivity per hour, and the resultant overtime labor needed to use equipment most effectively. Also included is compensation for some of the lost output.

The modified block cut technique combines mining and reclamation in an integral operation, in contrast to the other four techniques listed in Table 10-9. This method uses a large, 17-cubic-yard capacity front end loader instead of a dragline. The loader is used to move the overburden along the bench from one block to the next instead of depositing the overburden on the downslope. Bulldozers are used to remove surface vegetation and topsoil only before removing the overburden and shaping it back to its original contour. The coal is taken away with a 6-cubic-yard, front-end loader and transported off site by truck. The modified block cut method can produce more coal per hour than any of the other four methods listed in Table 10-9, because of the size of the front end loader.

Tables 10-10 through 10-15 show the cost estimates for each of the five mining techniques. The data represent direct production costs as well as overhead. However, they do not include any coal cleaning costs, railroad freight charges or profits. Depending on the technique, total costs range from $3.90 to $4.85 per ton. Therefore, reclamation costs may be increased by up to $.95 per ton over the non-reclamation case. Most Appalachian states have a minimum requirement that the spoil bank be smoothed. Therefore, it is more realistic to consider that $4.29 per ton is the base cost, with an incremental cost of $.30 per ton (see Table 10-9). Looking at it this way, it can be seen that terrace backfilling represents an increase of $.30 per ton over the base cost. Also, contour backfilling increases the cost of a ton of coal by $.56.

Table 10-10
Fixed Costs of Contour Strip Mining

	Dollars Per Ton
Union Welfare	$0.65
Administration, Marketing, Insurance	.35
Mining Rights (Royalty)	.25
Land rent	.05
Taxes	.20
Tippling Costs	.30
Total	$1.80

Source: Council On Environmental Quality, *Coal Surface Mining And Reclamation* (Washington, D.C.: Government Printing Office, 1973), pp. 101, 102, Tables E-1, E-2, E-3, E-4, E-5.

Table 10-11
Contour Strip Mining with No Reclamation

	Dollars Per Ton
Dragline	$0.45
Coal Loader, 6 Cu. Yard	.12
Drilling	.33
Blasting	.72
Trucking Off Site	.20
Bulldozer Operations	.28
	$2.10
Fixed Costs	1.80
Total	$3.90

Source: Council On Environmental Quality, *Coal Surface Mining And Reclamation* (Washington, D.C.: Government Printing Office, 1973), p. 101, Table E-1.

The modified block cut technique results in costs of $4.46 per ton, which increases the cost $.56 over the non-reclamation case. However, it only increases the cost of a ton of coal by $.17 over what is now required in many Appalachian states [7:100].

To conclude this section of the chapter, it is safe to say that although environmental damage from strip mining has been severe in Appalachia, it is not an unavoidable consequence of all forms of strip mining. Advanced reclamation practices can substantially reduce adverse environmental impacts in most areas. But certain reclamation practices appear to be essential in providing this protection. To provide a minimal level of protection, mining operations must be required: to backfill to the original or similarly appropriate contour; to prevent

Table 10-12
Contour Strip Mining with Smoothing of Spoil Bank

	Dollars Per Ton
Dragline	$0.45
Coal Loader, 6 Cu. Yard	.12
Drilling	.33
Blasting	.72
Trucking Off Site	.20
Bulldozer Operations	.61
Replanting	.06
	$2.49
Fixed Costs	1.80
Total	$4.29
Incremental Cost Over No Reclamation	.39
Incremental Cost Over Minimum Reclamation Required in 1972	.00

Source: Council On Environmental Quality, *Coal Surface Mining And Reclamation* (Washington, D.C.: Government Printing Office, 1973), p. 101, Table E-2.

Table 10-13
Contour Strip Mining with Terrace Backfilling of Bench

	Dollars Per Ton
Dragline	$0.60
Coal Loader, 5 Cu. Ft.	.15
Drilling	.40
Blasting	.73
Trucking Off Site	.25
Bulldozer Operations	.60
Replanting	.06
	$2.79
Fixed Costs	1.80
Total	$4.59
Incremental Cost Over No Reclamation	.69
Incremental Cost Over Minimum Reclamation Required in 1972	.30

Source: Council On Environmental Quality, *Coal Surface Mining And Reclamation* (Washington, D.C.: Government Printing Office, 1973), p. 101, Table E-3.

Table 10-14
Contour Strip Mining with Contour Backfilling of Bench

	Dollars Per Ton
Dragline	$0.74
Coal Loader, 6 Cu. Yard	.16
Drilling	.43
Blasting	.74
Trucking Off Site	.28
Bulldozer Operations	.64
Replanting	.06
	$3.05
Fixed Costs	1.80
Total	$4.85
Incremental Cost Over No Reclamation	.95
Incremental Cost Over Minimum Reclamation Required in 1972	.59

Source: Council On Environmental Quality, *Coal Surface Mining And Reclamation* (Washington, D.C.: Government Printing Office, 1973), p. 102, Table E-4.

Table 10-15
Contour Strip Mining Using Modified Block Cut

	Dollars Per Ton
17-Cu. Yd. Loader-Overburden	$0.65
Coal Loader, 6 Cu. Yd.	.13
Drilling	.35
Blasting	.73
Trucking Off Site	.22
Bulldozers (2 Required)	.55
Replanting	.03
	$2.66
Fixed Costs	1.80
Total	$4.46
Incremental Cost Over No Reclamation	.56
Incremental Cost Over Minimum Reclamation Required in 1972	.17

Source: Council On Environmental Quality, *Coal Surface Mining And Reclamation* (Washington, D.C.: Government Printing Office, 1973), p. 102, Table E-5.

the dumping of spoils down the slope, except as necessary to the original excavation of earth in a new mining operation; and to carry out reclamation as a concurrent part of the mining operation.

The modified block cut method, although not applicable to all states, incorporates reclamation as an integral part of the mining operation. Lands are reclaimed during mining by backfilling the previously worked area with newly removed overburden. Except for the initial cut, spoils are not deposited on the downslopes, and the land is almost immediately restored to its original contour. As a result, landslides, water pollution, esthetic blight, and other environmental effects are reduced. Although not widely used now, it offers one promising approach to reduce environmental effects in many, but certainly not all areas. Other mining techniques properly carried out on appropriate sites, can produce substantially similar levels of environmental impacts.

These strip mining techniques indicate that, in most cases, the incremental costs of adequate reclamation would not be significant and that integration of mining and reclamation operations offers the promise of significant cost savings over present practices which involve separate operations. Strip mined coal can be produced for $.75 to $2.50 per ton less than by underground mining. Therefore, the competitive position of strip mined coal would not worsen even at the highest range of reclamation costs.

Some areas, it appears, cannot at this time be successfully reclaimed because a combination of factors such as slope angle, type of overburden, and rainfall interact to produce environmental damages such as landslides, slumps, massive erosion, and acid mine drainage. Such areas should not be mined until adequate reclamation can be demonstrated.

Strip Mining Likely to Move West

In 1971, the United States got 56 percent of its strip mined coal from the Appalachian region and 12 percent from the western region. The Appalachian region provided for 68 percent of the total United States production and the western region provided 7 percent. However, the western region is the future energy reserve base of the United States. The Appalachian region has only 13 percent of the strippable reserves of the United States. The estimated strippable resources and reserves of coal and lignite in the western region is shown in Tables 10-16 and 10-17. Extensive deposits of subbituminous coal and lignite are located in the Northern Great Plains and Rocky Mountain regions. Almost 60 percent of the remaining strippable resource is considered strippable reserves in these areas. The largest concentration, 59 percent of the United States' 45 billion tons of strippable reserves, is in the Rocky Mountain and Northern Great Plains region. The Pacific Coast region has few reserves of coal. It has only 0.4

Table 10-16
Estimated Remaining Strippable Resources and Strippable Reserves of Coal, and Sulfur Category, for the Western Region (1968) (in millions of short tons)

Rank	Remaining Strippable Resources	Strippable Reserves	Strippable Reserves: Low Sulfur	Medium Sulfur	High Sulfur
Bituminous Coal					
Rocky Mt. and Northern Great Plains Regions:[1]					
Colorado	870	500	476	24	0
Utah	252	150	6	136	8
Total Bituminous	1,122	650	482	160	8
Subbituminous Coal					
Rocky Mt. and Northern Great Plains Regions:[2]					
Arizona	400	387	387	0	0
Montana	7,813	3,400	3,176	224	0
New Mexico	3,307	2,474	2,474	0	0
Wyoming	22,028	13,971	13,377	65	529
Total Subbituminous	33, 548	20,232	19,914	289	529

[1] Not estimated for Montana, New Mexico, Idaho, and Wyoming.

[2] Not estimated for Colorado.

Source: U.S. Congress, Committee On Interior And Insular Affairs, *Regulation Of Strip Mining*, Hearings, 92nd Congress, 1971, p. 623, Table II.

percent of the total strippable reserves. The Northern Great Plains region covers parts of Montana, North Dakota, South Dakota, and Wyoming. Parts of Arizona, Colorado, Idaho, Montana, New Mexico, Utah, and Wyoming are in the Rocky Mountain region.

As indicated in Tables 10-16 and 10-17, most of the deposits in the Western region have a low sulfur content. Low sulfur coal contains less than 1 percent of sulfur, by weight. Medium sulfur contains 1 to 2 percent sulfur. High sulfur coal is made up of over 2 percent sulfur, by weight.

Almost 75 percent of the United States' 31.8 billion tons of low sulfur reserve is subbituminous in rank, 17 percent is lignite, and 9 percent is bituminous coal. The Rocky Mountain and Northern Great Plains region contain over 80 percent of the total of 23.5 billion tons of low sulfur subbituminous coal. The state of Wyoming has by itself 13.4 billion tons.

Seventeen percent of the country's total of 2.8 billion tons of low sulfur coal

Table 10-17
Estimated Remaining Strippable Resources and Strippable Reserves of Lignite, by Sulfur Category, for the Western Region (1968) (in millions of short tons)

			Strippable Reserves		
Rank	Remaining Strippable Resources	Strippable Reserves	Low Sulfur	Medium Sulfur	High Sulfur
Lignite					
Rocky Mt. and Northern Great Region:					
Montana	7,058	3,497	2,957	540	0
North Dakota	5,239	2,075	1,678	397	0
South Dakota	399	160	160	0	0
Total Lignite	12,696	5,732	4,795	927	0

Source: U.S. Congress, Committee On Interior And Insular Affairs, *Regulation of Strip Mining*, Hearings, 92nd Congress, 1971, p. 623, Table II.

reserves is in the Rocky Mountain and Northern Great Plains regions. Of the total U.S. low sulfur lignite reserve of 5.5 billion tons, 85 percent is in Montana and North Dakota.

Montana has 224 million tons of medium sulfur subbituminous coal reserve, out of a U.S. total of 289 million tons. Wyoming has the remaining 65 million tons. Montana has 540 million tons and North Dakota has 397 million tons of the 1.6 billion tons of medium sulfur lignite in the United States.

In the United States there is a total strippable reserve of 9.2 billion tons of high sulfur coal. Bituminous coal accounts for 8.6 billion tons and subbituminous coal accounts for 529 million tons of the total. Only Utah in the Rocky Mountain region has any significant tonnage of high sulfur bituminous coal, 8 million tons. There are no high sulfur lignite reserves.

Out of all the states in the western region, Wyoming Montana, and North Dakota appear to offer the greatest potential for strip mining. Table 10-18 shows production by the different mining methods employed by the states in the western region.

Examining transportation costs of western strip mined coal to its consumption point is necessary, because it is an important element in the delivered cost of coal. Transportation usually accounts for over 50 percent of the total cost of coal in the West. The importance of transportation costs has new meaning in view of the concern over the environment. With the advent of new stricter air quality standards, there will be curbs on the use of high-sulfur fuel in the East and Midwest, and the need for low-sulfur fuels will become very significant. Coal from the western region is generally low in sulfur and is a likely source. However, western coal necessitates transportation over long distances.

Table 10-18
Production of Coal in the Rocky Mountain, and Great Plains Regions and Method of Mining (1969) (in millions of short tons rounded)

Rocky Mountain Great Plains, Regions	Underground	Strip	Total	Percentage Surface Mined
Colorado	3.6	2.0	5.6	34.6
Montana:				
Bituminous	.04	.7	.7	95.2
Lignite		.3	.3	100.0
New Mexico	.8	3.6	4.4	81.3
North Dakota:				
Lignite	–	4.7	4.7	100.0
Utah	4.7	–	4.7	–
Wyoming	.1	4.5	4.6	97.4
Total	9.24	15.8	25.0	63.8

Source: U.S. Congress, Committee On Interior And Insular Affairs, *Regulation Of Strip Mining*, Hearings, 92nd Congress, 1971, p. 245, Table 3.

Transportation costs for western coal being shipped to eastern and midwestern markets is illustrated in Tables 10-19 and 10-20. These tables compare the costs of western coal with the least cost high sulfur coal from specific areas. For example, how does the high sulfur coal produced in Illinois and shipped to Chicago compare with the delivered cost of western coal. Tables 10-19 and 10-20 were produced for illustration and there is a realization that it is quite unlikely that western coal will be transported into certain parts of the United States. For example, shipping coal from Wyoming to Boston is not probable [14:11].

Low sulfur coal from the West is more expensive than high sulfur coal in all cases in the East and Midwest. If there was a substitution of western coal for eastern or midwestern coal, the higher costs would be the result of increased transportation charges.

However, these increased costs would be offset somewhat by the less expensive mining costs in the West than in the East or midwest [14:13].

Table 10-21 makes a comparison on the cost of supplying coal from western producing districts and oil from the East Coast to eastern markets. Table 10-22 makes another comparison on the cost of supplying coal from western producing districts and oil from Gulf Coast to midwestern markets, in 1970.

Table 10-21 indicates that western coal is unable to enter into competition with foreign residual fuel oil in the Boston, Richmond, and New York fuel markets. The cost of supplying western coal to eastern markets ranges from 1.2 to 3.3 times higher than the cost of residual fuel oil.

Table 10-22 indicates that western coal may be a viable economic replace-

Table 10-19
Comparative Costs of Supplying High and Low Sulfur Coal from Selected Producing Districts to Selected Consuming Regions, 1970

	Consumer Region								
	East North Central (Chicago, Ill.)			West North Central (St. Louis, Mo.)			East South Central (Vicksburg, Miss.)		
	Cost of Coal Per Million BTU		Cost Difference	Cost of Coal Per Million BTU		Cost Difference	Cost of Coal Per Million BTU		Cost Difference
Producer District	Low[1] Sulfur	High[1] Sulfur		Low[1] Sulfur	High[1] Sulfur		Low[1] Sulfur	High[1] Sulfur	
Northern and Southern Colorado (Denver)	$ .45	$ .31	$ .14	$ .43	$ .31	$ .12	$ .51	$ .28	$ .23
Arizona and New Mexico (Albuquerque, New Mexico)	.34	.31	.03	.31	.31	.00	.30	.28	.02
Wyoming (Cheyenne)	.34	.31	.03	.33	.31	.02	.41	.28	.13
Utah (Salt Lake City)	.53	.31	.22	.52	.31	.21	.58	.28	.30
North and South Dakota (Fargo, North Dakota)	.34	.31	.03	.39	.31	.08	.53	.28	.25
Montana (Billings)	.34	.31	.03	.44	.31	.13	.43	.28	.15
Washington (Seattle)	.80	.31	.49	.82	.31	.51	.90	.28	.62

[1] Low-sulfur coal is from producer district given. High-sulfur coal given is least cost available high-sulfur coal for each consumer region.

Source: United States Department of the Interior, *Comparative Transportation Costs of Supplying Low Sulfur Fuels to Midwestern and Eastern Domestic Energy Markets*, by P.H. Mutschler, R.J. Evans, and G.M. Larwood (Washington, D.C.: Government Printing Office, 1973), p. 12, Table 2.

Table 10-20
Comparative Costs of Supplying High and Low Sulfur Coal from Selected Producing Districts to Selected Consuming Regions, 1970

	Consumer Region								
	South Atlantic (Richmond, Va.)			Middle Atlantic (New York, N.Y.)			New England (Boston, Mass.)		
	Cost of Coal Per Million BTU		Cost Difference	Cost of Coal Per Million BTU		Cost Difference	Cost of Coal Per Million BTU		Cost Difference
Producer District	Low[1] Sulfur	High[1] Sulfur		Low[1] Sulfur	High[1] Sulfur		Low[1] Sulfur	High[1] Sulfur	
Northern and Southern Colorado (Denver)	$ .60	$ .30	$ .30	$ .61	$ .32	$ .29	$ .63	$ .35	$ .28
Arizona and New Mexico (Albuquerque, New Mexico)	.45	.30	.15	.48	.32	.16	.50	.35	.15
Wyoming (Cheyenne)	.49	.30	.19	.50	.32	.18	.52	.35	.17
Utah (Salt Lake City)	.66	.30	.36	.66	.32	.34	.68	.35	.33
North and South Dakota (Fargo, North Dakota)	.59	.30	.29	.59	.32	.27	.62	.35	.27
Montana (Billings)	.50	.30	.20	.51	.32	.19	.53	.35	.18
Washington (Seattle)	.97	.30	.67	.97	.32	.65	1.00	.35	.65

[1] Low-sulfur coal is from producer district given. High sulfur coal given is least cost available high-sulfur coal for each consumer region.

Source: United States Department of the Interior, *Comparative Transportation Costs of Supplying Low Sulfur Fuels To Midwestern And Eastern Domestic Energy Markets*, by P.H. Mutschler, R.J. Evans, and G.M. Larwood (Washington, D.C.: Government Printing Office, 1973), p. 28, Table 6.

Table 10-21

Comparative Costs of Supplying Coal from Western Producing Districts and Oil from East Coast to Eastern Markets, 1970

	Consumer Region					
	South Atlantic (Richmond, Va.)		Middle Atlantic (New York, N.Y.)		New England (Boston, Mass.)	
	Cost Per Million BTU		Cost Per Million BTU		Cost Per Million BTU	
Coal Producer District	Coal	Oil	Coal	Oil	Coal	Oil
Northern and Southern Colorado (Denver)	$ .60	$ .29	$ .61	$ .38	$ .63	$ .34
Arizona and New Mexico (Albuquerque, New Mexico)	.45	.29	.48	.38	.50	.34
Wyoming (Cheyenne)	.49	.29	.50	.38	.52	.34
Utah (Salt Lake City)	.66	.29	.66	.38	.68	.34
North and South Dakota (Fargo, North Dakota)	.59	.29	.59	.38	.62	.34
Montana (Billings)	.50	.29	.51	.38	.53	.34
Washington (Seattle)	.97	.29	.97	.38	1.00	.34

Source: United States Department Of The Interior, *Comparative Transportation Costs Of Supplying Low Sulfur Fuels To Midwestern And Eastern Domestic Energy Markets*, by P.H. Mutschler, R.J. Evans, and G.M. Larwood (Washington, D.C.: Government Printing Office, 1973), p. 28, Table 6.

Table 10-22
Comparative Costs of Supplying Coal from Western Producing Districts and Oil from Gulf Coast to Midwestern Markets, 1970

	Consumer Region					
	East North Central (Chicago, Ill.)		West North Central (St. Louis, Mo.)		East South Central (Vicksburg, Miss.)	
	Cost Per Million BTU		Cost Per Million BTU		Cost Per Million BTU	
Coal Producer District	Coal	Oil	Coal	Oil	Coal	Oil
Northern and Southern Colorado (Denver)	$ .45	$ .48	$ .43	$ .63	$ .51	$ .48
Arizona and New Mexico (Albuquerque, New Mexico)	.34	.48	.31	.63	.30	.48
Wyoming (Cheyenne)	.34	.48	.33	.63	.41	.48
Utah (Salt Lake City)	.53	.48	.52	.63	.58	.48
North and South Dakota (Fargo, North Dakota)	.34	.48	.39	.63	.53	.48
Montana (Billings)	.34	.48	.44	.63	.43	.48
Washington (Seattle)	.80	.48	.82	.63	.90	.48

Source: United States Department of the Interior, *Comparative Transportation Costs Of Supplying Low Sulfur Fuels To Midwestern And Eastern Domestic Energy Markets*, by P.H. Mutschler, R.J. Evans, and G.M. Larwood (Washington, D.C.: Government Printing Office, 1973), p. 28, Table 7.

ment for foreign residual oil in the St. Louis, Chicago, and Vicksburg fuel markets. Coal from several western locations is potentially less expensive than oil, in all three midwestern markets. The average cost of residual fuel oil was 63 cents per million BTUs, while a low of 31 cents per million BTUs was the potential cost of coal in the St. Louis market [14:27].

In conclusion, the decisive factor of fossil fuel's potential market area is transportation costs. Foreign residual oil now dominates the eastern market. However, western coal would dominate the midwestern fuel market if pre-combustion air pollution regulations were put into effect nationwide.

Is Reclamation Possible in West?

Future strip mining will predominantly be in the West on a scale far larger than anything seen in the East. Western coal will be a boon to electric utilities because of its low sulfur content. Low sulfur coal is needed to meet new air pollution requirements. With increased strip mining in store for the West, it is necessary to look at the question of reclamation. Revegetation is not inherently difficult in most parts of Appalachia and in the Central coal regions. They both have adequate rainfall and sufficient topsoil. There are unanswered questions about the extent to which lands can be revegetated in the West, where there is little rainfall and the topsoil may be poor. The climate is not favorable for developing a layer of topsoil because there is little water and high winds. It is questionable whether any kind of reclamation will produce self-sustaining vegetation on the arid western range.

A stabilizing, sustained, useful vegetative cover is a common objective of a reclamation effort. A sustained forage cover is dependent on a soil medium that will support the plant cover. Natural soil development processes in the semiarid West are notoriously slow for a number of reasons, one of the most important being that excessive geological erosion is typical of the area. Another major limiting factor to soil development in most of the West is the fact that in areas like Eastern Montana 10 to 15 inches of annual precipitation is the rule, whereas in Appalachia the figure is nearly 50 inches. The lack of available moisture also plays a direct role in limiting plant reestablishment, especially during droughty periods. Soil alkalinity and salinity are properties associated with arid lands that will pose special problems in some locations of Montana for revegetation.

A study recently issued by the National Academy of Sciences tends to accentuate these general conclusions. Undertaken to determine whether or not western coal lands were likely to suffer especially seriously from the depredations of surface mining, the NAS study was neither alarmist nor precisely reassuring.

The bulk of the western coal resources lie in four Rocky Mountain and Great Plains states: Colorado, Montana, North Dakota and Wyoming. Over 87 percent of the more than 2 trillion tons of "estimated identified and hypothetical (coal) resources remaining in the ground as of January 1, 1972" in all nine of the coal bearing western states occurred in these four.[a] The importance of strip mining in these four states was suggested by the fact that about three-quarters of these coal resources lay "reasonably near the surface," that is with an overburden for the most part less than 1,000 feet [33:Table 3-2].

Surface mining is not as important in Colorado, where the bulk of the mining is carried on by conventional underground mining methods. However, Montana, North Dakota, and Wyoming had over 65 billion tons of coal estimated to be strippable resources or reserves as of January 1968, constituting 87 percent of the total in this category for the nine western states with sizable coal reserves.

Surface coal mining in the states of Montana, North Dakota, and Wyoming has been quite limited thus far. From 1920 to 1972, a cumulative total of about 230 million tons had been strip mined in these three states [33:Table 3-4]. To put this figure in context, we may note that in 1970 the surface mined output of Montana, North Dakota, and Wyoming amounted to only about 3 percent of the national output of bituminous coal in those years [33:Table 3-4] (also see Table 10-5 of this chapter).

The outlook, however, is for an immense acceleration in strip mining of coal in the states of Montana, North Dakota, and Wyoming, principally because the strippable coal resources in these states are low in sulfur content. Concern for air pollution has led to the adoption of legislative restrictions on the sulfur content of fuels, especially those consumed by public utilities, in order to attempt to reduce emissions of sulfur oxides. Of the 25 billion tons of coal having a sulfur content of less than 1 percent which is found in the nine western coal bearing states, 85 percent is found in Montana, North Dakota, and Wyoming [33:Table 3-3]. The outlook then, is for greatly increased strip mining in these states to meet the need for low sulfur fuels required to meet environmental restrictions on sulfur oxide emissions.

Accordingly, the National Academy of Sciences projects an increase in strip mining in Montana, North Dakota, and Wyoming such that output from 1972 to 1990 will exceed one trillion tons or about 4.7 times the cumulative output in these states from 1920 to 1972 [33:Table 3-6]. If this projected increase in mining activity occurs there will be a nearly corresponding increase in the acreage disturbed by surface mining activities. Thus, the cumulative acreage disturbed by strip mining in Montana, North Dakota, and Wyoming through 1971 was about 13.6 thousand acres. If the National Academy of Sciences' projections are realized, this figure will be increased threefold by 1990 to 41.9 thousand acres [33:Table 3-7].

Will this vast acreage remain as a permanent scar on the western hillsides? Or can it be restored to something approaching the pre-strip mining condition?

[a]The nine western states containing sizable coal reserves are: Arizona, Colorado, Montana, New Mexico, North Dakota, South Dakota, Utah, Washington, Wyoming [33:Table 3-2].

While the National Academy of Sciences project was set up with the basic purpose of answering these questions, it must be concluded that its answer is ambiguous. This is not attributable to any lack of expertise, but is rather intrinsic to the problem. In much of the arid and semi-arid west, the revival of vegetation after its removal is a matter not of decades but of centuries. Thus, the site of the ancient turquoise mines of New Mexico worked during the fourteenth and fifteenth centuries appear today much as they did when they were abandoned [33:60-1]. On the other hand, it is not at all certain that their appearance would have been much different if intensive efforts had been made for their rehabilitation. Having said that, it is well to note that more recent strip mining activities have left undeniable evidences of environmental despoliation [33:Figure 5-3].

Because there has been such limited experience with the rehabilitation of strip mined lands in the arid and semi-arid western states, the investigators of the National Academy of Sciences were understandably diffident with respect to conclusions concerning the rehabilitation potential in western coal fields. Specifically, the Academy scholars concluded: "We believe that those areas receiving ten inches (250 mm) or more of annual rainfall can usually be rehabilitated, provided that evapotranspiration is not excessive, landscapes are properly shaped, and techniques demonstrated to be successful in rehabilitating disturbed rangeland are applied. However, we must emphasize that this belief is not based on long-term, extensive, controlled experiments in shaping and revegetating western lands that have been surface mined" [33:2].

This clearly constitutes much less than a ringing declaration of faith that western coal lands that are strip mined can be restored to their original conditions. It should be noted, however, that some parameters argue in favor of a desirable outcome. Certainly, on any reasonable basis, western mine operators should be able to afford much more substantial expenditures for the rehabilitation of strip mined lands than would be feasible for coal mining companies in, for example, Appalachia. This is because mines in Montana, North Dakota, and Wyoming enjoy a favorable "stripping ratio," i.e., a low ratio of overburden that needs to be removed per ton of underlying coal (see, for example, Figure 6-6 in the NAS report). Accordingly, operating costs per ton of coal recovered tend to be low and this allows more funds for rehabilitation within a cost per ton still lying within competitive limits. Because coal seams, especially in Montana and Wyoming, are frequently thick with relatively thin overburdens, the NAS investigators concluded "that the mining operations can readily bear the full expense of rehabilitation in the western coal areas without the addition of more than a few cents per ton to the price of coal" [33:89]. It was noted earlier that transportation costs loom as a major factor in the delivered price of western coal. In view of that fact, there seems to be little reason to suppose that the relatively small costs of rehabilitation of western coal lands would constitute a decisive competitive handicap.

The United States has been unsuccessful in its reclamation programs thus far. Yet, there are examples of successful programs in West Germany and Great Britain. Why have they been successful and the United States hasn't? This chapter will now focus in on these two countries, to see if their programs have any relevance for the United States.

Reclamation Effective at British Strip Mines

Strip mining in Great Britain is called opencast mining. Opencast coal mining is very similar to area mining in the United States. The coal industry in Britain was nationalized in 1947, when the National Coal Board was formed to administer the mines in behalf of the government. The National Coal Board took over the open mining operations from the Ministry of Power in 1952. Since then all lands that were mined have been completely restored either for agricultural uses, for building land, or recreational uses. All strip mining is planned and initiated by the Opencast Executive, a division of the state owned National Coal Board. Although the mining work itself is performed by private contractors, all decisions concerning the amount of strip mining, the siting of particular mines, and environmental standards are directly under the control and initiative of the government.

The major goal of British policy since World War II has been to stabilize the underground mining industry, so that underground mines could meet planned long-term needs for coal. The government policy since 1967 has been to permit opencast sites only if the output would not prejudice production from deep mining. The main purpose of opencast mining is to meet shorter term and unexpected fluctuations in demand for coal. Presently 8 to 10 million tons are produced annually from strip mines, out of a total production of about 140 million tons [8:951].

Underground coal was produced for an average cost of $14.03 per ton and sold at an average price of $14.13 per ton, in 1970. Strip mine coal was produced at an average cost of $11.37 per ton and sold at an average price of $16.36 per ton. The cost of the strip mine coal included reclamation costs, which ranged from a low of $.63 per ton to a high of $4.24 per ton. The average cost is a $1.35 a ton [8:952]. Because of its planned marginal position in coal production, British strip mined coal sells at a premium above underground coal, thus providing ample overhead for reclamation.

The Opencast Coal Acts, building on the Town and County Planning Acts of 1962, clearly requires that the restoration of a given area to a usable condition be planned before mining begins, and that it remains under constant supervision from the beginning of coal stripping operations until five years after mining has been terminated. The planning of strip mining, initiated by the state's Opencast

Executive, is subject to review by local bodies, which have the power to modify the plans, and in some instances, to refuse them altogether.

The mine operator is responsible for the maintenance of a program of selective removal of all material overlying the coal to be mined. Topsoil and subsoil are removed and deposited in separate dumps. All remaining overburden must be stockpiled separately to prevent any admixture of topsoil, subsoil, or overburden at any time. Particular attention is given to possible disturbances caused by noise, dust, and blasting.

Agricultural restoration plans divide a mining project into two phases. The first phase of the program is a part of the total coal contract and is concerned primarily with backfilling and regrading. Once the contract is completed the sites are given over to the Ministry of Agriculture.

During the second phase the land is retained under Ministry management for five years. The Ministry's main concern during this time is to have proper management of grassland and grazing.

The next phase of restoration follows a plan generally similar to that employed for agricultural restoration. Forestry restoration usually takes place at a mining site which is located on a hillside or ridge type terrain. Restoration of the mining site must therefore incorporate the return to hill and dale contouring, permanently terraced and ditched to prevent soil erosion. Management is extended for a five-year period after regrading to insure proper planting and supervision.

The reclamation done by the strip contractors ranges from $944 to $7,179 an acre, with an average cost of $3,250 an acre. The reclamation includes backfilling, grading, rooting overburden and soils, and the respreading of subsoil and topsoil. However, the original cost of segregating the topsoil and subsoil is not included. Subsequent reclamation by the Ministry of Agriculture or by the Forestry Commission over a five-year period ranges from $368 an acre to $775 an acre, with an average of $456 an acre. Therefore, the total costs of reclamation range from $1,350 an acre to $7,542 an acre. The average cost of total reclamation is $3,976 an acre and the median cost is $4,110 an acre. As a percentage of total production costs, reclamation costs ranged from a low of 7 percent to a high of 33.6 percent [8:955].

There are a number of reasons why reclamation costs are so high in Britain and almost non-existent in the United States. First of all, strip mining in Great Britain is completely integrated into national and local land use planning, in a very strict manner. Great Britain is a country with a traditionally great concern for the conservation of its land. Contour strip mining is not allowed in Britain. Strip mining in Great Britain is state controlled, with private enterprise playing a subordinate role. Britain's high priced reclamation requirements discourage strip mining and encourage underground mining and purchases of low cost coal from the United States. On the other hand, strip mining in the United States is completely under the control of private enterprise. United States strip mining is not integrated into national or local land use planning. Finally, the United States does not have a tradition of conservation as Britain has.

The main difference between Britain and the United States is that strip mining in Great Britain is a supplementary process. Strip mining in Britain is not relied upon for basic long-term coal production needs. However, in the United States strip mining accounts for 50 percent of its coal production and this figure has been increasing.

There are many reasons why strip mining is gaining in popularity over deep mining in the United States. The operating costs of strip mining run 25 to 30 percent lower than that of underground mining. Also, another plus factor to the economy of strip mining is the enactment of the Coal Mine, Health and Safety Act of 1969, which raised underground mining costs still higher. There are two more distinct advantages that strip mining has over the underground method. First, it makes possible the recovery of deposits which for physical reasons cannot be mined underground. Secondly, it provides safer working conditions. In 1971 and 1972, there were 345 underground miners killed, compared to 49 in strip mining.

The United States would be ill-advised to switch to deep mining as its major source of coal and have strip mining as a supplementary process. Leaving aside the cost differential between underground mining and the strip mining of coal, there is another major factor. Switching to underground mining would be handicapped by the time needed to expand production from existing deep mines and to open new ones. Underground mining requires completely different equipment than does strip mining operations. Therefore, strip mining equipment cannot be used to expand underground mining production.

The opening of new underground mines and the expansion of old ones would require the purchase of highly specialized equipment such as loaders, track, cutters, ventilization devices, and roof jacks. It would take several years for delivery of the equipment. The lead time to open new mines runs from two to four years. It takes this amount of time to perform geological analyses, develop mining plans, provide railroad spurs, develop access roads and processing facilities, sink mine shafts to the coal seams, and put in the needed mining and safety equipment [7:57].

British strip mining provides coal at a premium price for special circumstances. This price structure provides ample economic leeway for high reclamation costs. On the other hand, strip mining in the United States characteristically produces coal at a below average price. It prospers by serving markets which demand coal at the lowest possible price. Therefore, there is little economic leeway for reclamation costs.

As already stated, the British spend an average of $3,976 an acre for reclamation, and five years of intensive soil treatment follow the strip mining. In the United States, on the other hand, reclamation–in those cases where it is undertaken–involves costs of no more than $200 to $500 per acre. U.S. strip mines treat the soil for one, or at the most, two growing seasons following mining [8:956].

There is no likelihood that the British experience will have substantial impact on strip mining practice in the United States. The fundamental difference between the two countries is that strip mining is practiced in Britain only for marginal output, whereas in the United States strip mining accounts for a large and growing portion of basic output. To put strip mining practice in the United States on the same level of reclamation as in Britain it could be necessary first of all to consign strip mining to a marginal role; thereafter, it would be essential to establish controls over strip mine practice in the United States as rigorous and sophisticated as those practiced in England. There is no reason to suppose that this is a realistic prospect.

The problem of meeting the ever-growing energy demands of society without needlessly destroying land, water, and forest resources is international in scope and must somehow be resolved. The land restoration policies which have been adopted in the Rhineland coal (lignite) fields of West Germany represent one possible way of achieving this goal. The German program for dealing with the social and environmental impacts from strip mining afford a valuable example to the United States. The Germans have developed an exemplary program for exploiting the mineral resources of a region without permanently impairing the quality of its environment. Presently, production of all lignite in Germany is from strip mining [8:1297]. As stated previously, the United States gets 50 percent of its coal from strip mining.

West German Strip Mining Preserves Land Values

In Germany, methods of mining and land reclamation have been developed and implemented in order to avoid serious environmental damage during the mining process. Each strip mine has a number of bucketwheel excavators for stripping off the overburden and removing the lignite without the use of explosives. Each machine costs up to $10 million. The bucketwheels are capable of excavating selectively, and because of this they are especially suitable for saving the topsoil for later use in land reclamation activity [8:1298]. The most successful use of bucketwheel excavators in the United States has been in the removal of unconsolidated overburden in North Dakota and in the removal of glacial drift in Illinois. However, it cannot be used for hard rock digging or for the handling of blasted materials. Therefore, it is inadequate for Appalachian contour mining because of the excavator's heavy weight and limited mobility. Nevertheless, the bucketwheel excavator might be used in the gently rolling plains of the western and northern coal states.

The strip mining of lignite necessitates that the water table in the proximity of the mines be lowered to prevent flooding of the pits. This is achieved by submersible pumps. On the average, some 14 tons of water must be pumped from the ground for each ton of raw lignite mined.

In Germany, reclaiming land to full agricultural productivity is the most expensive type of reclamation, with costs ranging from $3,000 to $4,500 per acre. Based on the average seam thickness of 20 meters, these reclamation costs amount to about $.05 per ton of raw lignite that is mined. This is equal to $.20 per ton of bituminous coal in a deposit yielding the same calorific value of fuel per acre [8:1299]. Thus even full reclamation costs appear to be modest in Germany.

Brown coal mining in Germany differs greatly from Appalachian strip mining economically, in the topography, the type of technology employed, and the degree of government regulation imposed upon the mining industry. In the German lignite fields, excavation is easier than in Appalachia because the coal beds are not covered with rock strata as in Appalachia. The terrain is relatively flat and sulfur bearing minerals are not present. Furthermore, the lignite fields are located in a rich agricultural area, providing a strong incentive for restoration of the land after mining is completed. In addition, almost all of the brown coal resources are located within a single state, permitting the economies of large-scale mining with low mining costs. This factor also makes it easier for the state to impose adequate land reclamation requirements because significant competition from neighboring states does not exist. As a consequence, nearly all government control of brown coal strip mining is by the state of North Rhine Westphalia rather than by the German federal government. Also, almost 90 percent of the lignite in Germany is produced by one firm. In contrast, the United States strip mining of coal is conducted by numerous private concerns spread throughout the entire country. Twenty-six of the fifty states produce coal by strip mining.

German reclamation costs compare favorably with the market value of the farmland restored to full agricultural productivity. However, reclamation was required in Germany long before an economic breakeven point was reached. The breakeven point was just recently accomplished through the development of more efficient mining and land restoration techniques. Therefore, the reclamation practices of the brown coal region cannot be explained as due entirely to economic efficiency in restoration. Rather, these practices seem to have been fostered by a thorough planning process which treats strip mining as a transitory land use [8:1299].

The land reclamation program starts with the creation of detailed plans for the relocation of villages and the specification of the type of reclamation, long before actual mining operations begin. Land use patterns are developed in conformity with guidelines from the regional planning commission and the new landscape is designed according to the topography, drainage, patterns, location of lakes, and the selection of areas designated for industry, forestry, agriculture, and public recreation. The thorough early planning is a very significant factor in lessening land reclamation costs because it enables mining operations to be coordinated with ongoing reclamation work. The broad outlines of the regional development plan are made known publicly some ten to fifteen years before the actual mining operations commence. This period of time allows the mining company and affected persons and businesses enough time to adapt to the requirements [8:1300].

In March 1950, North Rhine Westphalia passed Germany's first Regional Planning Law. This law established a Land Planning Commission which is charged with the responsibility of developing overall guidelines for land use within the region. The major purpose of the commission is to organize the social, economic, and industrial activities of the region and to plan for future developments. It also develops long-range plans for the preservation of historic sites, transportation networks, and the construction of recreational facilities to serve the whole region. In the United States there is no such land planning commission. Presently, there is no federal control over strip mining. As of 1972, the number of states that have passed laws for varying degrees of land reclamation is twenty-six. In many instances state standards and their enforcement have proved inadequate. There is no state enactment that approaches the strict, no-nonsense requirements of Germany and Great Britain.

In April 1950, the German legislature passed another law specifically for the brown coal producing areas. It is entitled The Law for Overall Planning in the Rhineland Brown Coal Area. This law created the very important Brown Coal Committee to develop detailed plans for exploiting the brown coal resources of the state within the framework of the regional planning law.

The main responsibility of the Brown Coal Committee is to protect land areas temporarily used for brown coal mining from long-term damage which would make the land unsuitable for more lasting uses. The makeup of the Brown Coal Committee, includes 27 members, is fixed by law. The committee members represent not only the mining interest, but other businesses, unions, and conservation groups as well [8:1301].

The main function of the Brown Coal Committee is to review proposals for extending mining operations to new land areas and to make appropriate recommendations to the state governor. The operation starts with the Rheinbraun Mining Company offering detailed mining and reclamation plans for new, unopened land areas. The committee inspects the proposals for conformance to regional planning guidelines and hears testimony from qualified experts, government officials, and representatives of the people who will be directly affected by the proposed mining. Based on the information presented, the mining company plan may be accepted, modified, or rejected. When final committee approval has been obtained, public hearings are held in which comments or objections to the plan are solicited. The mining plan and the public record are then given to the chief of the state land planning commission for final approval or rejection. As soon as he signs it, the plan becomes legally binding.

Once the plan is finally adopted, the State Mining Office is responsible for insuring that the mining and land reclamation activities are in accordance with its provisions. The State Mining Office enforces the explicit provisions of the plan. It also writes the supplementary regulations and largely sets the tone of the program.

Somewhat awkward because of its large size, the Brown Coal Committee has nevertheless operated effectively and has risen as an important element in the successful land reclamation of the Rhineland. The committee subjects the brown coal industry to public scrutiny and has helped produce the present conservation practices. The planning and enforcement process, with the participation of non-mining interests, gives flexibility in solving a wide range of social and environmental problems resulting from strip mining. Since discussions are held well in advance of the actual mining, there is sufficient lead time available for full consideration on all of the issues and problems.

In conclusion, the conservation accomplishments in the Rhineland indicate that a carefully planned land reclamation program can bring about impressive results. The United States might be better off if a planning institution similar to the Brown Coal Committee was formed. The United States committee would formulate land reclamation standards for varying conditions of terrain, climate, and land use patterns of the region being considered. In other words, the supplementary land restoration requirements would be flexible and would be based on the intended use of the land after mining is completed. They could prescribe results to be achieved rather than merely procedures to be followed. The committee should consider the broader range of economic and environmental issues associated with strip mining in a particular region. The committee should be given power to apply all measures needed for environmental protection and the preservation of land. This would include the complete banning of strip mining in terrain where adequate land restoration is impossible. Before actual strip mining operations are permitted, public hearings should be required to consider the adequacy of the proposed environmental safeguards. In order to gain full effectiveness, either the same agency or another should be given sufficient powers to insure that mining and land restoration are conducted according to approved standards.

In summary, the German program embodies four main principles which have contributed greatly to its success. First, the regulation of strip mining is embedded within an overall regional development plan. This makes it possible to protect the larger interests of the region as a whole. Second, the planning body is made up of representatives of different social and business interests. The group defines general mining and land restoration requirements long before the actual mining begins. Third, suggestions of the planning body are reviewed in public hearings before it is adopted and implemented. This allows for a useful check for the detection of possible adverse side effects that had gone unnoticed. Fourth, an enforcement agency is provided with the necessary powers to enforce the approved plan. The implementation of similar procedures in the United States could help to reduce the environmental impacts from coal strip mining.

Appendix 10A

Table 10A-1
United States Total Gross Consumption of Energy Resources by Major Sources (1940-1970) (in trillions of British thermal units)

Year	Anthracite	Bituminous & Lignite	Natural Gas Dry[1]	Petroleum[2]	Hydro-Power	Nuclear Power	Total Gross Energy
1947	1,224	14,600	4,518	11,367	1,326	–	33,035
1948	1,275	13,622	5,033	12,557	1,393	–	33,880
1949	958	11,673	5,289	12,119	1,449	–	31,488
1950	1,013	11,900	6,150	13,489	1,440	–	33,992
1951	940	12,285	7,248	14,848	1,454	–	36,775
1952	897	10,971	7,760	15,334	1,496	–	36,458
1953	711	11,182	8,156	16,098	1,439	–	37,586
1954	683	9,512	8,548	16,132	1,388	–	36,263
1955	599	10,941	9,232	17,524	1,407	–	39,703
1956	610	11,142	9,834	18,627	1,487	–	41,700
1957	528	10,640	10,416	18,570	1,551	1	41,706
1958	483	9,366	10,995	19,214	1,636	2	41,696
1959	478	9,332	11,990	19,747	1,591	2	43,140
1960	447	9,693	12,699	20,067	1,657	6	44,569
1961	404	9,502	13,228	20,487	1,680	18	45,319
1962	363	9,826	14,121	21,267	1,821	24	47,422

1963	361	10,353	14,843	21,950	1,767	34	49,308
1964	365	10,899	15,648	22,386	1,907	35	51,240
1965	328	11,580	16,098	23,241	2,058	38	53,343
1966	290	12,205	17,393	24,394	2,073	57	56,412
1967	274	11,982	18,250	25,335	2,344	80	58,265
1968	258	12,401	19,580	27,052	2,342	130	61,763
1969	224	12,509	21,020	28,421	2,659	146	64,979
1970	210	12,712	22,029	29,614	2,650	229	67,444

[1] Excludes Natural Gas Liquids

[2] Petroleum Products Including Still Gas, Liquefied Refinery Gas, And Natural Gas.

Source: U.S. Department Of The Interior, *United States Energy Through The Year 2000*, by Walter G. Dupree, Jr. and James A. West (Washington, D.C.: Government Printing Office, 1972), p. 40, Table 1.

Table 10A-2
U.S. Production of Crude Petroleum (1945-1971)

Year	Million Barrels
1945	1,714
1946	1,734
1947	1,857
1948	1,020
1949	1,842
1950	1,974
1951	2,248
1952	2,290
1953	2,357
1954	2,315
1955	2,484
1956	2,617
1957	2,617
1958	2,449
1959	2,575
1960	2,575
1961	2,622
1962	2,676
1963	2,753
1964	2,787
1965	2,849
1966	3,928
1967	3,216
1968	3,329
1969	3,372
1970	3,571
1971	3,454

Sources:

1. U.S. Department of the Commerce, Bureau of the Census, *Historical Statistics of the United States, Colonial Times to 1957* (Washington, D.C.: Government Printing Office, 1960), p. 362.

2. U.S. Department of Commerce, Bureau of Economic Analysis, *Survey of Current Business* (Washington, D.C.: Government Printing Office. Jan. 1974; Jan. 1972; Jan. 1970; Jan. 1968; Jan. 1966; Jan. 1964; Jan 1962; Jan. 1960; Jan. 1958), all on page 535.

Table 10A-3
U.S. Production of Residual Fuel Oil (1945-1972)

Year	Millions of Barrels
1945	469.4
1946	431.3
1947	447.7
1948	479.9
1949	424.9
1950	425.2
1951	469.3
1952	453.8
1953	449.9
1954	416.8
1955	420.3
1956	426.7
1957	415.7
1958	360.1
1959	348.0
1960	332.4
1961	315.6
1962	295.2
1963	275.9
1964	266.8
1965	268.6
1966	264.0
1967	276.0
1968	275.8
1969	265.9
1970	257.5
1971	274.7
1972	292.5

Sources:

1. U.S. Department of Commerce, Bureau of the Census, *Historical Statistics of the United States, Colonial Times to 1957* (Washington, D.C.: Government Printing Office, 1960), p. 362.

2. U.S. Department of Commerce, Bureau of Economic Analysis. *Survey of Current Business* (Washington, D.C.: Government Printing Office, Jan. 1974; Jan. 1972; Jan. 1970; Jan. 1968; Jan. 1966; Jan. 1964; Jan. 1962; Jan. 1960; Jan. 1958), all on page 535.

Table 10A-4
U.S. Production of Distillate Fuel Oil (1945-1972)

Year	Millions of Barrels
1945	249.2
1946	287.8
1947	312.1
1948	379.3
1949	340.8
1950	398.9
1951	475.8
1952	520.4
1953	528.1
1954	542.3
1955	603.1
1956	665.7
1957	668.6
1958	631.6
1959	679.2
1960	667.2
1961	697.2
1962	720.0
1963	765.1
1964	742.4
1965	765.4
1966	785.8
1967	840.8
1968	840.4
1969	848.4
1970	897.1
1971	912.1
1972	936.6

Sources:

1. U.S. Department of Commerce, Bureau of the Census, *Historical Statistics of the U.S., Colonial Times to 1957* (Washington, D.C.: Government Printing Office, 1960), p. 362.

2. U.S. Department of Commerce, Bureau of Economic Analysis, *Survey of Current Business* (Washington, D.C.: Government Printing Office, Jan. 1974; Jan. 1972; Jan. 1970; Jan. 1968; Jan. 1966; Jan. 1964; Jan. 1962; Jan. 1960; Jan. 1958), all on page 535.

Table 10A-5
U.S. Production of Bituminous Coal (1945-1971)

Year	Millions Short Tons	Price (Dollar per Ton)
1945	577.6	$3.06
1946	533.9	3.44
1947	630.6	4.16
1948	599.5	4.99
1949	437.8	4.88
1950	516.3	4.84
1951	533.7	4.92
1952	466.8	4.90
1953	457.3	4.92
1954	391.7	4.52
1955	464.6	4.50
1956	500.9	4.82
1957	492.7	5.08
1958	410.4	4.86
1959	412.0	4.77
1960	415.5	4.69
1961	403.1	4.58
1962	422.1	4.48
1963	459.0	4.39
1964	487.0	4.45
1965	512.1	4.44
1966	533.8	4.54
1967	552.6	4.77
1968	545.2	4.67
1969	560.5	4.99
1970	602.9	6.26
1971	552.1	7.07

Sources:

1. U.S. Department of Commerce, Bureau of the Census, *Historical Statistics of the U.S., Colonial Times to 1957* (Washington, D.C.: Government Printing Office, 1960), p. 356.

2. U.S. Department of Commerce, Bureau of Economic Analysis, *Survey of Current Business* (Washington, D.C.: Government Printing Office, Jan. 1974; Jan. 1972; Jan. 1970; Jan. 1968; Jan. 1966; Jan. 1964; Jan. 1962; Jan. 1960; Jan. 1958), all on page 535.

Bibliography

Public Documents

1. U.S. Department of Commerce, Bureau of the Census, *Statistical Abstract of the United States*. Washington, D.C.: Government Printing Office, 1972.
2. Environmental Protection Agency, *Legal Problems of Coal Mine Reclamation.* Washington, D.C.: Government Printing Office, 1972.
3. U.S. Congress. Committee on Interior and Insular Affairs. *Regulation of Strip Mining.* Hearings, 92nd Congress, 1971.
4. U.S. Congress. Committee on Interior and Insular Affairs. *Surface Mine Reclamation.* Hearings on S.217, 3126 and 3132, 90th Congress, 1968.
5. U.S. Department of the Interior. *Surface Mining and Our Environment.* Washington, D.C.: Government Printing Office, 1967.
6. U.S. Department of the Interior. *Stripping Coal Resources of the United States* (by Paul Averitt). Washington, D.C.: Government Printing Office, 1970.
7. Council on Environmental Quality. *Coal Surface Mining and Reclamation.* Washington, D.C.: Government Printing Office, 1973.
8. U.S. Congress. Committee on Interior and Insular Affairs. *Regulation of Surface Mining.* Hearings, 93rd Congress, 1973.
9. U.S. Congress. Committee on Interior and Insular Affairs. *Regulation of Surface Mining Operations.* Hearings, 93rd Congress, 1973.
10. U.S. Congress. Committee on Interior and Insular Affairs. *Surface Mining.* Hearings, 92nd Congress, 1972.
11. U.S. Congress. Committee on Interior and Insular Affairs. *Coal Surface Mining and Reclamation.* Hearings, 93rd Congress, 1973.
12. U.S. Congress. Committee on Interior and Insular Affairs. *The Issues Related to Surface Mining.* Washington, D.C.: Government Printing Office, 1971.
13. U.S. Department of the Interior. *Cost Analyses of Model Mines for Strip Mining of Coal in the United States.* Washington, D.C.: Government Printing Office, 1972.
14. U.S. Department of the Interior. *Comparative Transportation Costs of Supplying Low Sulfur Fuels to Midwestern and Eastern Domestic Markets* (by P.H. Mutschler, R.J. Evans, and G.M. Larwood). Washington, D.C.: Government Printing Office, 1973.
15. U.S. Department of the Interior. *United States Energy Through the Year 2000* (by Walter G. Dupree, Jr. and James A. West). Washington, D.C.: Government Printing Office, 1972.
16. U.S. Congress. Committee on Interior and Insular Affairs. *National Fuels Study.* Hearings, 87th Congress, 1961.

Other Publications

17. Kovarik, Thomas J. "Raped Earth of Strip Miners–Can it be Healed?" *Science Digest.* April 1972, pp. 64-69.
18. Kovarik, Thomas J. "Sharp Conflict on Strip Mine Reclamation." *Science News.* May 1, 1971, pp. 297-298.
19. Trawick, Jack. "Strip Mining." *National Parks and Conservation Magazine.* July 1971, pp. 11-14.
20. Caudill, Harry M. "Strip Mining–Coast to Coast." *The Nation.* April 19, 1971, pp. 488-490.
21. Faltermayer, Edmund. "Taming the Strip Mine Monster." *Life.* October 1, 1971, p. 21.
22. Faltermayer, Edmund. "Battle over Mining that Scars the Land." *U.S. News and World Report.* September 25, 1972, pp. 76-78.
23. Branscome, James. "Stripping for Pleasure and Profit." *Commonweal.* December 1971, pp. 229-231.
24. Mendes, Guy. "Peeling Back the Land for Coal." *Newsweek.* June 28, 1971, pp. 69-72.
25. Mendes, Guy. "The Coal Industry Makes a Dramatic Comeback." *Business Week.* November 4, 1972, pp. 50-58.
26. Mendes, Guy. "A Truce in the Western Strip Mining War." *Business Week.* November 18, 1971, p. 6.
27. Mendes, Guy. "The Surface Mining Issue: A Reasoned Response." *Coal Age.* March 1971, pp. 92-100.
28. Hechler, Ken. "TVA Ravages the Land." *National Parks and Conservation Magazine.* July 1971, pp. 15-16.
29. Sinclair, Ward. "Coals' Congressmen." *The New Republic.* January 15, 1972, pp. 9-10.
30. Sinclair, Ward. "Equipment Guidelines for Surface Mine Reclamation." *Coal Age.* July 1971, pp. 94-95.
31. Sinclair, Ward. "Surface Mining and Reclamation: A Boost with Bigger, Better Machines." *Coal Age.* March 1972, pp. 96-97.
32. Sinclair, Ward. "The Price of Strip Mining." *Time.* March 22, 1971, pp. 47-48.
33. National Academy of Sciences and National Academy of Engineering. *Rehabilitation Potential of Western Coal Lands* (A Report to the Energy Policy Project of the Ford Foundation). Cambridge, Mass.: Ballinger Publishing Co., 1974.
34. United States Federal Power Commission, *The 1970 National Power Survey.* December 1971. Washington, D.C.: Government Printing Office, p. 1-4-2, Figure 4.3.

11 Energy Under the Oceans: A Summary Report of a Technology Assessment of OCS Oil and Gas Operations

The Technology Assessment Group,
Science and Public Policy Program,
University of Oklahoma

Offshore oil is looked to as perhaps the most obvious solution to our newly discovered energy needs. What are the problems involved?

Energy consumption in the United States has been increasing at an annual rate of about 4 percent over the past ten years. An annual rate of 3.4 to 4.4 percent has been estimated for the period up to 1985. Given these rates of increase, U.S. energy consumption in 1985 will be between 112.5 and 130.0 quadrillion BTUs (the equivalent of 19.4 to 22.4 billion barrels of oil).

Demand for oil and gas accounts for almost 70 percent of this total, and, although energy from coal, nuclear, hydroelectric, or geothermal sources will increase, these sources are not likely to contribute enough to reduce oil and gas needs by very much. This point is emphasized by the two domestic production cases illustrated in Figure 11-1. The important question for the next twelve years, therefore, is where to get this supply of oil and gas.

Four areas from which it might be possible to increase domestic oil and gas production within this time period are:

1. onshore in the lower 48 states,
2. Alaska,
3. state lands offshore, and
4. the outer continental shelf (OCS).

Of the four, only Alaska and the OCS offer a potential for significant increases. With maximum development, the OCS may be expected to produce up to 2.6 million barrels of oil each day and 9.1 trillion cubic feet of gas each year; if the trans-Alaska pipeline system (TAPS) is available, daily Alaskan production can be as much as 2.6 million barrels of oil per day and 4.4 trillion cubic feet of gas per year.

But given even the lowest anticipated level of energy consumption, some portion of U.S. energy demand will have to be satisfied by imports. Assuming the lowest demand and highest domestic production situation shown in Figure 11-1, imports in 1985 would represent only 3 to 4 percent of supply, a level

This chapter is from *Energy Under the Oceans: A Summary Report of a Technology Assessment of OCS Oil and Gas Operations*, copyright 1973 by the Oklahoma Research Administration, published by the University of Oklahoma Press, Norman.

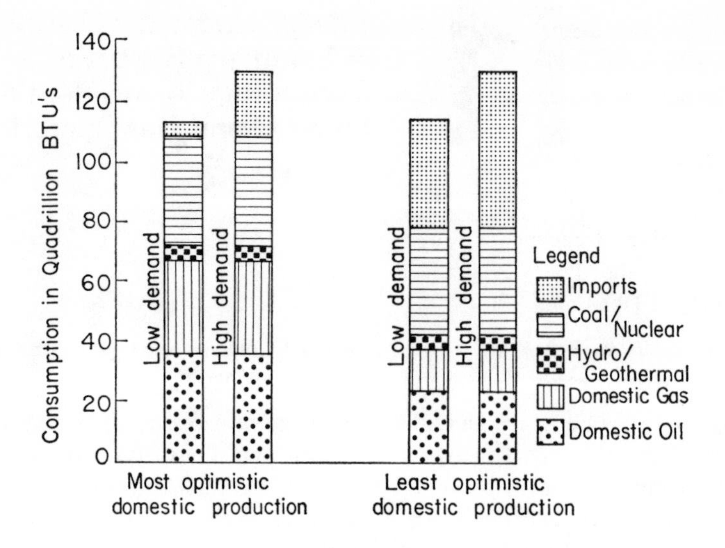

Figure 11-1. U.S. Energy Supply and Consumption in 1985

considered unrealistic by most experts. On the other hand, if it is assumed that demand will be at its highest while domestic production is at its lowest rates, imports will supply about 40 percent of total U.S. energy demand in 1985. The central question concerning energy supply, then, is to what extent will the U.S. be dependent upon imports. To answer this question requires that the costs and benefits associated with imports and the alternative means for increasing domestic production be compared.

There are objections to each of the domestic alternatives and to imports. The principal objections to Alaska and the OCS are environmental. Critics view developing either source as involving excessive, unwarranted environmental risks. Proponents of developing either Alaska or the OCS contend not only that imports are likely to be more of an environmental threat, but also that imports create national security, balance of trade, and economic problems.

Policymakers who have to choose from among these three alternatives must seek to balance demands for protecting the environment against demands for more energy. Technology assessment, an attempt to identify, analyze, and evaluate potential environmental, legal/political, and other social impacts, is one approach that can be used for informing policymakers of what the consequences of the choice or choices are likely to be.

This Study

This NSF-funded technology assessment of the OCS alternative has been conducted over a twenty-month period by an eight-man interdisciplinary team at the University of Oklahoma. The purpose of the study has been to:

1. assess a broad range of social impacts associated with the development of OCS oil and gas resources;
2. contribute to rational policy-making for the OCS;
3. contribute to the formulation of a social-technological system for the development of OCS oil and gas resources responsive to broad social concerns; and
4. make specific recommendations for changes in government policy and administration, industry management, and technologies which will contribute to optimal resources development on the OCS.

For purposes of the study, it has been assumed that environmental and quality-of-life concerns will continue to be a factor in making OCS policy. It has also been assumed that there will be no major changes from the present state of society, such as a major war or depression, for example; and that as a result of the limitations of other energy alternatives and other pressures, the OCS will continue to be developed. In scope the study is limited to the:

1. OCS off the lower 48 states and Alaska;
2. next fifteen years;
3. OCS oil and gas–consideration of alternatives has been limited to examining their feasibility as a replacement for OCS oil and gas during the next fifteen years; and
4. use of pipelines as the means for transporting oil and gas ashore–bulk carriers are considered only as a temporary storage and transport option.

Conduct of the Study

At the outset the assessment was designed to focus on existing and anticipated physical technologies and technological alternatives for finding, developing, producing, and transporting oil and gas. As the study progressed, it became clear that this accepted conception of technology assessment was inappropriate for this particular problem area. Technologies used on the OCS proved to be relatively stable and technological alternatives limited. This is not an area in which technological breakthroughs occur; technological change has been and will continue to be gradual unless there are major new initiatives by industry and/or government. How technologies are managed and regulated, what we came to call the social technologies, proved to be the critical element in this problem area. This is due primarily to changes in the social context within which OCS policies are made and administered: changes requiring that attention be paid to new concerns, especially for environmental quality and safety, and that participation be expanded to include others in addition to industry and government, particularly environmental and consumer interest groups.

Once it became clear that most of the changes required to provide for optimal development involve changes in the behavior and relationships of responsible persons in industry and government rather than changes in hardware, the focus of the assessment was shifted. Although physical technologies were still to be assessed, the major effort was redirected to become an assessment of social technologies associated with the present and future development of OCS oil and gas resources.

Specific policy issues related to particular negative or undesirable impacts were identified, alternative responses were defined and analyzed, and recommendations to achieve desirable changes were formulated. While limited quantitative data are available for defining certain undesirable impacts, such as blowouts during drilling, for example, the effect of most changes in rules, standards, and procedures cannot be measured in quantitative terms. As a consequence, the standards employed in the study are for the most part procedural; that is, the assessment is in terms of whether present or proposed rules, standards, management practices, and changes in decision-making and administrative processes, reduce or eliminate undesirable impacts.

The study is summarized beginning with our assessment of specific physical technologies and moving to our recommendations for an overall management plan for the OCS. This is followed by a brief summary comparison of the impacts of the OCS, TAPS, and imports.

The Issues and Our Recommendations

Although most physical technologies used on the OCS are generally adequate to permit oil and gas resources to be found, developed, produced, and transported safely with minimal adverse social impacts, our assessment identified a number of technological weaknesses.

Physical Technologies

Only three technologies were found to be inadequate—velocity actuated downhole safety devices,[a] well control technologies, and oil containment and clean-up devices.

1. *Downhole Safety Devices.* Although reliability data for velocity actuated downhole safety devices are limited, there are numerous indications of their inadequacy. For example, in recent major accidents in the Gulf of Mexico, 25 and 40 percent of them failed. The U.S. Geological Survey (USGS) now requires new wells to be equipped with a surface, rather than a velocity actuated, downhole safety device. However, this new requirement does not apply to wells presently producing until tubing has to be pulled for some other purpose, such

[a]This device is commonly called the "storm choke."

as a workover, for example. This may not occur for several years, if ever. Until there is a reliable replacement for "storm chokes" that can be installed in most producing wells without pulling tubing, the "storm choke" will continue to be a problem. Therefore, the "storm choke" must be made more reliable.

2. *Well Control Technologies.* The two principal approaches to re-establishing control over wells which are blowing out and/or burning are capping and drilling relief wells. (See the brief Glossary on page 285.) Capping is particularly difficult offshore because explosions and fires tend to destroy the platform requiring the cappers to provide their own work platform. Drilling relief wells consumes too much time. Alternatives for re-establishing well control must be developed. Possibilities include subsurface and/or above-the-mudline valves.

3. *Containment and Clean-up.* There was no effective capability for containing and cleaning up oil on the OCS before Union's blowout at Santa Barbara, California. Subsequent crash efforts by industry and government have produced only a limited capability even now. In fact, wave heights, wind velocities, and currents on much of the U.S. OCS exceed designed capabilities at least a third of the time.

Containment and clean-up on the OCS itself may be an illusory goal since, as a practical matter, there is an upper limit on sea conditions beyond which neither is possible. Although the primary effort should be to prevent accidents, it will never be possible to prevent all accidents and there must be some adequate means for responding when an accident does occur. Consequently, efforts should continue to be made to improve the performance of containment and clean-up devices. However, the primary development effort should be to achieve a capability to deal effectively with oil spills which threaten to come ashore.

Several other technologies being used on the OCS also require improvement:

1. *Drilling Bits.* Efforts to give drilling bits longer life have been a continuing research effort. These efforts should be continued and accelerated. Longer-lasting bits would reduce the number of trips made in drilling a well and would, thereby, reduce the risk of losing control of the well which is associated with this operation.

2. *Flaw Detectors.* Improved devices for detecting flaws in pipelines would make it possible to reduce chronic pollution from this source. Although usually dismissed as a minor source of polluting oil, in 1971 pipeline leaks and ruptures offshore accounted for 6 percent of all oil reported spilled into U.S. waters and 84 percent of all oil reported spilled from offshore facilities.

A number of technologies presently available for improving the quality of OCS operations are not being used by all operators.

1. *Sand Probes.* If sand probes, devices warning of excessive sand erosion, were used in all wells equipped with "storm chokes," these devices could be made more reliable.

2. *Mud Monitors.* Drilling safety could be enhanced if available devices for

monitoring small changes in the volume of drilling mud were used on all drilling rigs.

3. *Mass Flow Monitors.* Pipeline spills could be held to a minimum if available mass flow monitoring equipment were used on all pipelines.

4. *Additional Controls for BOP's.* Control might be regained more quickly on some blowouts if sufficiently remote and protected controls were provided for the blowout preventors (BOP's).

Operations on the OCS could be made safer and undesirable environmental impacts less likely if certain kinds of new technologies were developed and utilized. Among those are:

1. *Downhole Safety Devices.* Surface actuated downhole safety devices that could be installed in a producing well without having to pull tubing and without cutting production below an economic rate.

2. *Downhole Instrumentation.* A capability to measure pressure at the face of the bit to give faster and more reliable warning of potential blowouts.

3. *Event Recorders.* Event recorders designed to survive accidents and provide records of equipment malfunctions.

4. *Identification Devices.* Devices for identifying which wells on a burning multi-well platform are out of control.

5. *Multi-Phase Fluid Movement.* Pumps and pipeline coatings capable of efficiently moving oil and gas in the same pipeline to help reduce flaring, the number of pipelines required, and to simplify a move to subsea production systems.

6. *Automated Drilling.* More automated drilling in order to reduce accidents due to human error.

Why Do These Technological Weaknesses Exist? Most of these weaknesses in physical technologies exist because, until very recently, standards used for determining the adequacy of OCS technologies have been based largely on industry's judgment of what was economically feasible. Before events such as Union's Santa Barbara blowout attracted widespread public attention, continuous participation in policy-making and administration for OCS development had been pretty much limited to government and industry. The rules and regulations established by responsible government agencies had usually stated objectives rather than detailed specifications and standards. The most detailed rules, OCS orders issued for each USGS Area, had been and are the product of an institutionalized process of government-industry cooperation. Perhaps as a consequence, government regulation had tended to be heavily dependent upon industry's engineering and operational expertise when establishing OCS regulations.

In short, the system for managing and controlling OCS operations had been effectively closed to outside influences on a continuing basis. When it was subjected to close public scrutiny following Santa Barbara, some of the disadvantages of this closed decision-making system were identified. Many persons in responsible government agencies and in the petroleum industry recognize that the changed social context within which OCS development is now taking place necessitates changes in this system. Their goal is to respond to demands to improve the overall quality of OCS operations, and their efforts include taking steps to overcome some of the specific weaknesses identified here. For some of these weaknesses to be corrected, both industry and government agencies will have to change their past patterns of behavior. These should include changes in the way technologies are developed, maintained, and operated. Specifically:

1. Industry should modify its past pattern of incrementally adapting and linking components by making more extensive use of a systems design approach.[b]

2. Industry should expand its design criteria to focus explicit attention on human factors as a means for minimizing human error accidents.

3. USGS should establish equipment standards for all pieces of equipment affecting safety and the environment. Standards should be based on the objectives to be achieved and should not deter technological development.

4. USGS should appoint an independent and representative board of experts which would periodically review state-of-the-art OCS technologies and make recommendations concerning desirable changes, particularly changes in equipment and performance requirements and standards.

5. An improved system for recording and reporting equipment defects, malfunctions, and failures should be established. USGS should be responsible for insuring that these data are systematically analyzed and for issuing appropriate notices and directives for corrective action.

6. Investigative procedures to determine causes of major accidents and to provide data for improving safety should be strengthened and assigned to an independent investigative board within the Department of the Interior. This board should function within Interior as the National Transportation Safety Board does within the Department of Transportation. The Board should make appropriate recommendations for changes and additions to equipment, such recommendations to be available to the public.

7. USGS should undertake an expanded research, development, and testing program. This program should be aimed at identifying technological gaps and be designed to stimulate R&D. USGS initiatives should include involving organizations from outside the petroleum industry in order to promote the communication of perspectives from other technological communities. This program should also provide a means for USGS to develop and maintain a greater degree of technological independence from industry.

---

[b]Recent joint efforts by aerospace and petroleum companies to develop subsea production systems illustrate how this approach can be advantageously employed by the petroleum industry.

8. USGS should actively promote greater cooperation within industry in the development of safety, accident prevention, and environmental protection technologies. Industry should be assured that cooperation in these designated areas will not be subject to antitrust prosecution. This could be accomplished by having the Anti-Trust Division of the Department of Justice issue guidelines for cooperative efforts or by having the Division give an opinion on specific proposals.

9. USGS should establish uniform standards and certification requirements for personnel who perform inspection and test functions. As a first step, USGS should appoint a committee including representatives of the OCS operators and technical training specialists to recommend certification criteria and standards.

10. USGS should establish a program to develop improved and standardized procedures for operating personnel. This program should be developed in conjunction with technical experts and behavioral scientists who specialize in developing technical training programs.

11. USGS should appoint an advisory committee to assist its Area Supervisors in drafting and revising OCS orders. This committee should include representatives of parties of interest in addition to industry in order to broaden participation beyond the present pattern of government-industry cooperation.

12. USGS should review its sanctions for inadequate performance or nonperformance to insure that they are adequate to insure compliance with OCS orders and other regulations. The present sanctions system seems generally to be adequate and the principal need will be to extend it on the basis of the philosophy recommended here. This philosophy calls for more stringent enforcement of stricter regulations.

Each of these recommendations is aimed at bringing about desirable changes in the behavior of industry and its government regulators. The major thrust is to expand participation in policy-making and administration and to insure that both government and industry take advantage of developments within other technological communities.

Social Technologies

The public policy issues which arise in connection with government's management and control of the development of OCS oil and gas resources can be divided into four major categories: information and data, environmental quality, government management practices, and jurisdiction.

Information and Data.

Background Data. Some opponents of the OCS alternative have cited the paucity of background data against which to assess environmental consequences of a reason for either slowing or stopping altogether the development of OCS oil and gas resources. There is little, if any, disagreement concerning the lack of these data. The important issue for this study is what data are required adequately to inform policy-making for the OCS over the next fifteen years.

Current and programmed research will improve on present knowledge of the environmental effects of OCS development. However, acquiring a functional understanding of coastal environments as ecological systems is an extremely long-term and expensive goal. Consequently, at best, policy makers are likely to have only selective and incomplete background data upon which to base OCS policy during the next fifteen years. It is, therefore, particularly important that the allocation of research resources for this area be made so as to provide for the most essential data needs of policy makers. These include a special emphasis on acquiring a more complete knowledge of background levels of hydrocarbons in physical and biological components of the marine environment and the physiological effects of acute and chronic exposure to oil on marine plants and animals. Knowing those two kinds of data would make it possible to establish informed discharge and pollution regulations. Since this is the case, environmental research to acquire these data should be given a high priority by both government and industry. A single federal agency, either the Environmental Protection Agency (EPA) or the National Oceanic and Atmospheric Administration (NOAA), should coordinate and be a depository for the results for both kinds of studies. Background studies should be initiated by NOAA no later than when an OCS area is included on the Bureau of Land Management's (BLM) five-year lease schedule. In addition, when development activity actually begins on a tract, NOAA should be responsible for the continuous monitoring of physiological effects on marine plants and animals.

Exploratory Information. In addition to lacking background data, the Department of the Interior has limited geological and geophysical data to use in its management and control of OCS oil and gas development. This limits the Department's capability for long-range planning and has led to the pattern of OCS development being largely in response to industry's interest in specific OCS areas rather than according to an Interior plan for systematic, orderly development. Limited information also affects the ability of BLM and USGS to make economic, engineering, and geologic evaluations of each OCS tract considered for sale. Much of the data available to them is proprietary and cannot be publicly disclosed. Therefore, these data cannot be published in the environmental impact statements written for every OCS lease sale. Having USGS either gather all its own geological and geophysical data or purchase it without the proprietary restriction is too expensive to be either reasonable or feasible. However, USGS should be adequately funded to permit it to contract for exclusive seismic surveys in order to acquire adequate exploratory data for regional OCS development, including overall land use planning. And, before each lease sale, USGS should contract for both exclusive seismic and subsoil surveys to the extent necessary to acquire data for determining whether development can be carried out safely. Purchasing these data on an exclusive basis would permit their public disclosure, including their use in environmental impact statements.

Environmental Quality. Clearly, preservation and improvement of environmental quality are major concerns which must be accommodated if the public is to be persuaded that OCS oil and gas resources can be developed at an acceptable level of risk. Three major aspects of this issue have been addressed legislatively: environmental impacts (the National Environmental Policy Act [NEPA] of 1969); water quality (the Federal Water Pollution Control Act [FWPCA] Amendments of 1972); and development of a contingency plan for responding to oil spills (initially required by executive order, now contained in the FWPCA Amendments of 1972).

NEPA. A major impact of NEPA has been to open up OCS policy-making to much greater public scrutiny and much broader public participation. This is a consequence primarily of the section of NEPA which requires an environmental impact statement to be written whenever a contemplated major federal action may have significant impact on the human environment. Responsible agencies and other interested parties are still working out, largely in the courts, an acceptable interpretation of this provision of NEPA. One effect has been to delay the pace of OCS development.

The statement requirement has revealed or highlighted several problems, including: fragmented responsibilities for energy and land use programs; the inadequacy of existing partial and incomplete energy and land use policies; and an enormous amount of duplication in statements prepared for the same policy area. To resolve these problems, a regional programmatic statement should be written. Such a statement would help to eliminate unnecessary duplication, facilitate better planning for coordinated OCS development, and give a better basis for assessing overall impacts. These long-range statements would then be supplemented by statements for individual OCS lease sales.

To guard against an agency acting as an advocate in the statement process, the Council on Environmental Quality should constitute an ad hoc committee to review all OCS draft lease impact statements. This committee, which should represent a broad range of interests and expertise, should determine whether the draft statement is adequate and consistent with the regional plan.

FWPCA Amendments of 1972. A major approach to water quality during the past several years has been an attempt to develop enforceable discharge requirements. The FWPCA Amendments of 1972 is the latest effort to establish requirements and procedures. However, the applicability of some provisions of the Amendments to the OCS is ambiguous; and a lack of sufficient data on the effects of discharges may lead to OCS facilities being exempt from the permit requirements. The provisions of the Amendments should clearly be extended to cover OCS facilities.

National Contingency Plan. The National Oil and Hazardous Substances Pollution Contingency Plan was established to provide for efficient, coordinated, and effective action to minimize damage from oil and hazardous substances. This includes advance preparation for, as well as actually responding to, a spill.

One aspect of advance preparation has been the need to develop technologies for responding to spills on the OCS since no such technologies really existed prior to the Santa Barbara blowout. Coordinating government accident response R&D is the responsibility of the National Response Team established by the contingency plan. This is supposed to be achieved by a multi-agency R&D committee. This has not been very effective and as a consequence there appear to be gaps in government R&D efforts. EPA should be given responsibility for monitoring, coordinating, and filling gaps in R&D in this area.

The Coast Guard, with the advice of EPA and USGS, should establish equipment and performance standards to be met by the clean-up cooperatives which industry has established in its efforts to comply with lease requirements. This would assure coordination of government and industry operational response capabilities and help to eliminate overlaps and gaps.

Government Management Practices. Our assessment of the social technologies employed by government agencies overseeing OCS development identified three major management problem areas in addition to the management aspects of the problems already discussed. These three interrelated problems raise issues concerning leasing, planning, and cooperation and coordination.

Issues associated with all three arise in large part because of a fragmentation of responsibilities for energy and land use programs. Management of the lease system, for example, is affected by the lack of a policy specifying what portion of energy demand should be satisfied by OCS oil and gas. Planning to determine this is constrained in part by the lack of exploratory information, a problem which was mentioned earlier. And planning is inherently difficult within fragmented authority structures because of the extensive inter- and intra-agency cooperation and coordination which is required. Experience shows that responsible agencies tend to promote their own particular programs and respond to the interests and demands of their own constituency. This makes cooperation and coordination difficult.

The most straightforward approach for dealing with all three of these problems would be to establish a Department of Energy and Natural Resources. Alternatively, an administration official, either a departmental secretary of an officer in the White House, should be designated energy coordinator for the federal government. Either approach could facilitate the coordination of energy, environmental, and land use policies and planning. The objective, in any case, is to develop an administrative structure capable of formulating a ten-year energy, environmental, and land use plan. This plan would provide a basis for establishing the leasing pace required to produce OCS oil and gas at a specified rate.

Whatever the approach for dealing with the fragmentation problem, the functions of promoting and regulating OCS development should continue to be separated.

As we indicated in our discussion of environmental quality issues, environmental impact statements have, in effect, become a means for forcing development of more coordinated, longer-range energy and land use policies and plans. The series of programmatic or regional statements proposed here is intended to force changes which will bring greater stability and certainty to government management and control of OCS oil and gas development.

At the operational level, the management system could be made more effective by greater centralization of responsibility. Pipelines are a case in point. BLM, USGS, the Office of Pipeline Safety, and the Federal Power Commission either grant rights-of-way, approve easements, issue certificates of convenience, set design criteria, or measure production, for example. In general, operational oversight responsibilities of this sort should be assigned to USGS since it already has the bulk of these kinds of responsibilities and possesses the greatest expertise for being an effective overseer. In addition, when new requirements for OCS facilities are established, such as those required by the Occupational Safety and Health Act and the FWPCA Amendments of 1972, USGS should be made responsible for enforcing standards established for OCS facilities.

Jurisdiction. At least three kinds of jurisdictional questions arise concerning the OCS: gaps in federal jurisdiction, disputes between the states and federal government, and the definition of the shelf area under national jurisdiction.

Gaps in Federal Jurisdiction. There are two jurisdictional problems in addition to gaps in the FWPCA Amendments discussed earlier. These include: the ambiguity of federal jurisdiction under the Submerged Lands Act and the OCS Lands Act as they apply to the six-mile area between three and nine miles off the Gulf coasts of Texas and Florida; and the lack of certification requirements for some types of drilling rigs.

The courts have given both Texas and Florida jurisdiction over their adjacent submerged lands in the Gulf out to nine miles from their coasts. Ambiguity arises because the Submerged Lands Act applies only to the three-mile zone between the coast and the outer edge of the U.S. territorial sea; and the OCS Lands Act applies only to the portion of the shelf extending seaward from the outer edge of state jurisdiction. Those two Acts should be amended to provide unambiguously for clear federal authority in this six-mile zone.

As for certification of drilling rigs, Coast Guard authority for inspecting and certifying rigs is based on their being treated as vessels. Mobile bottom-standing rigs should also be classified as vessels and inspection and certification requirements clearly extended to cover them.

Federal-State. The principal problems between states and the federal governments involve jurisdictional and land use issues, and the environmental impact of OCS development on state lands. In the past, jurisdictional issues have been resolved in the courts, a slow and not altogether satisfactory approach. These disputes should be anticipated when areas are included in the long-term leasing plan, and a non-judicial agreement on jurisdiction negotiated. Even if this is only an interim agreement, it will permit development to proceed while jurisdictional problems are being resolved.

Environmental concerns, including those related specifically to land use, arise because OCS activities necessarily impact on the adjacent state. Oil and gas produced on the OCS have to be brought ashore, and facilities for processing or transshipping it are located ashore. A comprehensive federal land-use law should be enacted to provide for federal, state, and local coordination in land use planning, including the OCS and coastal zones, and the law should require that these plans include provisions for siting necessary onshore facilities essential to OCS operations.

National-International. At the present time, there is no clear international rule fixing an outer limit of national jurisdiction over adjacent submerged lands. Neither the 1958 Convention on the Continental Shelf nor the decision of the International Court of Justice in the *North Sea Cases* provides an adequate rule. The resulting uncertainty has become a significant international issue and attempts are being made to resolve it, primarily through negotiations within the United Nations. Two specific issues are being raised: who is to manage and control the development of seabed resources beyond 200 meters, and who is to benefit from their development?

A UN sponsored international conference to deal with this and other ocean space issues is to be convened in 1974. Proposals being considered range from extending national jurisdiction over the ocean and seabed some fixed distance (up to 200 miles) to retaining the present rule contained in the 1958 Convention and interpreting it to extend national jurisdiction to the outer edge of the continental margin. Most proposals now being discussed also provide for some sort of international authority to manage and control development of seabed resources outside national jurisdiction.

Participants engaged in working out new rules in this area represent a variety of values, perspectives, and ocean space interests.

Whatever the final solution to the seabed jurisdictional issue, if it is to accommodate the breadth of interest, values, and perspectives found within the international political system, it should fix an outer limit at some specified water depth or set number of miles from the coastline, establish an international authority to oversee development of resources beyond this limit, and accommodate other ocean space interests such as a right of transit and the special dependence of some countries on living resources in an extended adjacent ocean zone. The important point for development of oil and gas seabed resources adjacent to the U.S. is to provide for their orderly, safe development at an acceptable level of environmental risk.

A Management Plan for the OCS

Based on the assumption that OCS oil and gas resources will continue to be developed, this technology assessment was undertaken to determine how development of these resources could be optimized. In this chapter, we have summarized our major recommendations. While we have made recommendations for improving specific weaknesses in the physical technologies used on the OCS, our major effort has been to develop a management plan which will promote optimal development. An overall view of this plan and what it accomplishes is presented in Table 11-1. It should be kept in mind that the principal objectives intended to be achieved by the plan include:

1. effective, coordinated long-range planning and policy-making;
2. broadened participation in OCS policy-making and administration;
3. greater expertise within government to enhance the regulatory capabilities of responsible agencies;
4. more extensive, publicly disclosable information and data for making government management decisions;
5. greater centralization of responsibility and authority;
6. clarification of jurisdictional gaps and ambiguities; and
7. specifying the portion of the nation's energy demands to be satisfied by OCS oil and gas.

While recognizing that an ideal management system might look quite different if the existing system could be disregarded, our plan has been formulated on the assumption that constraints within our political system dictate modification rather than wholesale revision of the present system for managing OCS development. With this limitation in mind, we have used a hierarchy of environmental impact statements as a means for integrating land use and energy policies and planning, including provisions for long lead times and a formalized process for working out critical political accommodations. These are essential changes if OCS resources are to be optimally developed.

Table 11-1
A Plan for OCS Development

Present	Recommended	Changes Required to Implement Recommendations
	Jurisdictional Matters Affecting OCS Management	
Outer boundary of national jurisdiction uncertain	1. Fix outer boundary of national jurisdiction 2. Establish an international seabed authority	International agreement
Jurisdictional disputes between the federal government and state governments	Definitive agreement on state-federal boundary If not possible negotiate interim agreement	Legislative and/or Executive action
Ambiguous federal authority over 6-mile zone between territorial sea and OCS	Congress should clarify jurisdiction in this area	Amend Submerged Lands Act and/or OCS Lands Act
Uncertain	Top official designated federal energy coordinator on basis of organizational position	Presidential directive
	OCS Development Planning Responsibilities	
BLM prepares a 5-year tentative OCS leasing schedule in consultation with USGS	BLM formulate a 10-year OCS development schedule in consultation with USGS	Department action
BLM, as lead agency, is preparing a programmatic statement on the 5-year OCS leasing schedule	USGS, as lead agency, but with cooperation of Commerce, prepare a programmatic impact statement which will serve as a general development plan for each region included in the 10-year schedule which integrates energy, environmental, and land use components	
	BLM, in consultation with USGS, define by fixed coordinates areas (not tracts) included on 5-year lease schedule	Bureau action
USGS participation in group seismic shoots to collect data for tract evaluation; data is proprietary	USGS collect data to extent required to make tract evaluations. Data to be publicly available in lease sale impact statements	Increase appropriations for data collection
	NOAA initiate continuous hydrocarbon background studies of areas on 5-year lease schedule	Agency action
USGS grants exploration permits	Continue present responsibility, making all management decisions consistent with general development plan	

Table 11-1 (cont.)

Present	Recommended	Changes Required to Implement Recommendations
	OCS Development Planning Responsibilities (cont.)	
BLM receives and reviews lease nominations; BLM-USGS selects tracts BLM publishes list	Continue present responsibility, making all management decisions consistent with general development plan	
BLM prepares draft lease sale impact statement with inputs from other federal agencies	Based on the programmatic statement and nominations from industry BLM, with USGS, prepare draft lease sale impact statement to be available 15 months prior to sale and 3 months prior to public hearing; statement to supplement or amend programmatic statement as necessary	Department action
	CEQ constitute committees to review draft impact statements, review and report to be made public	Presidential directive
Interior publishes hearing notice	Continue present responsibility, making all management decisions consistent with the general development plan	
BLM prepares final lease sale impact statement; BLM publishes notice of lease sale 30 days after it is filed with CEQ	Continue present responsibility, making all management decisions consistent with general development plan	
BLM administers lease sale using the bonus bid-fixed royalty system	Continue present practice; consider selected experimentation with the staggered bonus-fixed royalty	Staggering bonus bidding requires legislative action
	Post-Lease Sale Management: USGS General Policy and Management	
General:	USGS prepare any post-lease sale impact statements judged necessary for especially sensitive areas	Department action
	USGS enforce all environmental quality standards	Inter-agency agreement
	USGS enforce OSHA on OCS	Inter-agency agreement
Drilling and Development: USGS requires lessee to submit exploratory drilling plan	Continue present responsibility, making all management decisions consistent with the general development plan	

Table 11-1 (cont.)

Present	Recommended	Changes Required to Implement Recommendations
Post-Lease Sale Management: USGS General Policy and Management (cont.)		
USGS requires field development plan	See recommended changes under Management of Technologies	
USGS requires "Application for Permit to Drill"	Continue present responsibility, consistent with general development plan	
USGS enforces regulations covering safety and equipment concerns not included in Labor, Coast Guard, or state and local government regulations	Continue present responsibility, consistent with general development plan	
Production: USGS requires monthly production report	Continue present responsibility, consistent with general development plan	
USGS requires monthly report of oil runs, gas sales, and royalties	Continue present responsibility, consistent with general development plan	
USGS requires reports and logs of well completions and recompletions	Continue present responsibility, consistent with general development plan	
USGS enforcement of safety and equipment regulations same as drilling and development phase	Continue present responsibility, consistent with general development plan	
Transportation: USGS authorizes rights-of-use for gathering lines	Continue present responsibility, consistent with general development plan	
Companies apply to FPC for approval of common carrier gas lines	USGS provide FPC estimates of recoverable gas reserves; USGS assess lines consistent with general development plan	Inter-agency agreement
BLM authorizes transmission lines rights-of-way	BLM grants rights-of-way only on recommendation of USGS	Departmental action
OPS responsible for common carrier design and performance standards	USGS enforce design and performance standards for offshore pipelines	Inter-agency agreement

Table 11-1 (cont.)

Present	Recommended	Changes Required to Implement Recommendations
	Management of Technologies	
USGS establishes and enforces general standards; trend toward more detailed standards	USGS establish equipment requirements in terms of objectives to be achieved; enforce by appropriate inspections and sanctions	USGS action
USGS requires limited reporting of failures and malfunctions	USGS improve reporting and analysis procedures for failures, malfunctions, and equipment defects; issue appropriate notices and warnings	USGS action
USGS investigates OCS accidents	Interior establish board similar to National Transportation Safety Board to investigate OCS accidents	Department action
	USGS appoint independent representative committee of experts to review state-of-the-art technologies periodically and recommend desirable changes	USGS action
	USGS undertake expanded R&D program	Increased appropriation
	USGS develop uniform standards and certification requirements for personnel who perform inspection and test functions	USGS action
	Industry develop program of improved and standardized training procedures for operating personnel	Industry action
	Industry expand its R&D programs	Industry action
	USGS promote greater cooperation within industry in development of safety, accident prevention, and environmental protection technologies	Interior and Justice action
	Industry increase use of systems design approach	Industry action
	Industry increase of human factors design criteria	Industry action
	USGS encourage development and use of subsea production systems	USGS and industry action
	USGS immediately compile a list of weak technological	USGS action

Table 11-1 (cont.)

Present	Recommended	Changes Required to Implement Recommendations
	Management of Technologies (cont.)	
	components; publish annual summary of progress in correcting weaknesses	
	USGS detail all specifications and regulations in OCS orders for each area	USGS action
USGS and industry groups review OCS orders	USGS appoint broadly representative committee for each USGS area to participate in review of OCS orders	USGS action
	Other Agencies' General Policy and Management	
Corps of Engineers authorizes placement of any permanent or floating structure in navigable waters	Continue present responsibility, making authorizations consistent with general development plan	
Coast Guard enforces regulations covering safety, equipment, vessel transportation, and accidents on the OCS; jack-up rigs are not certified since they are not defined as vessels	In addition to present responsibilities, Coast Guard establish formal certification and inspection requirements for jack-up drilling rigs	Coast Guard action
USGS establishes and enforces discharge standards; FWPCA not clearly applicable	EPA establish discharge standards for OCS; USGS enforce standards	Amend FWPCA
	With start of exploratory drilling, NOAA assure continuous monitoring of commercially useful and sensitive marine species	Agency action
	EPA or NOAA assume responsibility for monitoring, coordinating, and filling gaps in environmental research	Inter-agency agreement
National Response Team R&D committee coordinates cleanup, containment R&D	EPA assume responsibility for monitoring, coordinating, and filling gaps in R&D aimed at improving cleanup and containment technology	National Response Team agreement
	Coast Guard establish equipment and performance standards for cleanup cooperatives	Legislative action

A Comparison with Other Alternatives

Although this study was limited to a technology assessment of the OCS as a domestic source of oil and gas, we did compare the broad social impacts of developing the OCS, the North Slope of Alaska, and the increased levels of imports that will be required if either or both of these domestic alternatives are not developed. For purposes of illustration, comparisons of impacts are based on production and imports at a level of 2 million barrels per day. These overall comparisons, based in part on secondary sources, are summarized in Table 11-2.

OCS vs. Imports

As can be seen in Table 11-2, the OCS offers advantages over increased imports in each of the categories in which the two alternatives are compared except

Table 11-2
A Summary Comparison of Impacts

Categories of Impacts	OCS v. Imports	OCS v. TAPS
Economic		
Consumer costs	–	o
Balance of trade	+	o
Government revenue	+	+
Regional	+	+
Overall	+	+
Environmental		
Spill source (worldwide)	+	+
Spill rate (worldwide)	+	+
Location	+	+
Biological consequences	+	+
Overall	+	+
National Security (Overall)	+	o
Multiple Use		
Land Use	o	+
Coastal Use	–	–
Overall	–	?
OVERALL	+	+

Source: U.S., Department of the Interior, *Final Environmental Impact Statement: Proposed Trans-Alaska Pipeline*, Vol. I: *Introduction and Summary* (Washington: D.C.: Government Printing Office, 1972.

Note: Obviously many of the judgments made in this table are speculative; however, all entries are based on the quantitative and qualitative analyses made in this study.

+ indicates an advantage for the OCS; – indicates a disadvantage for the OCS; o indicates no discernable difference; ? indicates that no conclusion could be drawn.

multiple use. Land uses for both are about the same: both require refineries; the OCS needs additional onshore facilities; and imports require increased port facilities. But OCS development has a greater overall potential for interfering with other uses, particularly fishing and marine transportation. If subsea production systems are widely used, this potential conflict with other users will be reduced.

There are a number of economic advantages for the OCS; however, it appears now that imports offer some advantage in terms of consumer costs. This advantage is at best slight and subject to considerable uncertainty in the future.

Considering environmental impacts worldwide, the OCS is less of a threat to the environment than increased imports will be. In part, this is due to the harmful effects of refined products which have to be considered with imports. If only U.S. waters are considered, imports appear to have an advantage, but this fails to take into account differences in the kind of petroleum spilled and the fact that spills outside can produce effects within U.S. waters.

OCS vs. TAPS

Developing OCS resources offers some advantages over TAPS. In large part this is due to the greater complexity and exposure to a wider variety of potential risks associated with developing the North Slope, transporting the oil by pipeline to Valdez, and then by tanker to West Coast U.S. ports. There is no experience upon which to base an estimate of the risk involved with TAPS, but if anything, the risk is probably greater than in the lower 48 states. However, different impacts could be expected if an inland pipeline were to be developed. For example, this alternative to TAPS would eliminate marine impacts.

A beginning assumption of this study was that OCS development would continue. The overall objective set for the study was to find ways to insure that development was optimal in a broad social sense. On the basis of our comparison of alternatives, we have concluded that continued OCS development is socially preferable to increased imports or TAPS. This is on the basis of current policies, practices, and technologies. If changes result in more optimal OCS development, the advantages of developing the OCS should become even greater.

Glossary

Blowout. An uncontrolled flow of gas, oil, and other well fluids from a well to the atmosphere. A well blows out when formation pressure exceeds the pressure being applied to it by the column of drilling fluid.

Blowout Preventer (BOP). Equipment installed at the wellhead for the purpose of controlling pressures in the annular space between the casing and drill pipe, or in an open hole during drilling and completion operations.

Capping. Closing off a well to re-establish control after a blowout. If there is a fire, it must be extinguished before the well can be capped.

Continental Margin. The submerged prolongation of adjacent land extending to an average water depth of 200 meters (approximately 660 feet).

Downhole Safety Equipment. Valves or other devices installed below the Christmas tree in production wells to prevent blowouts.

Flaring. The disposal of unwanted gas by burning in the atmosphere.

Mass Flow Monitor. Device for metering flow through pipelines for the purpose of early identification of leaks.

Mobile Bottom-standing Rigs. Includes jack-ups, which have legs that extend to the ocean bottom and raise the hull to the water's surface, and barges which are used only in shallow water and are sunk after being towed to location.

Mud Monitor. Device for measuring sudden gain or loss of drilling mud in the well bore. Equipment which is capable of identifying the loss or gain of as little as one barrel is now available.

Multi-phase Pumping. A procedure for moving simultaneously through pipeline systems various combinations of oil, gas, and water.

Outer Continental Shelf (OCS). The submerged lands extending from the outer limit of the territorial sea to some undefined outer limit. In the U.S., this is the portion of the shelf under federal jurisdiction.

Relief Well. A well drilled to intersect another well at some point below the surface, used to regain control of wells that are out of control.

Sand Probe. A device used to warn of excessive sand erosion in wells containing velocity-actuated downhole safety valves.

Seismic Survey. A geophysical exploration technique in which generated sound waves are reflected or refracted from underlying geologic strata and recorded for later analysis.

Storm Choke. Common terminology for a velocity-actuated downhole safety device.

Subsea Production System (SPS). The complex of piping valves and related equipment used to produce oil and gas from individual or connected subsea completions.

Subsoil Survey. Investigation of shallow focus ocean bottom conditions, usually for the purpose of setting platforms or rigs.

Technology Assessment. An attempt systematically to identify, analyze, and evaluate the potential environmental, legal/political, and other social impacts of a technology.

Territorial Sea. The sea area immediately adjacent to a coastal nation within which it claims comprehensive jurisdiction.

Tubing. Conduit for routing oil or gas to the surface.

Acronyms

BLM	Bureau of Land Management
BOP	blowout preventer

BTU	British thermal unit
CEQ	Council on Environmental Quality
EPA	Environmental Protection Agency
FPC	Federal Power Commission
FWPCA	Federal Water Pollution Control Act
NEPA	National Environmental Policy Act of 1969
NOAA	National Oceanic and Atmospheric Administration
NSF	National Science Foundation
NTSB	National Transportation Safety Board
OCS	outer continental shelf
OPS	Office of Pipeline Safety
OSHA	Occupational Safety and Health Act of 1970
R&D	research and development
TAPS	trans-Alaskan pipeline system
UN	United Nations
USGS	U.S. Geological Survey

12 Oil Shale: A Huge Resource of Low-Grade Fuel

William D. Metz

Oil shale is a vast domestic source of petroleum. Does it represent an easy answer to our energy needs?

The rich oil shale deposits on the western slope of the Rocky Mountains constitute a potential source of fuel several times as great as the identified reserves of U.S. oil, and processes for extracting synthetic crude oil from the thick seams of brown-black rock have been ready to go for 15 years. Technologically, the production of synthetic crude oil from shale is a simple process. When the shale is crushed and heated to 480°C, raw shale oil is released. Because it does not require special mineral preparation, high pressures, or difficult catalytic procedures, the process of oil shale recovery is easier than either coal gasification or coal liquefaction. Until October 1973, the principal limitation to oil shale recovery was its price, which was projected by the National Petroleum Council to be about $5.50 per barrel.

Now that the price of domestic crude oil is at least $7 per barrel, oil shale appears to be economically viable. But "it's not simply a question of raising the price of conventional oil high enough and shale oil will automatically appear—although much of the material on the subject blithely makes this assumption," according to James E. Akins, U.S. Ambassador to Saudi Arabia and formerly chief State Department advisor on fuels and energy. "There are severe ecological and mining problems in extracting shale oil which cannot and must not be ignored." Under the close scrutiny that energy alternatives have received since the October 1973 boycott, it is clear that shortages of water will probably limit shale oil production to a few percent of the U.S. petroleum consumption, no matter what the crude oil price.

Water Shortage Limits Development

According to studies by the Department of the Interior and the Atomic Energy Commission, that limit will probably be 1 million barrels per day, a figure that pales beside the daily U.S. oil consumption, which is 18 million barrels. Not even

This chapter first appeared in *Science*, Journal of the American Association for the Advancement of Science, June 21, 1974, vol. no. 184, pp. 1271-1275.

oil shale enthusiasts seem to be proposing that shale oil can be squeezed out of the rock at a rate much higher than 1 million barrels per day, because 3 barrels of water will be required for each barrel of oil produced with the existing technology. The production rate could perhaps be doubled if in situ technology—the release of oil by cracking and burning the shale in place underground—were perfected, because water otherwise used to dispose of spent shale above ground could be saved. But in situ technologies have not been proved to be workable yet.

The actual amount of shale oil extracted will also depend heavily on government policies. The Department of the Interior, which controls 80 percent of the rich shale lands, is currently leasing six small (5120-acre) tracts for development. Estimates of the shale oil production from these tracts as well as from private lands range from a high of 300,000 barrels per day by 1980 to a low of 100,000 or 250,000 barrels per day by 1985. Private oil shale developers argue that production will not nearly reach 1 million barrels per day, even with an expanded leasing program, unless the government provides substantial economic incentives, such as guaranteed loans, rapid amortization of plants, import restrictions, or a price floor.

Oil shale is found in many areas of the contiguous 48 states and in Alaska, but almost all the shale that is rich enough to yield more than 15 gallons of oil per ton is in one geological formation, along the Green River in Colorado, Wyoming, and Utah (Figure 12-1). About 1800 billion barrels of shale oil are buried in the three-state region, but almost all the prime shales, with at least 30 gallons of oil per ton in seams 30 feet or more thick, are in the Piceance Creek basin in Colorado. Those are estimated at 117 billion barrels, about double the 52 billion barrels of oil that is identified and recoverable. Much of the high quality shale is in a formation called the Mahogany Zone, which is about 70 feet thick, and can be seen on the exposed faces of the canyon walls. The zone is a deposit from the sedimentation of Lake Uinta, a freshwater lake that covered the area during the Tertiary period. The burnable component of so-called oil shale, which is actually a marlstone rock, is an organic polymer called kerogen. Large quantities of nahcolite, $NaHCO_3$ and dawsonite, $NaAl(OH)_2CO_3$, are also found in parts of the Piceance basin, and they contribute to the natural alkalinity of the shale.

The people of western Colorado have heard recurring rumors that oil shale developers would start mining the region since before World War I. Oil shale was mined in Scotland for about 100 years after 1860, and 25 million tons of shale are now mined each year in Estonia, more than half for burning at mine-mouth generating stations rather than for conversion to oil. In 1957 the Union Oil Company of California tested a pilot plant that was built on privately held land along Parachute Creek, south of the major deposits in the Piceance Creek basin, which is predominantly federally owned. The Union plant successfully extracted oil from 300 to 1000 tons of shale per day, and was closed in 1958 because market prices for crude oil were too low to make the operation profitable. Union recently announced plans to build a full-size plant by 1979 that will

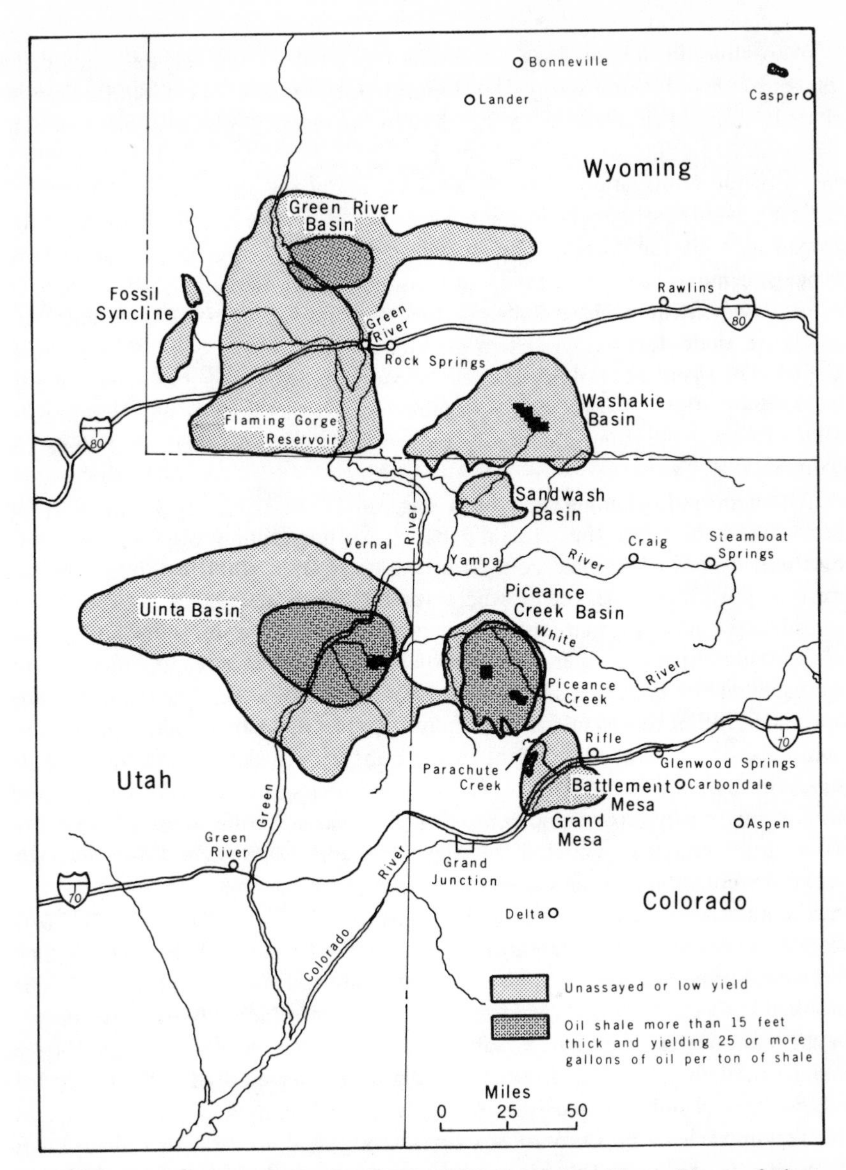

Figure 12-1. The Major U.S. Oil Shale Deposits. Note: Most of the high-grade ore is found in the Piceance Creek basin in western Colorado. Black areas indicate tracts of federal land leased for private development.

produce 50,000 barrels per day with an improved version of the earlier process. The original version employs a large piston to continually pump rock upward through a retort where hot gases release shale oil; the oil drains out the bottom and spent shale is forced out the top in the form of large chunks or clinkers.

Also along Parachute Creek, only about 75 miles from the well-known ski center of Aspen, the Colony Development Corporation[a] owns 8000 acres of shale land, and has spent up to $55 million for research with a pilot plant that has processed 1000 tons of shale per day. The Colony operation is the source of much of the information currently available about the environmental effects of shale recovery, particularly in the environmental impact statement filed by the Department of the Interior for its oil shale leasing program. In the Colony process, ceramic balls are heated and then mixed with finely crushed oil shale to break down the kerogen to shale oil. Heat transfer is very efficient because two solids are in contact rather than a gas and a solid, and the yield of shale oil is virtually 100 percent. Colony has nearly completed the design of a large plant, to produce 50,000 barrels per day, which is scheduled to begin operation in 1977. Colony will almost certainly be the first company to market shale oil commercially.

Although strip mining has been suggested, the most likely method of removing shale from the Piceance basin is underground mining, since the overburden above the shale zone is generally at least 1000 feet thick. In the western portion of the basin, where tract C-a was leased, the overburden is considerably less, so that tract is a possible candidate for strip mining. A 100,000-barrel-per-day operation is projected. But for most of the region strip mining would be too expensive, so the shale will probably be removed by carving out underground rooms in the Mahogany Zone, leaving behind pillars of shale to hold up the roof. Since the height of the mine ceiling, determined by the shale zone thickness, will be 50 to 70 feet, an underground shale mine could accommodate very large equipment, such as trucks and front loaders. According to a study recently prepared by the Cleveland Cliffs Iron Company, the operation would be very similar to one in a large surface mine.

The alternative to mining would be in situ conversion of shale to oil, but true in situ processes appear to require at least 15 more years of development, if they ultimately prove feasible, in spite of much private and government research. The problem is to create enough void space in the shale, which is quite impermeable in its natural state, so that a flame front will burn evenly through a large underground region. The Bureau of Mines has tested hydraulic fracturing followed by chemical explosives. After the shale is ignited, air is pumped into the region at high pressure to sustain burning, and the resulting shale oil is pumped out. Although this process was promising in shallow shale beds, it has not yet worked in deeper zones.

A hybrid process, often called in situ, has been proposed by the research arm of the Occidental Petroleum Corporation. Up to the point of igniting the shale, the Occidental process requires conventional underground mining. A stope mining technique is used to tunnel into the shale, hollow out a low room, and

[a]Colony is a joint venture by the Atlantic Richfield Company, the Oil Shale Corporation (TOSCO), Ashland Oil and Refining, and Shell Oil Company.

then blast down the ceiling. Then the room is sealed off and ignited. Shale oil is drained out through a trough previously cut in the floor. From 20 to 35 percent of the shale would have to be removed from the mine to make the tunnels and rooms, and that shale would be processed above ground with the same problems of any shale extraction operation. Under ground, the main uncertainties about the Occidental process are how large the rooms can be, how uniform the broken shale will be, and whether the process will work in shale that is susceptible to groundwater. Once ignited, a room would need to burn for several months to extract all the shale oil.

Occidental has successfully produced shale oil from a room 30 feet square and 70 feet high, and is now lighting a second room the same size. The yield is much lower than in aboveground processing. Experiments that simulated the Occidental in situ retorting with a large steel retort of the Bureau of Mines at Laramie, Wyoming, obtained a maximum of 60 percent shale oil recovery. After testing a third small room, Occidental is proposing to mine a much larger room, about 250 feet high, to test the process on a commercial scale. To extract 30,000 barrels per day would require completing at least one such room each week, according to knowledgeable observers. Even if all four test rooms are successful, Occidental would probably require at least 4 years to begin commercial operation, in the opinion of Gerald Dinneen, director of the Bureau of Mines station at Laramie. Others, particularly competitors, predict that 10 years will be required for development.

Unless a true in situ process can be found, oil shale development will require a mining effort that can only be called gargantuan. Oil shale has less energy value per ton than practically any substance that has been used for commercial fuel. Even for the high-grade deposits, about 1.5 tons of shale will be required to produce each barrel of oil. With coal, only about 0.5 ton would be needed to produce a barrel of synthetic crude oil, if the technology for coal liquefaction were available. To get 1 million barrels of oil per day from shale would require mining, transporting, crushing, and retorting 1.5 million tons of oil shale, then disposing of 1.3 million tons of residue. Altogether, 2.8 million tons of material would be handled each day, or about 1 billion tons per year. Last year, the total coal production in the United States was 570 million tons. Even if the Occidental in situ process were used, about 250 million tons of rock would be mined and moved each year. In order to obtain 3 million barrels of oil per day from shale, these figures would have to be trebled. In the course of 3 months, as James Akins likes to note for perspective, the tonnage would be equivalent to the weight of earth and rocks excavated in constructing the Panama Canal. "This does not mean that extraction of oil from shale is impossible," says Akins, "It is just that it is very difficult."

Because of the low energy value of shale, which necessitates such massive mining, and the aridity of the region where shale naturally occurs, its development will inevitably alter the environment and has the potential for extensive damage. The growth of a mature oil shale industry could present problems with the disposal of spent shale, revegetation of affected areas, disturbance of natural habitats, increase in the salinity of the Colorado River, and release of dust and sulfur dioxide to the air. A true in situ process would eliminate some problems, but by no means all. According to Charles Prien, at the University of Denver, Colorado, in situ methods may just hide the environmental problems underground. Specifically, there are two major aquifers for return of water to the Piceance Creek–which eventually flows into the Colorado River. In the northern half of the basin, one aquifer is above the shale zone and one is below. Prien points out that in some areas of the basin mining could create communication between these two aquifers, and that water passing through the spent shale could leach salts out of it. He thinks that the problem could probably be better controlled in the course of conventional mining than by in situ methods. In the following discussion of environmental effects, conventional underground mining is assumed.

Environmental Changes

After processing, most companies plan to dispose of waste shale above ground, probably in nearby canyons, rather than store it back underground in the mines at a higher cost. Spent shale from the original Union Oil Company process is composed of large chunks at least 10 cm across, shale from a Bureau of Mines process is about 1 cm in average size, and spent shale from the TOSCO process is a fine powder about 0.07 mm in size. Characteristics such as permeability and alkalinity also vary significantly, so the shale disposal problems with one process may be different from the problems with another. The spent shale from a 50,000-barrel-per-day plant will fill a typical canyon in the region, such as Davis Gulch where Colony will dispose of its shale, to a depth of 250 feet in 7 years, leaving about 700 acres of bare shale exposed on top. To minimize the volume, each layer of shale will be compacted as the pile is built up, and the pile will be stable against sliding if the slope is less than 3:1, according to Colony. However, tests by John C. Ward at Colorado State University, Fort Collins, have shown that snowfall destroys the compaction of spent shale to a depth of at least 2 feet. So sliding could occur if the spent shale isn't protected with topsoil and vegetation, according to Ward.

How readily the native grasses and shrubs can grow on spent shale is a question that will largely determine the environmental effects of the shale industry. Of course, the aesthetic appeal of a canyon cannot be regained by growing grass on the false floor, but the success of revegetating spent shale will affect the quantity of salts leached out of the shale by rain and snowfall, and the reduction of the wildlife population in the area, as well as the stability of the pile against erosion. The Piceance basin is the winter range for one of the largest herds of migratory deer in the world, 30,000 to 60,000 mule deer, and is the home of at least a dozen nests of golden eagles. The Department of the Interior estimates that a mature shale industry would disturb 80,000 acres of land over 30 years.

Spent shale must be heavily watered to remove salts before most grasses will grow. The spent shale from the TOSCO process has a significant alkalinity (pH about 9) and contains essentially no nitrogen or phosphorous. In many small test plots, Colony has gotten at least three native grasses to grow profusely, after watering heavily for the first year, using commercial fertilizer, and mulching. Two native shrubs have also grown, but so far none of the woody species used as browse by deer. Colony does not readily distribute data on how much watering was necessary to achieve this. The annual rainfall in the Piceance Creek region is only 12 to 15 inches per year, and one of the major unanswered questions is whether vegetation growing on shale can survive several dry seasons. A test plot of spent shale from the Union Oil process grew grasses naturally, however, according to Harry Johnson at the Interior Department. Tests on a plot of TOSCO spent shale by Ward showed that after 41 inches of water had been applied, with no fertilizer or mulch, only tiny weeds appeared 2 years later. Ultimately, it may be necessary—and cheaper—to cover spent shale with topsoil than to revegetate it. Besides a shortage of natural water, the industry faces another problem for artificial seeding of shale piles: where can you buy four-winged saltbush seeds?

The residents of the Rocky Mountain region and the Far West could also experience a problem with increased salinity in the Colorado River. A mature shale industry (1 million barrels per day) would deplete the quantity of fresh water flowing into the Colorado enough that the salinity at Hoover Dam would increase by 1.5 percent, according to the Department of the Interior. But many observers think that the 1.5 percent effect could be dwarfed by the contributions of salts added to the Colorado River from saline underground aquifers and by leaching of spent shale. The environmental impact statement for prototype leasing did not try to estimate such salt loading, which has been estimated to increase the salinity at Hoover Dam as much as 50 percent.

There is no doubt that the runoff from bare shale is extremely salt-laden. Ward found that the concentration of dissolved inorganic solids to be as high as 5000 mg/liter, about five times the salinity of the Colorado River in the region. Colony plans to install a catch basin at the toe of the shale embankment to keep the runoff from reaching Parachute Creek and thence the Colorado. To evaporate water fast enough so that it didn't overflow, such a basin would have to be quite large. Further increases in the salinity of the Colorado River might require considerable expenditures for desalinization downstream, where there is heavy demand for municipal and agricultural water.

Besides the potential for polluting the Colorado River, an oil shale industry could pollute the air with dust from mining, crushing, and disposal operations, and with sulfur dioxide emission from the retorting process. Some local impact on plants would occur, and there is considerable doubt whether shale operations could meet the recent court ruling that air quality not be degraded when it is purer than environmental standards.

Environmental degradation is certain to occur with a mature oil shale industry, and there is the potential for a very serious impact, according to Harry Johnson, although only local degradation will occur with the prototype program. All parties, including the environmental groups, agree that not enough is known from small test plots to assess the environmental damage to be expected on a larger scale, and the Department of the Interior intends to monitor the prototype program closely and use the information to decide about expanding the industry.

Since most of the rich oil shale is on federal lands, the industry cannot grow to 1 million barrels per day without additional public land leasing. Critics of the prototype leasing operation argue, however, that 5120-acre leases were certainly not needed to determine the environmental effects, that there are many loopholes in the stipulation that all affected lands should be rehabilitated, and that the enforcement provisions are inadequate.

Perhaps one of the most important concerns is that the shale industry will grow up, prove unsuccessful, and be abandoned, leaving the western slope of the Rockies in somewhat the same condition as Appalachia. There is no doubt that a 1 million-barrel-per-day oil shale industry would alter the character of the region. The industry would bring in 115,000 people, more than double the population that now lives in the counties where oil shale is found. Towns would have to expand their municipal services, mobile home settlements would be wheeled in, and many rural aspects of life in the region would disappear. If the industry failed, it could leave the region environmentally desolated and economically broke.

Although oil shale is a bounteous reserve compared to oil, it is clear that it cannot be extracted from the earth without paying a far greater environmental cost. Even so, the rate at which oil can be fired out of shale will be more dependent on the water reserves than on shale reserves. Even the estimated ceiling of 1 million barrels per day may be high, because much of the available water in the region has reportedly been cornered for surface coal mining. One would think that the major reason for urgently developing shale oil would be to utilize its portability as a liquid fuel. But, more likely than not, the first use of raw shale oil will be to burn it, in place of sulfureous coal, for generating electricity in the far Southwest.

13 The Energy Picture: A Summary

Alfred J. Van Tassel

Looked at in a world perspective it was perhaps inevitable that the United States would react most sharply to the world shift in the relationships of supply and demand for energy. Accustomed almost throughout its history to the lavish use of energy, the slightest tightness in supply was bound to feel like a painful squeeze. The main facts were almost tiresomely familiar: with 5.8 percent of the world's population, the United States managed to consume 32.9 percent of the world's energy.[1] There was no need to detail the roots of this disproportion; most importantly it derived from America's devotion to the private automobile, a worship in process of rapid transfer to other capitalist nations, but it was compounded as well of a restless desire to be somewhere else, however transported, and when there to alter the climate from what it had been.

For the conservationist the signs of energy shortage were a partial fulfillment of frequent forecasts from the past. But to many who had been irked by restrictions on pollution advanced in the name of improving the environment, it was a time to do battle against all measures that would impede in the slightest the abundant flow of energy.

Where were energy savings possible? A glance at Table 13-1 showing consumption by sector suggests the answers. The purveyors of energy had been quick to adjust prices upward in response to signs of shortage and there would presumably be a more or less automatic response in the industrial sector accounting for 31 percent of national consumption. Not only were there opportunities for energy economies in the industrial sector, but the output mix might also be altered in favor of less energy intensive products. Each such change would be so small in itself as to be almost undetectable, but the cumulative effect might be tremendous. Where there was a profit to be made, industry was likely to respond with alacrity. Thus, the overwhelming bulk of such genuine recycling of materials as took place in America was carried on by industry.

An appeal to patriotism was relied upon to reduce fuel consumption by the ordinary citizen and the small businessman. A prime target for such exhortation was the 45 percent of national energy consumption represented by (a) transportation, and (b) household and commercial. The 24 percent for transportation was mostly accounted for by gasoline for the family car, and the exhortation to drive less and more economically could be and was supplemented by on-again, off-again speed limits which apparently effected considerable savings when in force.

Table 13-1
Total U.S. Consumption of Energy by Consuming Sector (1970)

Sector	Percentage
Electric Utilities	25
Transportation	24
Industrial	31
Household and Commercial	21
	100[a]

[a]May not add to 100 due to rounding.

Source: Resources for the Future, *Energy, Economic Growth, and the Environment* (Baltimore, Md.: Johns Hopkins University Press, April 1971), Table 5, p. 176.

Much of the 31 percent of national energy consumed by the household and commercial sector went for space heating and cooling and all stops were out on the public relations programs to gain popular cooperation. These voluntary measures were apparently sufficient to avoid a full-scale crisis in the winter of 1973-74.

In many respects, the most interesting response was from the electric utilities, which account for 24 percent of national energy consumption. We can leave aside the widespread and entirely predictable effort to get approval from state utility commissions for upward adjustments in rate schedules to compensate for any increase in fuel costs. By far the most spectacular reaction was from the American Electric Power Company, which used the occasion to mount a massive advertising and public relations attack on efforts of the Environmental Protection Agency to reduce sulfur oxide pollution. Not all utilities took this line; many cooperated with EPA in experimental efforts to clean stack gases, and there were numerous promising developments as a result. The Pacific Gas and Electric Company tapped a little used source–geothermal power–and contrived in this way to supply half the electricity needs of the city of San Francisco. From the viewpoint of the national energy problem, the principal virtue of geothermal power was that it was necessarily a domestic source. Supporting an ingrained habit of free-handed use of energy might prove disastrous for the balance of payments if foreign sources made up all the deficit.

The major domestic source of energy was coal, as shown by Table 13-2. Actually the proven reserves of coal were more abundant than its 20 percent share of the national energy supply might suggest. But there were more major obstacles to the expanded utilization of coal. Underground mining was fraught with dangers and was expensive despite intensive mechanization. Strip mining had left what threatened to be enduring scars in Appalachia without bringing prosperity, hardly an example likely to inspire acceptance of stripping in the wide open spaces of the Far West where the scenary was an important economic asset. Nevertheless much of the immense reserves of the west was within reach of strip mining, whose economies of production would probably be necessary if western coal were to overcome the handicap of high costs of transportation to the principal markets–all some distance east.

Like coal, natural gas was basically a domestic fuel resource mostly produced and consumed at home; but unlike coal, reserves of natural gas were as scarce as coal's were abundant. Many felt that the natural gas shortage was contrived by producers to free them from price constraints, but for whatever reason there was little expectation that natural gas would play much part in easing an energy shortage.

This was not so in the case of petroleum–the source of 43 percent of the nation's energy supply, as shown in Table 13-2. The role of petroleum was not only large, it was pervasive, since it was the source of fuel for automobiles (to which coal and gas made no contribution), for space heating (where coal had lost its place almost entirely), and was as well an important fuel for utilities and industry. To most people the answer to a petroleum shortage seemed to lie in increased domestic output, and since the contribution of conventional wells was slowing down this seemed to mean intensified offshore drilling or exploitation of the immense western reserves of shale oil. Chapters 10, 11, and 12 examined the environmental consequences of intensified use of strip mining, offshore oil, or shale oil, respectively.

As shown in Table 13-2 hydro and nuclear power contributed only 4 percent of the nation's energy consumption. Of these, hydro might be looked upon as belonging entirely to the past since almost all usable sites had been put to use, and nuclear as belonging to the future inasmuch as the use of fission atomic power was believed to be in its infancy. Many projections had envisioned nuclear power becoming the principal source of energy by the end of the century. Opponents and proponents alike agreed that such an outcome would be fraught with profound social consequences.

Table 13-2
Total U.S. Consumption of Energy by Primary Source (1970)

Source	Percent
Coal	20
Natural Gas	33
Petroleum	43
Hydro and Nuclear	4
	100

Source: Resources for the Future, *Energy, Economic Growth, and the Environment* (Baltimore, Md.: Johns Hopkins University Press, April 1971), Table 5, p. 176.

Waste Heat from Power Plants

As indicated in Table 13-1, the electric utilities of America accounted for about one-fourth of the nation's energy consumption, but there is a disproportionate share of opportunities for energy saving in this sector. This is because this 25 percent of the nation's energy consumption is concentrated in a relatively few, generally sizable, generating plants. Thus, the waste heat generated by power plants occurs at concentrated localities and in very large quantities, and this increases the possibility of economically feasible recovery. In Chapter 2, John Bandel described and analyzed a number of efforts to recover for beneficial purposes some part of the enormous amount of waste heat rejected to the environment by electric generating stations. If not utilized, such rejected heat is the cause of a variety of problems covered by the general term "thermal pollution."

Of the total heat energy produced by the burning of fuel at electric generating stations, only about 40 percent is converted to power, and more than this, or 45 percent, is rejected to the environment. Thus, the possible opportunities for substantial savings are large and growing. They are growing particularly because the amount of rejected heat in electric power generation is much greater for plants fueled by nuclear power than by conventional stations employing fossil fuels. Nuclear plants account for less than 4 percent of present power production, but this figure is expected to increase to at least 39 percent by 1990.

Several possible uses of waste heat from power plants are being explored currently. Most interesting of these is the possible hookup with industry to utilize the exhaust steam from power plants as process steam in industry. At least one such linking of a power station and a manufacturing plant is under construction at Midland, Michigan. There, a new nuclear plant will generate 1300 megawatts of electricity, and will simultaneously supply 4,050,000 pounds per hour of process steam to a Dow Chemical plant in Midland. The Midland linkage is especially attractive because it makes it possible to withdraw steam from the power generating cycle at a higher temperature than would otherwise be feasible. One of the greatest limitations on the use of exhaust steam from power turbines is that the steam is usually at a relatively low temperature, and its usefulness is limited accordingly. By extracting steam at higher temperatures from the power generating cycle, the manufacturing plant can make highly efficient use of the steam, and the overall system efficiency of the electric generating plant plus the manufacturing plant can approach almost complete efficiency.

The other major use for low-temperature heat is for space heating and cooling. At its most sophisticated, this would involve heat-electric systems which would fill most of the energy needs of the communities in which they are located. Such installations have been designated as energy utilization systems or

EUS and envisage the supply of all the community's need for electricity plus energy for space heating and cooling, domestic hot water, water distillation, process steam for industry, and heat for snow and ice melting.

Such interfaces between power stations and industries or communities are beset by many difficulties that still await solution particularly of the social and political problems involved. The waste heat from power stations is not a fully dependable source because it may be lost due to power shutdowns, and there is a severe limit on the radius within which such waste heat may be transported economically.

For these reasons, many feel that the best opportunities for the use of waste heat lie in agriculture. Agricultural use of waste heat has taken two forms. One is the use of heated water from power generating facilities to maintain soil temperatures in agricultural fields, which stimulates plant growth and permits an extension of the growing season. The other approach has been to substitute the waste heat from power stations for conventional heating by fossil fuels in greenhouses which are devoted to intensive agriculture such as the growing of lettuce, onions, tomatoes, and radishes out of season. Somewhat similar is the possible use of waste heat in aquaculture. Intensive farming for fish is made enormously more productive by the use of waste heat, which stimulates the growth of aquatic life. As in possible industrial and community uses of waste heat, the uses in agriculture and aquaculture are severely impeded by the difficulties associated with possible power shutdowns and the limited radius of economically feasible transfer of waste heat. Perhaps more serious than these technical problems are the social and financial obstacles. All these uses of waste heat, however desirable socially, represent untested areas of investment that would be looked upon with caution under the best of circumstances. Bandel is no doubt right when he maintains that the first requisite for encouraging such investment would be a clear statement of the operating rules upon which utilities could depend for the future.

There should be no tendency to underestimate the obstacles that stand in the way of effective utilization of any large fraction of the heat rejected to the environment by electric power plants. On the other hand, there should be no disposition to minimize the potential for energy saving in this direction. Table 13-1 indicates that electric utilities account for 25 percent of gross U.S. energy consumption and Bandel has shown that nearly half of this is rejected to the environment. If half of this rejected heat could be put to useful ends, there would be a saving of 5 percent in gross U.S. energy consumption.[a] Perhaps most importantly, finding effective use for a substantial part of the heat presently rejected to the environment by electric utilities would help to solve one of the industry's most pressing problems. Bandel points out that much of the industry and some in regulatory agencies appear to have concluded fatalistically that the

[a]If such a "saving" were accomplished, it would not necessarily be reflected in a 5 percent reduction in gross U.S. energy consumption.

only acceptable solution to the thermal pollution problem is to build cooling towers. But these cooling towers are expensive, frequently unsightly, and sometimes associated with undesirable environmental side effects of their own. How much better it would be to achieve the positive solution of waste heat utilization, which would incidentally demonstrate that there is no inevitable conflict between balancing the energy budget and protecting the environment!

Opportunities for Recovery of By-Products from Stack Gases of Power Plants

Opportunities for recovery of waste heat from electric power plants are still largely potential and experimental, but to the extent that they exist at all, it is because electric generating plants are points of concentrated conversion of energy. For much the same reason, there are opportunities for the recovery of by-products from fuel combustion that would be totally infeasible at less concentrated points of usage. Thus far, utilities have generally regarded whatever materials were recovered from power plants as solid wastes representing a disposal problem, but A.A. Ferrara adduces evidence in Chapter 3 that suggest that this situation might change.

Henry Adams chose the dynamo as the symbol of modern science and efficiency. Certainly, the power-generating stations of the United States, linked together as they are in a vast nationwide interlocking power grid, represent the epitomy of our industrialization. But actually, as we saw in Chapter 2, power generation is a process in which only about 40 percent of the heat energy consumed is converted into electrical energy.

For generating stations burning fossil fuels, some of the loss of energy is manifested visibly in the plumes of smoke issuing from the chimneys of power plants. This emission of smoke-designated technically as particulates–will not exceed tolerable limits so long as the level of operations is not too high and air currents are sufficient to disburse and dilute it. Such dilution is enhanced by increasing the height of the smoke stack, and the impact of the smoke nuisance may be lessened generally by careful site selection. For many years, utilities contented themselves with such "solutions," but the mounting demand for power and heightened environmental consciousness has led to governmental regulation of the permissible volume of particulate emissions from power plants. Probably more hazardous to human health than particulates are the sulfur oxides that form part of the emissions from those stations burning sulfur-bearing fuel. The sulfur oxides formed in combustion may, after further atmospheric oxidation, fall to earth in raindrops as diluted sulfuric acid.

Although removal of 99 percent of the particulates is quite feasible technically, employing a combination of devices, the cost is substantial, both in terms of capital investment and operating costs. Moreover, there are additional costs involved in disposal of the captured fly ash. Ferrara estimates the present cost of fly ash disposal is more than $40 million. These costs are expected to climb in the coming decades, exceeding $100 million by the end of the century. However, it should be noted that these costs are dwarfed by the cost of fuel and capital charges for generating equipment. The fact that fly ash recovery costs are a relatively small part of total power generating costs may account for the very subdued interest utilities have shown in offsetting particulate suppression cost by sale of the recovered fly ash. Ferrara demonstrates that Detroit Edison can recover over $2 million annually by the aggressive marketing of fly ash. For all U.S. utilities the potential return from fly ash could amount to about $100 million annually. In its most important possible uses, fly ash would replace aggregates—sand and gravel. In addition, fly ash may be used as a partial substitute for portland cement. The aggregates and portland cement industries have fought the use of fly ash as a substitute for their products by resisting changes in highway standards and building codes that would permit their use.

The recovery of particulates is accomplished principally by physical means, but the removal of sulfur oxides necessitates the use of chemical reactions. If captured in alkaline substances, the sulfur oxides may be subsequently reduced to sulfur, or, alternatively, the sulfur oxide gases may be further oxidized with the aid of catalysts to sulfur trioxide and dissolved in water to form sulfuric acid. Both by-product sulfur and sulfuric acid find their major but by no means exclusive outlets in the production of fertilizer.

The problem of holding sulfur oxide emissions within legally permissible limits may be lessened by input controls which restrict the maximum permissible sulfur content in the fuels consumed. However, this is accomplished at the expense of less efficient operation of fly ash collection equipment, the efficiency of which is positively correlated with the sulfur content of the fuels consumed.

The costs of sulfur oxide control are substantial, but not staggering. Operating costs are likely to be a higher percentage of total cost than in the case of fly ash collection. With the exception of the catalytic oxidation system, it is necessary to supply a chemical to absorb the sulfur oxides. In throwaway systems, all such absorbents must be disposed of and even where sulfur is recovered, there is much solid waste to be discarded. Total costs including solid waste disposal of sulfur oxide removal systems are nevertheless likely to be less than the premium that must be paid for low sulfur fuels to achieve the equivalent lowering of the level of sulfur oxide emissions. Thus Ferrara says: "Typically, the cost of these (sulfur oxide removal systems) has been found to be in the range of about two to four mils per kwh in comparison to a premium low sulfur content fuel cost of approximately three to eight mils per kwh."

The American Electric Power System, a holding company for a group of mid-American power companies, rejects stack gas scrubbers as a means of removing sulfur oxides and has mounted a massive advertising campaign directed against the EPA, (see e.g., *Time Magazine*, May 13, 1974, p. 81), which advocates instead the "release of the enormous reserves of U.S. Government-owned low sulfur coal in the West." This latter option is considered in Chapter 10 of this volume. Meanwhile, it may be noted that although the cost of pollution control is small as a percentage of total cost of electricity, the difficulty that utilities are experiencing in raising capital in recent years is stiffening resistance to adoption of new sulfur oxide control technology.

Phillip Sporn, a former president of American Electric Power, contends that neither low-sulfur coal or stack gas scrubbers are necessary to hold sulfur oxide emissions from fossil fuel burning power plants to acceptable levels. Rather, he holds that very tall smokestacks will effect sufficient dispersion.

There is no way of foretelling with assurance what the responsible legislative and regulatory bodies will do about setting standards for emissions from fossil fuel burning power plants. However, assuming no substantial change from present standards, it would appear desirable to do everything possible to maximize returns from the recovery of by-products. Two specific recommendations may be made:

1. Research should be encouraged to find additional use for sulfur since increasing amounts are likely to be generated;
2. Laws should be enacted to encourage the by-product use of fly-ash in concrete and in paving materials, or at least to remove obstacles to such use.

Desulfurization of Fuel Oil

We have been concerned thus far with the possible recovery of waste heat from power plants (Chapter 2) or of by-products recovered from the stack gases of these same generating stations (Chapter 3). Table 13-1 shows that the "household and commercial" sector accounts for 21 percent of U.S. gross energy consumption, and most of this is consumed as fuel oil, principally for space heating. This consumption is spread over so many millions of structures that economic recovery of waste heat or by-products of combustion is out of the question, and control of sulfur oxide emissions from this source is accomplished by input restrictions. Many states and municipalities specify the maximum sulfur content permissible for fuel oils consumed in their jurisdiction, either in space heating or power generation. These legislative restrictions have meant that low sulfur fuel oils have commanded a premium price and this price differential has made it commercially feasible to set up plants to desulfurize fuel oils to the extent necessary to bring them within the acceptable fuel oil input standards established by environmental agencies.

The desulfurization process and the costs associated with it are described and analyzed in Chapter 4 and the details need not occupy us in this summary chapter. The important point is that this desulfurization process yields elemental sulfur as a by-product and this confronts, in the marketplace, the by-product sulfur recovered from the stack gases of power plants. It may be noted also that insofar as stack gas recovery of sulfur oxides results in production of sulfuric acid this enters into what is basically the same competitive arena since the principal use of sulfur is as a raw material for production of sulfuric acid. The sulfuric acid in turn is used mainly in the conversion of phosphatic rock into super phosphates for use as fertilizer. This restriction on the available market has meant that much of that which might be recovered would have no nearby market. The sulfur market has been burdened as well in recent years by large quantities derived from the desulfurization of Canadian natural gas.

Existing markets are hardly likely to be able to absorb all the by-product sulfur that could be produced from desulfurization processes or by recovery from stack gases of power plants. However, Ferrara cites studies by the Sulfur Institute which suggest that the market for sulfur might be greatly expanded.

One of the major problems facing producers of desulfurized fuel oil has been the difficulty in finding suitable locations for their plants convenient to water transportation, and in areas not denied them by effective opposition of local citizenry. Thus, a measure to improve the environment generally by reducing air pollution may be thwarted by the opposition of local groups organized in the name of protecting the environment.

Garbage Power

Chapter 5, by Stewart Winnick, examines the problems and possibilities in the disposal of solid waste. Disposal by landfill has been the preferred method of dealing with municipal solid waste in the overwhelming majority of American cities. At its best this was a planned procedure known as "sanitary landfill" in which each day's addition to the dump was leveled, covered with a layer of dirt, and further compacted by rolling. In this way some cities had carried on a program of converting undesirable areas to parks and playgrounds according to a planned program. At its worst, however, landfill operations amounted to no more than open dumping, leading to noxious, noisome, and rodent ridden eyesores. Whether well or poorly conducted, with urban and suburban expansion and growth it has become increasingly difficult to find land for disposal areas sufficiently close to the area of collection to hold down hauling costs. Since collection and hauling costs are about four-fifths of overall solid waste disposal costs, this question of land availability was crucial.

A relatively few large cities have employed incinerators, usually as a supplement to sanitary landfill. Chicago employed relatively sophisticated incinerators to handle all solid waste collected by the municipal collection service. The Chicago incinerators generated steam and also recovered unburnable metals as a by-product.

With the energy shortage of recent years, Winnick points out that several municipalities are undertaking to use garbage as a fuel for power generation, simultaneously seeking to recover metal scrap and other by-products from the residue. Plans are underway for such projects in the state of Connecticut, in Westchester County, and in New York, the town of Hempstead, Long Island. William C. Kasper, principal economist in New York State's Office of Economic Research, has estimated that the refuse generated annually by the state's urban population has a caloric content sufficient to generate 13 billion kilowatt hours of electricity "equivalent to the amount that would be produced by burning 5.7 million tons of low-sulfur coal or 23.8 million barrels of low-sulfur residual fuel oil."[2]

The Connecticut experiment is especially interesting because it involves political as well as technical pioneering. Such energy and by-product recovery projects are only economically feasible where a sufficient volume of refuse is generated to permit scale economies. Political fragmentation of areas sometimes stands in the way of achieving sufficient volume. The state of Connecticut is bypassing this difficulty by combining the solid waste outputs of several areas.

Many environmentalists feel that the most satisfactory use of solid waste would be to compost it, making a soil conditioner and thereby achieving the ideal of recycling. One difficulty has been the cost of marketing the resulting compost material. Winnick points out that such efforts have not been successful from an economic standpoint, but adds that this should not always rule out composting if no other socially useful application of solid waste can be found.

Power Saving Through Redesign of Refrigerators and Color TV

Residential customers expanded their use of electric power faster than the economy as a whole during the 1960s, and residential sales of electric utilities went up from 196 billion kwh in 1960 to 402 billion in 1969. Figure 6-4 shows the distribution of electrical energy use as of 1967 and the indications are that this distribution held well into the 1970s. The expanded residential consumption of electricity, then, has been a consequence principally of expanded usage of electric lighting and of appliances which were already widely employed during the 1960s. The extent of usage of such appliances is a function principally of economic forces and efforts to restrict such usage is likely to be an ineffective way of slowing the increase of electrical consumption. More promising might be the redesign of major appliances to conserve electricity.

The Center for Policy Alternatives of the Massachusetts Institute of Technology examined the potentialities for energy savings in the redesign of refrigerators and color television sets in a study, a summary of which is reproduced in this volume as Chapter 6. Color TV was selected for study because it ranked first among appliances as an object of consumer expenditures in 1972, and also because color TV was representative of the category of "brown goods," which consists of "home electronic products characterized by their high technology which are subject to more rapid technological change than white goods." "White goods," represented in the MIT study by refrigerators, consists of "such major household appliances as ranges, clothes washers and dryers, freezers and refrigerators." In 1972, refrigerators ranked third among appliances as a subject of consumer expenditures, just behind black and white television sets [1:6].

In their study, the MIT group made use of the concept of "life-cycle costs" in which both original purchase cost *and discounted* future expenditure for servicing and power requirements are calculated over the expected life of the appliance. "Discounted" future expenditures represents the amount of money one would have to put in the bank at the time of purchase in order to have enough money to pay service and power expenses in the future. "Life-cycle costs" are calculated in constant dollars and do not take account of future inflation. The MIT investigators found that the life-cycle costs of color television sets declined by about one-third from 1954 to 1972, due principally to substantial decline in purchase costs *even in constant dollars* and a modest expenditures for power (see Figure 6-2). Both drops reflected the rapid technological progress in the television industry expressed mainly in the application of principles of solid state physics. Power accounted for about 12 percent of life-cycle costs in 1972.

Quite different trends in life-cycle costs were found for refrigerators, where the total declined only slightly and irregularly over the period 1954 to 1972 despite a substantial decline in original purchase cost. Negating the declines in life-cycle purchase and service costs was the near tripling of power costs from 1954 to 1972, accounting for 58 percent of the total in the latter year (see Figures 6-2 and 6-3).

If the slight reduction in color television life-cycle power cost is attributable to technical progress, the substantial increase in life-cycle power costs for refrigerators are traceable mainly to product changes. The average size of refrigerators has been greatly increased concomitantly with the introduction of "frost-free" freezer areas.

The refrigerator-freezer accounts for 14.1 percent of total residential electrical energy use and leads all other appliances in this respect. All types of TV account for only 4.5 percent (see Figure 6-4). Thus the refrigerator is a good target for design changes to reduce life-cycle energy costs. The MIT study estimates that improved insulation of the refrigerator could effect an overall reduction in total life-cycle costs of $85, or about 10 percent, despite a small increase in initial purchase price.

Heavier wiring in the refrigerator motor could result in a discounted net saving of $35 or about 5 percent over the life of the refrigerator, again despite an increase in initial purchase price.

How Much Should We Rely on Nuclear Power?

Resources for the Future, in a projection made in 1971, foresaw a 67-fold increase in the input of nuclear energy to electric utilities from 1969 to 1980. It was expected that nuclear energy would account for 30 percent of the total energy input of electric utilities in the latter year [1:190]. In the expectation of a continued upward surge in electric generation (such as that pictured in Figure 2-1), RFF envisioned a nuclear input to electric utilities 217 times as great as that of 1969 by 1990 and 366 times as great in the year 2000. The same source foresaw that nuclear energy would account for 54 percent of electric utility energy inputs in 1990 and a slightly higher percentage in 2000 [1:191].

The RFF projections tend to justify Weinberg's conclusion, in Chapter 7, "that nuclear energy is on the verge of becoming our dominant form of energy." His finding leads Weinberg to undertake a penetrating thoughtful analysis of the "new and peculiar demands mankind's commitment to nuclear energy may impose on our human institutions." Only time will tell whether the RFF forecasts will be realized in practice, and whether therefore problems of the magnitude envisioned by Weinberg will arise. For the near term there is reason to suppose the combination of the opposition of environmental groups and the cost of investment capital may well mean a shortfall in the 1980 estimates.

Whether the opposition of environmental groups is inspired by irrational fears of atomic accidents or the sensible prudence of concerned citizens doesn't affect the outcome much, which is likely to be delays in construction.

For the longer run most students see an expanded use of nuclear power substantially in line with RFF projections. In the first place there is no clear indication yet of any sharp slowdown in the growth of electric power production. If the rate of increase continues at anything like the rate that has prevailed since World War II, it will provide powerful impetus for increased use of all types of fuels in the generation of electricity. Moreover, nuclear energy may be in a favorable competitive relationship to other fuels. Table 7-1 indicates that estimated unit costs for nuclear plants are 3 to 5 percent less than for comparably equipped coal plants without sulfur dioxide removal system, and 16 to 17 percent less if equipped with the latter. Moreover, Weinberg points out that technical progress in nuclear generation of electricity is continuing tending to reduce nuclear fuel costs and permit larger and stronger reactor vessels which in turn make for economies of scale.

Hannes Alfven, in Chapter 8, raises the question of "whether atomic energy can be considered a cheap and competitive source of energy without the more or less hidden subventions" which have led to a "great deal of development work," basically for military purposes. Certainly, it appears undeniable that the rapid worldwide spread of atomic technology has been principally motivated by military considerations rather than the adoption of the most economical fuel.

Certainly the military origins of the use of atomic energy has colored the public perception of nuclear power as a competitive energy source. Despite assurances of safety from prestigious scientific sources much of the public has continued to see nuclear energy shrouded by the mushroom cloud that accompanied the first awful atomic explosions.

But this public fear of nuclear power has motivated nuclear scientists and technologists to extraordinary efforts to assure safety in the operation of atomic reactors, and in the subsequent disposal of atomic wastes.

Weinberg makes a persuasive case for the position that all the significant sources of error and accident have been anticipated and dealt with successfully. Such vigilance, he feels, has reduced the probability of a serious accident in the generation of atomic power to such a miniscule figure that the risk is more than balanced by the benefits forthcoming. It is, of course, evident that the possibility of a serious accident cannot be excluded with certainty. Weinberg says: "When we say that the probability of a serious reactor incident is perhaps 10^{-8} or even 10^{-4} per reactor per year, or that the failure of all the safety rods simultaneously is incredible, we are speaking of matters that simply do not admit of the same order of scientific certainty as when we say it is incredible for heat to flow against a temperature gradient or for a perpetuum mobile to be built." The probability of an accident is extremely small, but certainty is beyond the grasp of science–is trans-scientific. Also trans-scientific, says Weinberg, is proof of some of the genetic effects attributed to radioactivity by opponents of nuclear power. If it requires experimentation on 8 billion animals to establish a given genetic effect of radioactivity with 95 percent confidence, we have passed beyond the boundaries of scientific proof.

Actually, Weinberg says, "a *properly* operating nuclear power plant and its subsystems . . . are, except for the heat load, far less damaging to the environment than a coal-fired plant would be." The nuclear plant emits more waste heat per unit of power generated, but it emits none of the sulfur oxides or particulates which create problems for coal burning plants.

Hannes Alfven thinks it is a mistake to depend as heavily on nuclear energy as we seem likely to do. Plutonium and several of the waste products of nuclear fission are among the most poisonous elements known to man. Even very minute leakage may be dangerous because "there are in nature a number of complicated biologic processes which enrich some of the radioactive waste products by a factor of 1,000 or 100,000."

There is general agreement that the problem of disposing of nuclear wastes is principally one of scale. Alfven says that "a few reactors which are carefully controlled are not likely to constitute a very serious ecological threat," meaning no doubt that safe disposal of the relatively small amounts of resulting waste can be dealt with safely. Weinberg demonstrates, in Table 7-2, the enormous scale on which atomic wastes will be generated if present projections on the use of nuclear energy are realized. He shows no disposition to dismiss the resulting disposal problem as trivial. On the contrary, his findings inspire him to undertake an extraordinarily thoughtful analysis of the enormous problems of dealing with the consequences of nuclear electrical generation deployed on the scale usually projected for output. This will require depositories for tremendous quantities of highly poisonous, long lasting wastes which must be guarded against accidental intrusion and leakage. This will demand, he says, "both a vigilance and a longevity of our social institutions that we are quite unaccustomed to." If the necessary discipline seems unattainable, it may be well to recall that the atomic bomb "has imposed an additional burden on our social institutions," which we have supported for more than a generation however precariously.

For his part, Alfven opts for less reliance on nuclear energy and a more active exploration of the steps needed to expand output of energy from fossil fuels, atomic fusion, the sun and geothermal sources. The remaining chapters to be summarized here will deal with problems associated with some of these sources.

Geothermal Energy

Geothermal power, which is discussed in Chapter 9 by Ronald Axler, is a source of power that, like nuclear power, is still largely a potential source. Geothermal power is not regarded by its proponents as being as promising, nor by its opponents as being as menacing as nuclear power.

At present, in its only major commercial application, geothermal power provides roughly half the electricity needs of the city of San Francisco, which shows that its contribution to energy needs is not negligible. At the same time, it is evident that geothermal power is no answer to the national energy crisis. Even its most optimistic backers do not foresee geothermal power equaling hydro's 4 percent contribution to the national energy budget [1:164-166] in the next ten years (see Table 9-10). Moreover, geothermal power is similar to hydro in that its potential lies largely in the western states and its generation is limited to particular localities, unlike thermal generation which can occur wherever fuel can be shipped. Like hydro power also, geothermal power can usually be generated without major adverse effects on the environment.[b]

Despite its limitations, it is entirely possible that geothermal power could make a significant contribution to the nation's energy with a minimum adverse impact on the environment. What is required most of all is research and exploration, areas in which the federal government is especially well equipped to make a contribution.

---

[b]Of course, many environmentalists regard the hydro power reservoirs with their fluctuating levels as a form of pollution, and as noted in Chapter 9 geothermal waters have been found in California's Imperial Valley that are so corrosive as to be unusable.

Strip Mining

The increased demand for electricity after 1960 was reflected especially in increased demand for coal. Earlier in the period since the 1920s coal had lost ground steadily to oil in powering railroads and in home heating and later to natural gas for domestic purposes. However, as such devices as air conditioners, radios, television sets, and refrigerators began to account for more and more of the household energy demand, the relative demand for electricity likewise rose. Since coal was the preferred fuel for electric power stations, it benefited accordingly.

However, this favorable position for coal was not achieved without an intense competitive struggle with other energy carriers such as oil and natural gas. To achieve the lowest cost per unit of energy, coal users turned increasingly to strip mined coal rather than that derived by conventional underground mining methods. Some coal could only be reached by underground methods, but deposits that were strippable were increasingly preferred.

Chapter 10 records some of the damage that was wreaked on coal mining states in the United States, especially in Appalachia, by the widespread application of strip mining. There is little hope that the harm done to Appalachia by strip mining will be redressed and the subject is predominantly of historical interest. However, strip mining attains unquestioned contemporary importance when considered in relation to the vast coal deposits of the western states. Much of the traditional wide open spaces of the Great Plains and Rocky Mountain states is underlain by coal within comparatively easy reach of the bulldozers and power shovels of the strip miners. Moreover, much of this coal is of sufficiently low sulfur content to make it acceptable to power stations in major metropolitan areas from the standpoint of sulfur oxide emissions. The energy companies holding federal leases to the mineral rights for these deposits have been impatient to get at the strip mining of low-sulfur western coal. Action has been delayed, however, because of the recognition of special difficulties likely to be encountered in the rehabilitation of strip mined lands in the west. Whereas there is abundant rain to provide vegetation on strip mined lands in the east, much of the western coal occurs in arid or semi-arid regions where the restoration of vegetation would be slow and uncertain. The National Academy of Sciences was asked to conduct a study of the prospects for rehabilitation of strip mined western coal lands. The Academy scholars reached a carefully hedged conclusion, which is quoted in Chapter 10, that suggested very tentatively that strip mined western coal lands could probably be rehabilitated in those areas receiving ten inches or more of annual rainfall. Much of the western coal land received less than this amount and successful rehabilitation of strip mined lands might be impossible for other reasons even where watered by 10 inches annually.

Shale Oil

The development of another huge resource counted on to realize the goals of Project Independence (the campaign to make the U.S. self-sufficient in energy) will also be handicapped by lack of water. Oil shale is found in immense amounts in the Rocky Mountain states of Colorado, Wyoming, and Utah, but the deposits are very low grade and will require the processing of tremendous quantities of rock–about 1.5 tons for each barrel of oil produced. Moreover, the existing technology requires the consumption of about 3 barrels of water for each barrel of oil produced. For these reasons studies of the Interior Department and the Atomic Energy Commission have concluded that the limit on yield from oil shale will probably be about 1 million barrels per day, or a miniscule contribution to the U.S. daily consumption of 18 million barrels.

Metz's summary, in Chapter 12, of the oil shale outlook makes it evident that the environmental price would be enormous for even a 1 million barrel per day industry. "Perhaps one of the most important concerns," says Metz, "is that the shale industry will grow up, prove unsuccessful, and be abandoned, leaving the western slope of the Rockies in somewhat the same shape as Appalachia." Moreover, he feels the ceiling of 1 million barrels per day may be high "because much of the available water in the region has reportedly been cornered for surface coal mining."

It is interesting to note that most of the best oil shale deposits are on federal lands. The devastation wrought by strip mining has frequently been attributed to weak and compliant state officials who fail to enforce state laws concerning strip mining. The oil shale situation is clearly almost entirely under federal control and responsibility for environmental damage will be clearly borne by federal officials.

Energy Under the Oceans

Few economic and political issues have been more hotly debated than the question of whether or not there should be extensive exploitation of offshore oil deposits as a way of meeting the nation's demand for energy. The Science and Public Policy Program of the University of Oklahoma undertook a study financed by the National Science Foundation to examine this question intensively, and their findings were published by the University of Oklahoma Press. Chapter 11 consists of a summary report of this study.

The authors point out that the projected energy consumption of the United States in 1985 "will be between 112.5 and 130.0 quadmillion BTUs (the equivalent of 19.4 to 22.4 billion barrels of oil). Demand for oil and gas accounts for almost 70 percent of this total" and other sources of energy "are not likely to contribute enough to reduce oil and gas needs very much."

The study envisioned four areas as possible sources of expanded domestic oil and gas production: "(1) Onshore in the lower 48 states, (2) Alaska, (3) State lands offshore, and (4) the Outer Continental Shelf (OCS)." Of these, "only Alaska and the OCS offer a potential for significant increases. With maximum development the OCS may be expected to produce up to 2.6 million barrels each day and 9.1 trillion cubic feet of gas each year; if the Trans-Alaska pipeline system (TAPS) is available, daily Alaskan production can be as much as 252.6 billion barrels of oil per day and 4.4 trillion cubic feet of gas per year.

"But given even the lowest anticipated level of energy consumption, some portion of U.S. energy demand will have to be satisfied by imports, . . . if it is assumed that the demand will be at its highest while domestic production is at its lowest rates, imports will supply about 40 percent of total U.S. energy demand in 1985."

The Oklahoma study was designed to supply policymakers with some guidelines to aid in deciding whether to seek to hold down imports to a comparatively low level by developing to the fullest new domestic sources, in general, and Alaskan and OCS sources in particular. Since the decision had already been made to build the Trans-Alaska pipeline system, the principal focus of the study was on the OCS.

The general methodology of the study was to make an assessment of the technological problems involved in exploiting the resources of the OCS with minimum harmful impact on the environment. Initially, it was expected that the assessment would focus on physical technologies for finding, developing, producing, and transporting oil and gas, but it soon became apparent that it was essential to consider the social context in which such physical technologies were managed and regulated. It was found that the physical technologies used in the OCS were relatively stable and technological alternatives limited, with the prospect that technological change would continue to be gradual. "How technologies are managed and regulated, what we came to call the social technologies, proved to be the critical element in this problem area." It was necessary to take account of new concerns, especially for environmental quality and safety and to take account in the decision-making process of environmental and consumer interest groups in addition to industry and government.

To provide an adequate summary of Chapter 11, which is itself a highly compressed version of the lengthy, original study–*Energy Under the Oceans*–is a next to impossible task. Chapter 11 will provide the reader with a guide to the original study so he may read further in subjects of special interest. Here only a few selected findings will be noted.

The Oklahoma investigators found only three technologies to be inadequate–"velocity activated downhole safety devices, well control technologies, and oil containment and clean-up devices." The downhole safety device known as a "storm choke" is the first line of defense against a "blowout," i.e., an uncontrolled flow of gas and oil to the atmosphere. Between 25 and 40 percent of these "storm chokes" were found to have failed in recent accidents in the Gulf of Mexico and the studies first recommendation is that "the 'storm choke' must be made more reliable."

Once a blowout is underway, well control technologies must be brought into play and these include capping and drilling relief wells. The study found that capping was "particularly difficult offshore" and relief well drilling too time consuming. The studies second recommendation was: "Alternatives for re-establishing well control must be developed."

Once a blowout that cannot readily be brought under control occurs in a well, containment and cleanup devices enter the picture. Here again the Oklahoma study found existing technology inadequate. "There was no effective capability for containing and cleaning up oil on the OCS before Union's blowout at Santa Barbara, California. Subsequent crash efforts by industry and government have produced only a limited capability even now. In fact, wave heights, wind velocities, and currents on much of the U.S. OCS exceed designed capabilities at least a third of the time." The Oklahoma study accordingly advances as its third recommendation that "efforts should continue to be made to improve the performance of containment and clean-up devices" with special attention to dealing "effectively with oil spills which threaten to come ashore."

Two other technologies were identified as requiring improvement: (1) Drilling Bits: The continuing research effort to develop longer lasting bits could indirectly reduce frequency of oil spills; (2) Flaw Detectors: Improved devices for detecting flaws in pipelines could reduce pipeline leaks and ruptures which accounted for "84 percent of all oil reported spilled from offshore facilities."

Finally, the study points to four "technologies presently available for improving the quality of OCS operations (which) are not being used by all operators" and six areas where new technologies should be developed and utilized.

Why do these technological weaknesses exist? The Oklahoma study answers its own question by noting that "standards used for determining the adequacy of OCS technologies have been based largely on industry's judgment of what was economically feasible. Before events such as Union's Santa Barbara blowout attracted widespread public attention, continuous participation in policy-making and administration for OCS development had been pretty much limited to government and industry. . . . Perhaps as a consequence, government regulation had tended to be heavily dependent upon industry's engineering and operational expertise when establishing OCS regulations." Having thus stated the problem, the study lists twelve specific recommendations for changes in the patterns of operations to correct the situation.

This summary of a summary has been concerned thus far with the adequacy of physical technologies for dealing with the OCS. But the Oklahoma study also considered the requirements for social technologies capable of meeting the needs of oil discovery and recovery from the OCS.

A primary task under this heading would be provision for research and investigation which would make it possible for the environmental agencies charged with regulating operations on the OCS to establish informed discharge and pollution rules. This will involve fundamental "knowledge of background levels of hydrocarbons in physical and biological components of the marine environment and the physiological effects of acute and chronic exposure to oil in marine plants and animals."

Another major data lack on the part of the federal regulatory agencies is the absence of adequate fundamental geological and geophysical data in the areas being offered for lease sale. The Oklahoma study points out that: "This limits the (Interior) Department's capability for long-range planning and has led to the pattern of OCS development being largely in response to industry's interest in specific OCS areas rather than according to an Interior plan for systematic orderly development." Accordingly, the study recommends that: "USGS should be adequately funded to permit it to contract for exclusive seismic surveys in order to acquire adequate exploratory data for regional OCS development, including overall land use planning."

Although the Oklahoma study does not make this point relative to the OCS, it has been widely observed that federal regulatory agencies have a tendency to become overly accommodating to the industries whose operation they are supposed to oversee. Certainly dependence for the information necessary to regulate upon industry sources would encourage tendencies toward federal subservience to the industry.

Public support for federal actions to minimize the adverse environmental impacts of OCS operations is essential for success of the program. An avenue for enlisting informed public support is provided in the National Environmental Policy Act (NEPA) of 1969 "which requires an environmental impact statement to be written whenever a contemplated major federal action may have significant impact human environment." This NEPA provision has opened the way to broader public participation in policy making.

However, the number of agencies with some measure of responsibility in the field has led to duplication in statements relating to the same policy area. The Oklahoma study suggests that this difficulty might be met by preparing broad regional programmatic statements which would give a better basis for assessing overall impacts. These long-range statements would provide a context for judging the environmental impact statements relating to individual lease sales.

The fragmentation of authority among federal agencies, which has led to duplication of environmental impact statements, also has profound consequences for the day-to-day administration of environmental programs in general and the OCS program in particular. The Oklahoma study accordingly recommends the establishment of a Department of Energy and Natural Resources to assure a coordinated approach. Failing that the study suggests the designation of a high administration official to serve as coordinator of energy related programs.

The need for a coordinator reflects the existence of both overlapping and gaps in the jurisdiction of federal agencies. Similar jurisdictional questions arise in relations between federal and state agencies, and national and international bodies. Mechanisms should be developed for the resolution of such matters.

On the basis of its study the Oklahoma group recommended an overall management plan for the OCS. Too detailed for coverage here, a digest of the plan is given in Table 11-1.

Conclusions

In commenting on a massive federal report on energy, Leonard Silk recently said: "The remarkable aspect of the current decade is the way that the long run has suddenly come to seems like the very short run."[3] Mr. Silk was referring to the speed with which intensely difficult energy problems have descended upon us as if from nowhere, but the remark is pertinent to many other environmental problems as well. The proximate cause of the current energy crisis may appear to lie in Arabian policy concerning their oil reserves, but the real root difficulty is simply the staggering magnitude of the energy consumption pattern that has developed in the United States. Similarly, the central thrust of the Hofstra environmental studies published by Lexington Books has been that the scale of output has raised pollutional effects to a new order of magnitude that called for urgent ameliorative action.

The occasion for Silk's remark was the issuance of a 762-page report on Project Independence by John C. Sawhill, the former Federal Energy Administrator.[4] The Sawhill report painted a generally bleak picture of the energy outlook, but despite its tone of urgency did not envision or call for sweeping changes in the sources of energy or the ways in which they were employed. This was wise because the situation is not one in which rapid change is likely. To a large extent, both businesses and individuals are locked into fuels and methods of consuming them by heavy fixed investments in such things as oil burners and automobiles. There is therefore not likely to be a rapid increase in the use of coal at the expense of oil although the United States is blessed with abundant domestic supplies of the latter and is increasingly dependent on foreign supplies of petroleum.

Conservation by reducing consumption seemed like the path best suited to some immediate easing of the energy shortage. But here as well the outlook was for slow response—voluntary conservation measures would have limited effect and political timidity would probably bar more effective steps.

One predictable reaction was to use the energy crisis as an occasion for an attack on the measures that had been proposed to improve the environment by reducing pollution. If automobile emission controls reduce gas mileage or pollution control devices for power plants add to costs already swollen by rising fuel prices, it might be expected that public reaction would sweep environmental constraints aside. But here again inertial forces were found to be operative and politicians were not prepared to conclude hastily that the environment could be sacrificed.

Another line of attack was to charge that environmental protection made a major contribution to inflation. Russell W. Peterson, chairman of the Federal Council on Environmental Quality, pointed out that "1973 expenditures to satisfy the requirements of Federal air and water pollution control legislation amounted to approximately 1 percent of our gross national product,"[5] if all environmental outlays were included the total would still not exceed 2 percent of the GNP.[6]

There were many non-issues raised, then, as some used the energy crisis to mount an attack on the environmentalists. But it must be recorded that environmental groups had brought this upon themselves by arguing in many instances for a standard of perfection.

There were, for example, those who advocated standards of zero population growth which had never been achieved since mankind entered the scene. Less extreme but nonetheless mischievous were those self-appointed protectors of the environment who were unwilling to use or give appropriate weight to the dispersing effects of currents in air and water. Such a perfectionist approach was likely to discourage efforts to improve the environment by making them appear too expensive. In fact dilution by dispersing, if not overloaded, could do much of the job of maintaining a livable environment.

A good example of the sort of perfectionist demand raised periodically by some environmentalists was the call for elimination of overhead telephone and electrical lines. The advantages of service uninterrupted by storms were obvious, but the expense and disruption involved in a shift to underground lines were sufficient overweighting factors.

This sort of environmental perfectionism was especially difficult to understand in the face of the very real threats to the environment posed by the strip mining of coal in the west and the exploitation of shale oil. Large-scale development of either seemed likely to leave devastation of the wide open spaces in its wake. The wry words of Archibald MacLeish might yet be realized:

> Everything sticks to the grease of a gold note–
> Even a continent – even a new sky![7]

For the inhabitants of the sprawling megalopolises of the coastline, especially those of the northeast, the sea represented the last easily accessible "wide open spaces." To this last frontier, large-scale offshore oil drilling represented the most immediate threat. Certainly it was not too much to ask that all the technical problems be solved before such a program was launched.

Finally, the evidence seemed clear that it was possible to effect genuine gains for society such as savings of energy while also protecting the environment. What was needed was the will.

Notes

1. Sam H. Schurr (ed.). *Energy, Economic Growth, and the Environment* (Baltimore, Md.: Johns Hopkins University Press, 1972), p. 178.
2. *New York Times*, May 22, 1974, p. 42.
3. Leonard Silk, "Missing: Energy Policy," *New York Times*, November 13, 1974, p. 68.
4. *New York Times*, November 13, 1974, p. 1, column 6.
5. Gladwin Hill, "Environmental Outlays and Inflation: Is There a Link?" *New York Times*, November 9, 1974, p. 14.
6. Ibid.
7. Archibald MacLeish, "Burying Ground by the Ties," in Selden Rodman (ed.), *A New Anthology of Modern Poetry* (New York: The Modern Library, 1938), p. 313.

Index

About the Editor and Contributors

Dr. Alfred J. Van Tassel, as part of a varied career in economics, served as staff studies director of a committee of the United States Senate and as Executive Secretary of the United Nations Scientific Conference on the Conservation and Utilization of Resources. He has been a member of the faculties of New York University, Columbia University, the University of Illinois and Hofstra University. He was editor of *Environmental Side Effects of Rising Industrial Output* and *Our Environment: The Outlook for 1980*, both published by Lexington Books, D.C. Heath and Company.

Dr. Hannes Alfven is a member of the Royal Institute of Technology in Stockholm and Professor of Physics at the University of California, San Diego. In 1970 he shared the Nobel Prize in physics with Louis Neel of France.

Ronald Axler who holds a Bachelor of Science degree from Ithaca College and a masters degree from Hofstra University is an auditor for Abrams, Meresman and Company in New York City.

John M. Bandel, Jr. graduated from Hofstra University which also awarded him a Master of Business Administration degree. He is a businessman in New York City.

Angelo A. Ferrara who is a project engineer for Grumman Aerospace Corporation has bachelors and masters degrees in engineering from Polytechnic Institute in Brooklyn and the MBA from Hofstra University.

Dr. J. Herbert Hollomon is the director of the Institute of Policy Alternatives of the Massachusetts Institute of Technology in Cambridge, Mass. He served as Assistant Secretary of Commerce from 1962 to 1967.

Professor Don E. Kash is Director of the Science and Public Policy Program of the University of Oklahoma, Norman, Oklahoma.

Dr. William D. Metz has the Ph.D. in nuclear physics from Yale University and is on the staff of *Science*. He is coauthor of *Energy and the Future*, published by the American Association for the Advancement of Science.

Gerald L. Neway is Assistant Controller of Shaween Natural Resources Co., Inc. and holds bachelors and masters degrees from Niagara and Hofstra Universities.

Dr. Alvin M. Weinberg is Director of Energy Research and Development of the Federal Energy Administration and before that was director of the Oak Ridge National Laboratory of the Atomic Energy Commission.

Stewart Winnick is senior accountant in the International Division of Mobil Oil Corporation and holds bachelors and masters degrees from Hofstra University.

Irwin Zurkowski is a New York businessman who has a Master of Business Administration degree from Hofstra University.

DATE DUE
